PCR for Clinical Microbiology

Margret Schuller · Theo P. Sloots ·
Gregory S. James · Catriona L. Halliday ·
Ian W.J. Carter
Editors

PCR for Clinical Microbiology

An Australian and International Perspective

Springer

Editors
Dr. Margret Schuller
Profitable Business Development
1499 Sutherland NSW
Australia
mschuller@profitbd.com

Gregory S. James
Centre for Infectious Diseases
and Microbiology
ICPMR
2145 Westmead NSW
Australia

Ian W.J. Carter
Prince of Wales Hospital
SEALS Microbiology Department
2031 Randwick NSW
Australia

Dr. Theo P. Sloots
Sir Albert Sakzewski Virus Research Centre
Queensland Children's Medical
Research Institute
Herston Road
4029 Herston QLD
Australia

Catriona L. Halliday
Centre for Infectious Diseases
and Microbiology ICPMR
2145 Westmead NSW
Australia

ISBN 978-94-017-8336-1 ISBN 978-90-481-9039-3 (eBook)
DOI 10.1007/978-90-481-9039-3
Springer Dordrecht Heidelberg London New York

Printed on acid-free paper

Springer is part of Springer Science+Business Media (www.springer.com)

Foreword

The Polymerase Chain Reaction is a technique developed over 20 years ago. Like the preceding tools of nucleic acid hydridisation and DNA sequencing, it represented a breakthrough method that was almost instantly adopted as an indispensable tool in the biomolecular sciences. It is probably true to say that there is no organism that has been investigated by humans that has not had PCR applied to it to solve some basic aspect of its biology. The potential practical applications of the PCR were also recognized very early and its penetration into some industries is well known. Perhaps the most notable of these is in its application in forensic biology, a fact obvious in popular culture as can be seen from many prime time TV crime shows.

The potential of PCR as a medical diagnostic tool was also recognized in the early days of PCR. Its uptake in this field has been relatively slow however and the early expectations are still to be realized. There are a number of reasons for this. Most of these however centre around the fact that its reliability and reproducibility to a standard required for clinical diagnosis is difficult to achieve. In a research laboratory, PCR is highly tailored to address specific research questions and time and resources are generally available to identify inconsistent data and trouble shoot where necessary. Clinical priorities however are driven by the need to have rigorously standardized protocols that can be consistently applied across diverse laboratories by personnel whose primary training is not necessarily in PCR. Apart from the technical aspects of carrying out a PCR, it is also the case that diagnostic tests have to be carefully designed with respect to primers to be used and to how test samples are collected and the DNA therein purified. Improper design or execution of any of these steps can lead to adverse outcomes for a patient if misdiagnosis occurs. On top of this of course is the fact that PCR is extraordinarily sensitive so that laboratories dedicated to the method have to be meticulously designed to minimize contamination of test material.

This book is an attempt to bridge the gap from "bench to bedside" and is notable for a number of reasons. Firstly, while the book is written with an international audience in mind it very much has an Australian identity. This is not just because it is written by a panel of Australian experts in the field of basic and diagnostic biology generally and medicine specifically. It is also because it includes advice on diseases that are emerging or actual important pathogens in Australia and the nearby Asia/Pacific region region including dengue fever and malaria. Apart from providing defined methods for the identification of a diverse range of viral, bacterial and parasite pathogens it also describes protocols for susceptibility screens that encompass important resistance and virulence genes. This is however far more than just a "recipe" book. Rather, it is designed to educate the reader in the intricacies of PCR and includes an introductory chapter on PCR basics, PCR methodology and laboratory accreditation standards. The following chapters are individually and collectively of educational value to the novice. Chapters are structured into bacteriology, virology, fungal and parasitic as well as susceptibility screens. Fundamentals are covered in Part I and the medical diagnostic perspective for disease states is introduced by a clinician in Part II. The inclusion of medical criteria broadens the appeal of the compendium and will help the reader understand, not just the strengths of PCR as a tool, but its limitations in specific contexts in relation to other diagnostic tests.

This book, like most great ASM initiatives, originates from the efforts of many ASM members at a local level, in this case the NSW ACT Branch of the ASM. The Antimicrobial Special Interest Group of this branch has previously published compendia "in-house" that were marketed locally but this effort is more ambitious and comprehensive. The book actually grew beyond the scope originally intended in the early design stage as the importance of the topic was realized. In the end over 50 authors, most ASM members, made primary contributions. I would also like to acknowledge the efforts of Margret Schuller the overall coordinating Editor along with the four scientific editors, Greg James and Catriona Halliday from Westmead Hospital, Ian Carter from the Prince of Wales Hospital and Theo Sloots from the Sir Albert Sakzewski Virus Research Centre. The chapters in the book underwent rigorous peer review and for this I would like to acknowledge the efforts of the ASM Standing Committee in Clinical Microbiology. ASM is also indebted to Springer for agreeing to act as publisher. I have no doubt this book will quickly become a standard reference in clinical microbiology laboratories.

ASM President Hatch Stokes

Contents

Contributors

Sam Abraham School of Biological Sciences, University of Wollongong, Gwynneville, NSW 2500, Australia

Rafat Al-Jassim School of Animal Studies, University of Queensland, Brisbane, QLD 4072, Australia

Snehal Anuj Queensland Paediatric Infectious Diseases Laboratory, Sir Albert Sakzewski Virus Research Centre, Queensland Children's Medical Research Institute, Children's Health Service District, Brisbane, QLD 4029, Australia, snehal_anuj@health.qld.gov.au

G. Barber Office of Health Protection SSBA Regulatory Scheme, GPO Box 9848 MDP 54 Canberra ACT 2601, Australia

John Bates Public Health Microbiology, Forensic and Scientific Services, Clinical and Statewide Services Division, Queensland Health PO Box 594, Archerfield, QLD 4108, Australia, John_Bates@health.qld.gov.au

Seweryn Bialasiewicz Queensland Paediatric Infectious Diseases Laboratory, Sir Albert Sakzewski Virus Research Centre, Queensland Children's Medical Research Institute, Children's Health Service District, Brisbane, QLD 4029, Australia, seweryn@uq.edu.au

Cheryl Bletchly Molecular Diagnostic Unit, Microbiology Division, Pathology Queensland Central Laboratory, Brisbane, QLD 4029, Australia, Cheryl_Bletchly@health.qld.gov.au

Scott Bowden Victorian Infectious Diseases Laboratory, VIDRL, Wreckyn St, Melbourne, VIC, Australia, scott.bowden@mh.org.au

James Branley Microbiology and Infectious Diseases, Nepean Hospital, University of Sydney, Sydney, NSW, Australia, branlej@wahs.nsw.gov.au

James Buttery Enteric Virus Research Group, Murdoch Childrens Research Institute, Royal Children's Hospital, Parkville, VIC 3052, Australia, jim.buttery@mcri.edu.au

Shane Byrne Molecular Pathology Department, Sullivan Nicolaides Pathology, 134 Whitmore Street, Taringa, QLD 4068, Australia, Shane_Byrne@snp.com.au

Ian Carter SEALS Microbiology Division, Prince of Wales Hospital, Randwick, NSW 2031, Australia, ian.carter@sesiahs.health.nsw.gov.au

Julia A. Cattell Division of Microbiology, PathWest Laboratory Medicine, QEII Medical Centre, Nedlands, WA 6909, Australia

Glenys R. Chidlow Division of Microbiology, PathWest Laboratory Medicine, QEII Medical Centre, Nedlands, WA 6909, Australia, glenys.chidlows@health.wa.gov.au

James Chin School of Veterinary Science, University of Queensland, Brisbane, QLD 4072, Australia, xion.biotech@gmail.com

Geoff Coombs Department of Microbiology and Infectious Diseases, PathWest Laboratory Medicine, Royal Perth Hospital, Perth, WA, 6847, Australia

Dominic Dwyer Centre for Infectious Diseases and Microbiology, Westmead Hospital, Westmead, NSW 2145, Australia, dominic_dwyer@wmi.usyd.edu.au

Cassandra Faux Queensland Paediatric Infectious Diseases Laboratory, Sir Albert Sakzewski Virus Research Centre, Queensland Children's Medical Research Institute, Children's Health Service District, Brisbane, QLD 4029, Australia, cassyfaux@uq.edu.au

Rebecca Fonte Department of Microbiology and Infectious Diseases, PathWest Laboratory Medicine, Royal Perth Hospital, Perth, WA 6847, Australia

Janet Fyfe Mycobacterium Reference Laboratory, Victorian Infectious Diseases Reference Laboratory (VIDRL), 10 Wreckyn Street, North Melbourne, VIC 3051, Australia, janet.fyfe@mh.org.au

Suzanne M. Garland Microbiology and Infectious Diseases Department, Royal Women's Hospital, Parkville, VIC 3052, Australia; Department of Obstetrics and Gynecology, University of Melbourne, Melbourne, VIC, Australia; Murdoch Children's Research Institute, Parkville, VIC 3052, Australia, Suzanne.Garland@thewomens.org.au

Gwendolyn Gilbert Centre for Infectious Diseases & Microbiology Laboratory Services (CIDMLS), Institute for Clinical Pathology and Medical Research (ICPMR), Westmead, NSW, Australia

Philip M. Giffard Menzies School of Health Research, Institute of Advanced Studies, Charles Darwin University, Casuarina, NT, Australia, Phil.Giffard@menzies.edu.au

Maria Globan Mycobacterium Reference Laboratory, Victorian Infectious Diseases Reference Laboratory (VIDRL), 10 Wreckyn Street, North Melbourne, VIC 3051, Australia, maria.globan@mh.org.au

David L. Gordon Department of Microbiology and Infectious Diseases, Flinders University, Bedford Park, SA 5042, Australia, d.gordon@flinders.edu.au

Stephen Graves Microbiology, Hunter Area Pathology Service (HAPS), Newcastle, NSW 2310, Australia; Australian Rickettsial Reference Laboratory (ARRL), Barwon Biomedical Research, The Geelong Hospital, Bellarine Street, Geelong, Victoria 3220, Australia, stephen.graves@hnehealth.nsw.gov.au

Catriona Halliday Centre for Infectious Diseases and Microbiology, Westmead Hospital, Westmead, NSW 2145, Australia, Catriona.Halliday@swahs.health.nsw.gov.au

Gerald B. Harnett Department of Microbiology and Infectious Diseases, PathWest Laboratory Medicine, QEII Medical Centre, Nedlands, WA 6009, Australia, Gerry.Harnett@health.wa.gov.au

Jonathan Howard SEALS Virology Division, Prince of Wales Hospital, Randwick, NSW 2031, Australia, jonathan.howard@sesiahs.health.nsw.gov.au

Leonard Izzard Australian Rickettsial Reference Laboratory (ARRL), Barwon Biomedical Research, The Geelong Hospital, Bellarine Street, Geelong, VIC 3220, Australia

Greg James Centre for Infectious Diseases and Microbiology Laboratory Services (CIDMLS), Institute for Clinical Pathology and Medical Research (ICPMR), Westmead, NSW 2145, Australia, Greg.James@swahs.health.nsw.gov.au

Neisha Jeoffreys Molecular Biology, Centre for Infectious Diseases and Microbiology, Institute for Clinical Pathology and Medical Research, Westmead Hospital, Westmead, NSW 2145, Australia, Neisha.Jeoffreys@swahs.health.nsw.gov.au

Ian D. Kay Department of Microbiology and Infectious Diseases, PathWest Laboratory Medicine, Royal Perth Hospital, Perth, WA, 6847, Australia, Ian.Kay@health.wa.gov.au

Timothy M. Kidd Queensland Paediatric Infectious Diseases Laboratory, Sir Albert Sakzewski Virus Research Centre, Queensland Children's Medical Research Institute, Children's Health Service District, Brisbane, QLD 4029, Australia; Microbiology Department, Pathology Queensland – Central Laboratory, Herston Hospitals Campus, Brisbane, QLD 4029, Australia, Tim_Kidd@health.qld.gov.au

Carl Kirkwood Enteric Virus Research Group, Murdoch Childrens Research Institute, Royal Children's Hospital, Parkville, VIC 3052, Australia, carl.kirkwood@mcri.edu.au

Tuckweng Kok Microbiology-IMVS, SA Pathology, Adelaide, SA 5000, Australia, Tuckweng.Kok@adelaide.edu.au

Stephen B. Lambert Queensland Paediatric Infectious Diseases Laboratory, Sir Albert Sakzewski Virus Research Centre, Queensland Children's Medical

Research Institute, Children's Health Service District, Brisbane, QLD 4029, Australia, sblambert@uq.edu.au

Marcel Leroi Department of Microbiology, Nepean Hospital, Penrith, NSW 2750, Australia, LeroiM@wahs.nsw.gov.au

Michelle Lockhart Australian Rickettsial Reference Laboratory (ARRL), Barwon Biomedical Research, The Geelong Hospital, Bellarine Street, Geelong, VIC 3220, Australia

Ian M. Mackay Queensland Paediatric Infectious Diseases Laboratory, Sir Albert Sakzewski Virus Research Centre, Queensland Children's Medical Research Institute, Children's Health Service District, Brisbane, QLD 4029, Australia, ian.mackay@uq.edu.au

Christopher McIver SEALS Microbiology Division, Prince of Wales Hospital, Randwick, NSW 2031, Australia, christopher.mciver@sesiahs.health.nsw.gov.au

John Merlino Department of Microbiology and Infectious Diseases, Concord Hospital, Hospital Rd, Concord, NSW 2139, Australia; Department of Microbiology and Infectious Diseases, PathWest Laboratory Medicine, Nedlands, WA 6909, Australia, john.merlino@sswahs.nsw.gov.au

Graeme R. Nimmo Department of Microbiology, Pathology Queensland Central Laboratory, Herston, QLD 4029, Australia, Graeme_nimmo@health.qld.gov.au

Michael D. Nissen Queensland Paediatric Infectious Diseases Laboratory, Sir Albert Sakzewski Virus Research Centre, Queensland Children's Medical Research Institute, Children's Health Service District, Brisbane, QLD 4029, Australia, theniss@uq.edu.au

Silvano Palladino Department of Microbiology and Infectious Diseases, PathWest Laboratory Medicine, Royal Perth Hospital, Perth, WA, 6847, Australia

Tracy M. Perris Division of Microbiology, PathWest Laboratory Medicine, QEII Medical Centre, Nedlands, WA 6909, Australia

Todd M. Pryce Department of Microbiology and Infectious Diseases, PathWest Laboratory Medicine WA, Royal Perth Hospital, Wellington Street Campus, Box X2213 GPO, Perth, WA, 6847, Australia, Todd.Pryce@health.wa.gov.au

Mala Ratnamohan Molecular Biology, Centre for Infectious Diseases and Microbiology, Institute for Clinical Pathology and Medical Research, Westmead Hospital, Westmead, NSW 2145, Australia, Mala.Ratnamohan@swahs.health.nsw.gov.au

William Rawlinson SEALS Virology Division, Prince of Wales Hospital, Randwick, NSW 2031, Australia, w.rawlinson@unsw.edu.au

Jason A. Roberts Poliovirus Reference Laboratory, Victorian Infectious Diseases Reference Laboratory, North Melbourne, VIC 3051, Australia, Jason.Roberts@mh.org.au

Jennifer M.B. Robson Sullivan Nicolaides Pathology, Cnr WhitmoreSt & Seven Oaks St, Taringa, QLD 4068, Australia, jrobson@snp.com.au

Rebecca Rockett Queensland Paediatric Infectious Diseases Laboratory, Sir Albert Sakzewski Virus Research Centre, Queensland Children's Medical Research Institute, Children's Health Service District, Brisbane, QLD 4029, Australia, r.rockett@uq.edu.au

Bronwen Roy Department of Microbiology, Nepean Hospital, Penrith, NSW 2750, Australia, royb@wahs.nsw.gov.au

Leanne Sammels Department of Microbiology, PathWest Child and Adolescent Health Service, Princes Margaret Hospital, Perth, WA 6008, Australia, Leanne.Sammels@health.wa.gov.au

Gillian M. Scott Virology Division, Department of Microbiology, SEALS, Prince of Wales Hospital and School of Biotechnology and Biomolecular Sciences, Randwick, NSW 2031, Australia; University of New South Wales, Sydney, NSW, Australia, gillian.scott@sesiahs.health.nsw.gov.au

Theo P. Sloots Queensland Paediatric Infectious Diseases Laboratory, Sir Albert Sakzewski Virus Research Centre, Queensland Children's Medical Research Institute, Children's Health Service District, Brisbane, QLD 4029, Australia, t.sloots@uq.edu.au

David W. Smith Division of Microbiology, PathWest Laboratory Medicine, QEII Medical Centre, Nedlands, WA 6909, Australia; Faculty of Life and Physical Sciences and Faculty of Medicine, University of Western Australia, Nedlands, WA 6009, Australia, David.Smith@health.wa.gov.au

Greg Smith Australian Animal Health Laboratory, CSIRO Livestock Industries, Geelong, VIC 3220, Australia, Greg.A.Smith@csiro.au

Helen V. Smith Public Health Microbiology Communicable Diseases Forensic and Scientific Services, Queensland Health, 39 Kessels Rd, Coopers Plains, QLD 4108, Australia, Helen_Smith@health.qld.gov.au

Ina Smith Australian Animal Health Laboratory, CSIRO Livestock Industries, Geelong, VIC 3220, Australia, Ina.Smith@csiro.au

Damien Stark Microbiology Department, St Vincents Hospital, Darlinghurst, NSW, Australia, dstark@stvincents.com.au

John Stenos Australian Rickettsial Reference Laboratory, Barwon Biomedical Research, The Geelong Hospital, Bellarine Street, Geelong, VIC 3220, Australia; Division of Verterinary and Biomedical Science, Murdoch University, Murdoch, WA 6150, Australia, jstenos@unimelb.edu.au, JOHNS@BarwonHealth.org.au

Sepehr N. Tabrizi Microbiology and Infectious Diseases Department, Royal Women's Hospital, Parkville, VIC 3052, Australia; Department of Obstetrics and

Gynecology, University of Melbourne, Parkville, VIC 3052, Australia, Sepehr.Tabrizi@thewomens.org.au

Bruce R. Thorley Poliovirus Reference Laboratory, Victorian Infectious Diseases Reference Laboratory, NM 3051, Australia, Bruce.Thorley@mh.org.au

Sarah J. Tozer Queensland Paediatric Infectious Diseases Laboratory, Sir Albert Sakzewski Virus Research Centre, Queensland Children's Medical Research Institute, Children's Health Service District, Brisbane, QLD 4029, Australia, s.tozer@uq.edu.au

David M. Whiley Queensland Paediatric Infectious Diseases Laboratory, Sir Albert Sakzewski Virus Research Centre, Queensland Children's Medical Research Institute, Children's Health Service District, Brisbane, QLD 4029, Australia, d.whiley@uq.edu.au

Ms Robyn Wood MASM, Consultant to the Department of Essential Health Technology, World Health Organisation, Switzerland, woodro4@gmail.com

Ren Zhang School of Biological Sciences, University of Wollongong, Gwynneville, NSW, 2500, Australia

Abbreviations

5′UTR	5 prime untranslated region
AFLP	Amplified fragment length polymorphism
AIDS	acquired immune deficiency syndrome
AMV	avian *myeloblastosis* virus
anti-HBc	antibodies to hepatitis B core antigen
anti-HBe	antibodies to hepatitis B "e" antigen
anti-HBs	antibodies to hepatitis B surface antigen
ARTG	Australian Register of Therapeutic Goods
ATCC	American type culture collection
bfpA	bundle forming pilus
BKV	human polyomavirus BK virus
BL	Burkitt's lymphoma
bp	Base pair
BSC	Biological safety cabinet
BSI	British Standards Institute
BVCC	Tryptose blood agar supplemented with vancomycin-cefixime-cefsulodin
C	C structural protein of dengue virus
CAB	Conformity Assessment Branch
CCID	Cell Culture Infective Dose
cDNA	complimentary DNA
CFU	colony forming units
CGH	Comparative genome hybridization
CHEF	Contour clamped homogenous electric fields
CLSI	Clinical and Laboratory Standards Institute
CMV	Cytomegalovirus
CNS	central nervous system
CNS	Coagulase negative Staphylococci
Cp	Crossing point
CS	Citrate synthase
CSF	Cerebrospinal fluid
Ct	Cycle threshold
DAEC	diffusely adherent Ec

DEC	Diarrheagenic Ec
DEN-1	Dengue virus type 1 serotype
DEN-2	Dengue virus type 2 serotype
DEN-3	Dengue virus type 3 serotype
DEN-4	Dengue virus type 4 serotype
DENV	Dengue virus
DEP	Diarrheagenic Ec Pathotypes
DHF	dengue hemorrhagic fever
DIG	digoxigenin
DIG-11-dUTP	digoxigenin-11-uridine 5'-triphosphate
DIG-POD	digoxigenin-11 peroxidase
DMSO	dimethyl sulphoxide
DNA	Deoxyribose nucleic acid
DNase	Deoxyribonuclease
dNTP's	Deoxynucleotide triphosphates
ds	double stranded
DTT	dithiothreitol
E	E structural protein of dengue virus
eaeA	intimin
EAEC	enteroaggregative Ec
EBV	Epstein-Barr virus
Ec	Escherichia coli
EDTA	Ethylenediaminetetraacetic acid
EHEC	enterohemorrhagic Ec
EHV	Equine herpesvirus
EIA	enzyme immunoassay
EIEC	enteroinvasive Ec
ELISA	Enzyme-linked immunosorbent assay
EPEC	enteropathogenic Ec
EQAS	External quality assessment schemes
ETEC	enterotoxigenic Ec
FDA	Food and Drug Administration
FRET	Fluorescence resonance energy transfer
G1	Genotype G1
G2	Genotype G2
G3	Genotype G3
G4	Genotype G4
G8	Genotype G8
G9	Genotype G9
gB	glycoprotein B gene
GLP	Good laboratory practice
GMO	Genetically modified organisms
GMP	Good manufacturing practice
HAV	Hepatitis A virus
HBcAg	hepatitis B core antigen

HBeAg	hepatitis B "e" antigen
HBsAg	hepatitis B surface antigen
HBV	Hepatitis B virus
HCMV	Human cytomegalovirus
HCV	Hepatitis C virus
HDV	Hepatitis D virus or delta agent
HEPA	High efficiency particulate air
HEV	Hepatitis E virus
HGT	Horizontal gene transfer (exchange)
HHV	human herpesvirus
HHV6	Human herpes virus type 6
HHV7	Human herpes virus type 7
HHV8	Human herpes virus type 8
HIV	Human immunodeficiency virus
hnRNA	Heterogeneous nuclear RNA
HPLC	High pressure liquid chromatography
HPV	Human papillomavirus
HRM	High resolution melt
HRP	horse radish peroxidase (substrate)
HSV	Herpes simplex virus
HSV	Herpes simplex virus
HSV1	Herpes simplex virus type 1
HSV2	Herpes simplex virus type 2
Ial	invasion associated locus
IgG	immunoglobulin G
IgM	immunoglobulin M
IGS	Intergenic spacer region.
IPC	Internal positive control
ITS	Internal transcribed spacer
IVD	In-vitro diagnostic device
JCV	human polyomavirus JC virus
LB	Luria-Bertani broth
LED	Light emitting diode
LGV	lymphogranuloma venereum
LiPA	Line Probe Assay
LNA	Locked nucleic acid
lt	heat labile toxin
MBL	Metallo-β-lactamases
mEC	modified EC broth
MGB	Minor groove binding
MIE	the major immediate early
MLST	Multilocus sequence typing
MLVA	Multilocus variable number tandem repeat (VNTR) analysis
MM	Master mix
MOMP	Major outer membrane protein

MOTT	Mycobacteria other than tuberculosis
MRAB	Multi-resistant *Acinetobacter baumannii*
mRNA	Messenger RNA
MRO	Multi-resistant organism
MRSA	Methicillin resistant *Staphylococcus aureus*
MRSE	Methicillin resistant *Staphylococcus epidermidis*
MSM	Men who have sex with Men
MSSA	Methicillin sensitive *Staphylococcus aureus*
MSSE	Methicillin sensitive *Staphylococcus epidermidis*
MTBC	*Mycobacterium tuberculosis* complex
MV	Measles virus
NA	Nucleic acid
NAAT	Nucleic acid amplification test
NAT	nucleic acid testing
NATA	National Association of Testing Authorities
NBSC	newborn screening cards
NDC	No DNA control
NHS	National Health and Security
NPA	nasopharyngeal aspirate
NPAAC	National Pathology Accreditation Advisory Council
NPV	negative predictive value
NRL	National Serological Reference Laboratory
NS1	NS1 nonstructural protein of dengue virus
NS2a	NS2a nonstructural protein of dengue virus
NS2b	NS2b nonstructural protein of dengue virus
NS3	NS3 nonstructural protein of dengue virus
NS4a	NS4a nonstructural protein of dengue virus
NS4b	NS4b nonstructural protein of dengue virus
NS5	NS5 nonstructural protein of dengue virus
NTC	no template control
O157:H7	EHEC serotype strain
O165:HNM	uncommon EHEC serotype
OGTR	Office of the Gene Technology Regulator
OMP	Outer membrane protein
ORF1	open reading frame 1
ORF2	open reading frame two
PBS	Phosphate buffered saline
PCR	Polymerase chain reaction
PCRS	PCR sequence
PFGE	Pulsed field gel electrophoresis
PhHV	phocine herpes virus
PHLN	Public Health Laboratory Network
PHRANA	Phylogenetic hierarchical analyses using nucleic acids
PML	progressive multifocal leukoencephalopathy
PNA	Peptide nucleic acid

PPV	positive predictive value
prM	prM structural protein of dengue virus
PVL	Panton-Valentine Leukocidin
PVPP	polyvinylpolypyrolidone
QAP	Quality assurance program
QC	Quality control
QCMD	Quality Control for Molecular Diagnostics
Q-PCR	(qPCR) Quantitative PCR
RAPD	Random amplification of polymorphic DNA
RC	relaxed circular (DNA)
RCPA	Royal College of Pathologists Australia
RFLP	Restriction fragment length polymorphism
RNA	Ribonucleic acid
RNase	Ribonuclease
rRNA	Ribosomal RNA
RSV	Respiratory syncytial virus
RT	Real time
RT-PCR	reverse transcription-PCR
SARS	Severe acute respiratory syndrome
SFG	Spotted fever group
SMAC	Sorbitol-MacConkey.
SNPs	Single nucleotide polymorphisms
SSBA	Security sensitive biological agents
ssRNA	single stranded ribonucleic acid
SST	Serum separator tubes
SSU	small-subunit
st	heat stable toxin
ST	Sequence types
STEC	shiga toxin-producing Ec
STI	Sexually transmitted infection
stx1	shiga toxin1
stx2	shiga toxin2
TAE	tris acetate edta (buffer)
TB	Tuberculosis
TBE	Tris Borate EDTA
TG	Typhus group
TGA	Therapeutic Goods Administration
Tm	Melting temperature
TMB	tetramethylbenzidine
tRNA	Transfer RNA
TTV	transfusion transmitted virus
UC	ultracentrifugation
UN	United Nations
UNG	Uracil-*N*-glycosylase
US	United States

USA	United States of America
UV	Ultraviolet
VMM	viral maintenance medium
VNTR	Variable number tandem repeat
VP	viral protein
VRE	Vancomycin-resistant enterococci
VTEC	verocytotoxin-producing Ec
VZV	Varicella-zoster virus
WHO	World Health Organization

Introduction

Gwendolyn Gilbert

In the 56 years since James Watson, Francis Crick, Maurice Wilkins, Rosalind Franklin, Maurice Gosling described the DNA double helix in five articles in *Nature* in 1953 [3] – for which three of them subsequently won the Nobel Prize (Physiology or Medicine, 1962) – molecular biology has progressed at an ever accelerating pace. Some other important milestones have been descriptions of nucleic acid hybridisation (1960) and DNA amplification using primers (1971) and the discovery of a heat stable polymerase produced by *Thermus aquaticus* (*Taq* 1976), which made multiple cycles of amplification practicable. Collectively they set the scene for the development of the polymerase chain reaction (PCR) by Kary Mullis, in 1983, for which he also won a Nobel Prize (Chemistry, 1993). In an article in Scientific American [1], Mullis described a process which, at the time, must have seemed almost magical to the general reader: "Beginning with a single molecule of the genetic material DNA, the PCR can generate 100 billion similar molecules in an afternoon. The reaction is easy to execute. It requires no more than a test tube, a few simple reagents, and a source of heat...".... "The DNA sample ... can be pure, or ... a minute part of ... complex biological mixture ... from a hospital tissue specimen, a single human hair, a drop of dried blood, ... tissues of a mummified brain or ... a 40,000-year-old woolly mammoth frozen in a glacier". PCR became the basis for the new and rapidly evolving discipline of molecular biology research, including molecular microbiology.

Molecular diagnosis in clinical microbiology has evolved at a relatively stately pace. However, for nearly 20 years it has been slowly changing the speciality from a cottage industry, which often produced results long after the patient could benefit, to one with the potential to compete with other branches of clinical pathology for short turn-around times and immediate clinical relevance. In the 1980s, it was not difficult to create a wish-list for nucleic acid amplification tests (NAATs), once their potential was recognised. Herpes simplex encephalitis, for which treatment had just become available, required a brain biopsy for diagnosis; diagnosis of tuberculous meningitis required prolonged CSF culture and, often, seriously delayed or inappropriate therapy. Suspected intrauterine infection with *Toxoplasma gondii* was diagnosed, if at all and only after several weeks, by mouse inoculation of amniotic fluid; diagnosis of genital chlamydial infection in men was rarely attempted because it required deep and painful urethral scrapings to harvest epithelial cells, culture was not widely available and antigen tests were insensitive and prone to false positive results. Increasing numbers of fastidious or non-culturable respiratory and enteric pathogens were being identified, without suitable diagnostic tests being available.

Initially, most diagnostic NAATs were developed in larger laboratories, using a variety of different reagents, equipment, test methods, quality controls and evaluation procedures – and therefore widely variable quality and clinical accuracy. Inter-laboratory quality assurance studies often showed alarming variations in sensitivity and specificity of assays in different laboratories. What to use as a gold standard for comparison was a source of debate and controversy, because of the ever-present risk of nucleic acid contamination, the presence of inhibitors in specimens and the fact that the NAAT was often more sensitive than conventional methods, leading to false positive or negative (or wrongly interpreted true positive) results. Clinical criteria for interpretation were difficult to apply, even if enough cases were available since the use of NAAT often uncovered previously unrecognised asymptomatic acute or latent infections. A positive or negative NAAT result does not necessarily mean the same as a positive or negative culture result. This was a source of confusion and, often, either scepticism or naïve trust, among clinicians.

Slowly, the introduction of quality assurance programs for molecular diagnostics revealed and improved our understanding of these discrepancies and, in the process, led to improvement in standards and consistency, among the limited numbers of laboratories using in-house tests. At the same time, commercial kits started to became available, which provided much wider access for all laboratories. However, they also were not without problems – some had had very limited clinical evaluation and there were limited data about their predictive values; most were expensive, which led to the temptation to limit the use of controls; some were withdrawn from sale without explanation and sometimes, they were not performed or interpreted by the users according to the manufacturer's instructions. It became apparent that different skills and knowledge are required, from those needed for conventional microbiology, which sometimes take time to acquire.

Over the past 10 years or so there has been an enormous improvement in the quality of commercially available molecular microbiology diagnostic assays, but their range is still limited and quality variable. Some have been so widely accepted that there is little need for laboratories to use in-house tests. For example, one of the most widely used has revolutionised the diagnosis of chlamydial infection, which, unlike culture or antigen tests, is sensitive enough to detect *C. trachomatis* DNA in voided urine – a boon to patients and clinicians alike. Development of quantitative molecular assays for HIV has allowed rational prescribing of antiretroviral agents and early detection of resistance, thus contributing to the survival of people living with HIV infection. The availability of treatment for hepatitis C infection makes it essential that antibody tests are supplemented with testing for the presence of HCV RNA, to confirm the presence of persisting, rather than past, infection.

There is still a continuing demand for in-house assays, even for some high volume tests. Recognition of the limitations of serology for the diagnosis of pertussis led to increasing use of PCR and an almost complete reversal in proportions of diagnostic methods used for laboratory notifications – from predominantly serology (high level IgA), especially in older children and adults, to predominantly PCR in all age-groups, during the recent widespread pertussis epidemic in New South Wales. This not only improved the timeliness and accuracy of clinical diagnosis but increased the confidence of public health officials in notification data.

Similarly, the need for high volume molecular testing for influenza and other respiratory viruses, has led to development of in-house multiplex assays, which are flexible enough to be rapidly modified as new viruses are identified and new pandemic strains of influenza A emerge. This on-going development and real-time evaluation is not for the faint-hearted and, to be clinically, epidemiologically and medico-legally acceptable, must be done by scientists with appropriate skills using accepted quality standards.

The publication of "Laboratory Accreditation Standards and Guidelines for Nucleic Acid Detection Techniques 2006" by the National Pathology Accreditation Advisory Committee, was a major step towards improving the quality of molecular diagnostics in Australia. It has provided the basis for laboratory accreditation of laboratories developing and using NAATs. As the clinical need and demand for more widespread application of molecular microbiological diagnostics continues, there will be an on-going need for education of microbiology laboratory scientists, clinical microbiologists, clinicians and public health physicians about the methods, limitations and interpretation of microbiological NAATs [2]. Whether laboratories are using commercial kits or in-house methods developed in their own laboratories or adopted from published methods or those of other laboratories, all clinical microbiology laboratories need to be able to understand, critically evaluate, perform and interpret these tests according to rigorous and clinically appropriate standards. The cost and effort of development and evaluation of in-house tests is considerable and many laboratories do not have the resources to do so. If we are to maintain and improve the clinical relevance and high quality of diagnostic microbiology there will be an increasing need for sharing of expertise, methods and evaluation results. This compendium is an important vehicle for such sharing and I warmly congratulate those who have worked so hard to produce it and ASM for making its publication possible.

References

1. Mullis K (1990) The unusual origin of the polymerase chain reaction. Sci Am 262 (4): 56–61, 64–65
2. National Pathology Accreditation Advisory Committee. Laboratory Accreditation Standards and Guidelines for Nucleic Acid Detection and Analysis, 2009. Department of Health and Aging. Available at http://www.health.gov.au/internet/main/publishing.nsf/Content/health-npaac-docs-nad.htm (accessed December 2009)
3. Nature Archives Double Helix of DNA: 50 Years. http://www.nature.com/nature/dna50/archive.html

Part I
PCR Fundamentals

Chapter 1
PCR Basics

G. James

Abstract The polymerase chain reaction (PCR) is used as a routine tool in diagnostic microbiology laboratories to detect and/or characterize disease causing organisms from patient specimens. The implementation of PCR testing must be planned by experienced scientists, with appropriate laboratory areas designated and good molecular biology laboratory practices diligently adhered. The inclusion of controls to alert to the possibility of false negative and false positive results and to monitor the performance of the PCR assay provides reassurance as to the quality and accuracy of results.

Keywords PCR basics · Workflow · Nucleic acid extraction · Storage · Conventional PCR · Real time PCR · Quantitative PCR · Multiplex PCR · Molecular sub-typing · Quality control · Accreditation

1.1 Introduction

The polymerase chain reaction (PCR) has become a routine tool in diagnostic microbiology laboratories to detect and/or characterize disease causing organisms from patient specimens. PCR assay formats have evolved from the simple conventional assays to Real-Time (RT) formats. In addition, advances in quality and performance of reagents, understanding of the dynamics of PCR and evolution of complementary products and methods such as fluorescent probes and DNA sequencing have provided increased scope for multiplexing different nucleic acid (NA) targets in a single PCR tube to provide rapid diagnosis of disease states and characterization of organism subtype and potential antimicrobial susceptibility.

G. James (✉)
Centre for Infectious Diseases and Microbiology Laboratory Services (CIDMLS), Institute for Clinical Pathology and Medical Research (ICPMR), Westmead, NSW 2145, Australia
e-mail: Greg.James@swahs.health.nsw.gov.au

M. Schuller et al. (eds.), *PCR for Clinical Microbiology*,
DOI 10.1007/978-90-481-9039-3_1,

Conventional PCR principally uses tube formats and detection of amplified DNA product is most often visualized following separation of DNA segments within an agarose or polyacrylamide gel under the influence of an electric current (electrophoresis). Addition of a dye such as; ethidium bromide or Sybr Safe, that binds to DNA, either during electrophoresis or in a bath following electrophoresis, and excitation of the bound dye by irradiation with UV light will show the length of the PCR product DNA. The identity of the PCR product is confirmed by comparison against a size ladder of DNA fragments of known length. A photograph can be taken for analysis of product and storage of the result.

Conventional PCR assays typically take 4–5 h from PCR setup to analysis of product following electrophoresis. This method is simple and relatively cost effective and has been widely used by laboratories since the earliest examples of published PCR assays. It is also used during the development of RT-PCR assays for initial confirmation of product amplification and when trouble-shooting. However as the amplified DNA is taken from the PCR tube and placed into a gel for electrophoresis the PCR product can be easily dispersed during manipulations of the gel post electrophoresis and particularly during cleanup of the apparatus as PCR product DNA will be present in large number in the electrophoresis buffer. DNA fragments are very easily aerosolized and dispersed on air currents and on the clothes and skin of laboratory staff and can contaminate subsequent PCR assays resulting in the production of False Positive results.

In addition, DNA segments less than 300 bp measure under 0.1 μm and pass through HEPA filters (0.3 μm pore size) in biological safety cabinets (BSC) and are not eliminated by UV light or many disinfectant solutions. If contamination is extensive, development of another PCR assay with a different target NA sequence may be required.

Conventional PCR reagent master mix (MM) may use a fluorescent probe such as a TaqMan probe (described in more detail in Chapter 2) that enables the assay to be completed in a closed tube format greatly reducing the potential for PCR product contamination events. Assays detecting product by either gel electrophoresis or fluorescent probe systems provide a qualitative result.

RT-PCR combines PCR DNA product amplification and detection in one closed tube within the thermal cycler. PCR formed DNA segments are detected by bound fluorescent dyes such as SYBR Safe or through use of fluorescence generating probes and are said to be detected in 'Real-Time' as they are produced in each PCR cycle. RT-PCR assays can be designed to provide a quantitative result important for establishing disease state and treatment efficacy in infectious diseases such as human immunodeficiency virus (HIV) and hepatitis C.

The size of DNA segments formed in RT-PCR are usually designed to be not more than 100 bp and as the PCR tubes used have excellent thermal conductivity, assays that can detect a single organism can be completed in less than an hour and often in around 40 min. PCR product is retained within the reaction tube and so cannot contaminate subsequent PCR assay samples and the potential for false positive results is greatly reduced but not eliminated as any tube breakage following amplification could result in subsequent contamination events.

Laboratories wishing to employ PCR assays for patient diagnosis must invest in; staff experienced in molecular biology and particularly PCR development and evaluation, equipment, procedures and processes required to maintain good molecular biology laboratory practices and in space needed to allow physical separation of crucial steps in the process. Each of these elements must be in place prior to the commencement of PCR development work so that the potential for contamination of the work environment by PCR amplified DNA is minimized. Further, laboratory management and staff must be knowledgeable and have achieved requirements described in standards and guidelines for the use of NA tests to enable testing of patient specimens and issuing of results. In Australia, the National Pathology Accreditation Advisory Council (NPAAC) has published, 'Laboratory Accreditation Standards and Guidelines for Nucleic Acid Detection and Analysis, 2006' [2] and 'Requirement for the Development and Use of In-House In Vitro Diagnostic Devices, 2007' [3] that form the basis for laboratory accreditation under the *Health Insurance Act 1973* and assessed by National Association of Testing Authorities (NATA). The Therapeutic Goods Administration (TGA) accredits laboratories performing tests that are classified as Class 4. These tests are; used to screen the blood supply, detect transmissible agents that cause serious disease with a risk of propagation in the Australian population (influenza A H5N1) or are used in response to a potential act of bioterrorism.

1.2 Facility Layout, Airflow and Workflow

The ability of PCR assays to detect even a single copy of a target sequence provides laboratories with an exquisitely sensitive tool for the diagnosis of infectious disease, however this extreme sensitivity also presents PCR Achilles heel in that any contamination from other highly positive specimens or previously produced PCR amplified DNA will produce a false positive result and potential change in patient management and patient outcome.

The relatively few studies of the rate of specimen carry over contamination during processing and inoculation of cultures in microbiology laboratories have mostly been conducted by laboratories cultivating tuberculosis as incidents can be detected by epidemiologic information and molecular subtyping. The rate varies widely but in a review by Burman and Reves [1] was found on average to be 3.1%. Factors that contribute to carry over contamination include; mislabeling of specimens, aerosols produced during processing, splashes, loop and pipette transfers, safety cabinet airflows, contaminated solutions and equipment and the number of positive specimens processed within the laboratory. All of these factors also occur during the processing of specimens for PCR. Further, microorganisms can be cultivated on plates to greater than 10^6 organisms per colony or 10^{10} organisms per milliliter of broth. Cultivated organisms also provide a source for carry over contamination during specimen processing for PCR if specimen processing, cultivation and examination of cultures are performed in the same area.

With current methodology PCR assays must produce approximately 10^9–10^{10} copies of a 100 bp target DNA segment for that segment to be detected by the RT-PCR instrument or visualized in the gel. When a single target is present in the sample this is theoretically reached after 30 cycles of RT-PCR and 34 cycles of conventional PCR.

The number of DNA segments produced in a typical PCR assay of 50 cycles starting with a single copy of the target and 100% amplification efficiency is 1.1×10^{15}. If the PCR target is 100 bp long and each segment was placed end to end it would stretch 37.2 km. This number of DNA segments is not produced in a PCR due to saturation of the assay with product and exhaustion of reagents. Although each positive PCR assay tube will contain between 10^9 and 10^{12} copies of the target DNA segment and if dispersed into the laboratory could result in on-going contamination events. Staff performing activities such as cultivation of PCR target microorganisms and PCR amplified DNA analysis provides a vehicle for transport of DNA to contaminate future assays.

Due to the potential for carry over and PCR amplified product contamination, laboratories performing PCR need to have in place good molecular biology practices and appropriate laboratory layout, work flow and air flow to minimize the potential for contamination and generation of false positive results.

Laboratories performing PCR require at least three physically separate areas with known air flows to reduce the risk of contamination. These areas can be separated by; walls, distance, strict laboratory practice and by performing tests within the working space of an instrument such as an automated specimen processing instrument or liquid handling robot. Laboratories performing nested PCR require a fourth laboratory for the transfer of PCR amplified product DNA from the first round of PCR into the second round MM [2].

Laboratory areas can be divided into three classes;

1. Super clean area used for the preparation of reagents and PCR assay MM. This area is not used for specimen processing or cultivation of microorganisms and is ideally distant to laboratories performing those tasks and PCR.
2. Clean area where dedicated PCR specimens are processed to liberate NA and PCR assays are set up. Processing of specimens for culture or the cultivation of microorganisms must not be performed in this laboratory.
3. Dirty areas are defined as areas where microorganisms are cultivated or where PCR product is produced and analyzed.

Figure 1.1 depicts an ideal laboratory layout and air flow. A one-way workflow from clean pre-PCR areas to dirty post-PCR areas, compartmentalized equipment and consumables and good molecular biology practices are essential for minimizing the potential for assay contamination. Using Fig. 1.1 as an example, PCR work begins in the super clean Reagent Preparation room that is distant to all other laboratories performing PCR work and also to laboratories cultivating microorganisms. PCR MM and other reagents are prepared in this room. A clean gown and gloves must be worn and all equipment and consumables must have arrived in this room

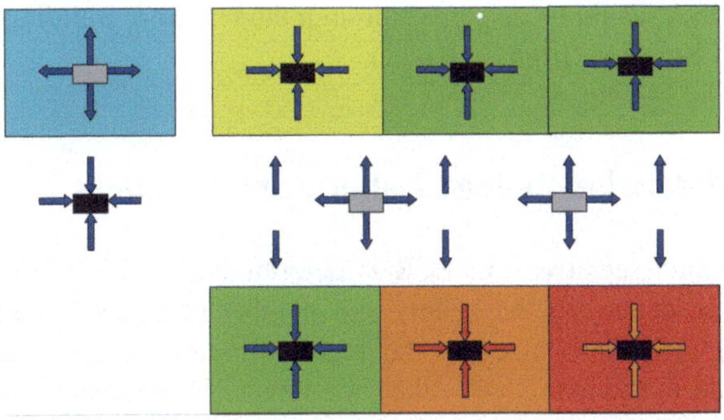

Fig. 1.1 Ideal laboratory layout and airflow. The reagent preparation room is coloured *blue* and is distant to other laboratories involved in the PCR process and cultivation of microorganisms. This room ideally has positive pressure and air flows out of this room from the air vent (coloured *grey*) to the exhaust vent in the corridor depicted as *black*. NA extraction and sample addition is performed in the *yellow* laboratory that has normal airflow from the corridor to the exhaust vent in the laboratory. This laboratory is separated by distance and airflow to the laboratory in which PCR amplification is achieved and product analyzed (depicted as *red*). This laboratory ideally has a negative pressure in relation to surrounding laboratories and is most likely contaminated with PCR product DNA. Laboratories performing nested PCR require a fourth laboratory, depicted as orange, for the transfer of amplified PCR DNA product from the first round of PCR to the second round MM. This room is also likely to be contaminated with PCR product DNA

new or not used in any other laboratory cultivating microorganisms or involved in PCR. Reagents and accompanying racks travel from this laboratory to the NA Extraction and PCR set-up laboratory (clean room) for sample preparation and addition of NA to PCR MM. Equipment and consumables used in this room must have arrived new or come from the Reagent Preparation room. They must not travel back to the Reagent Preparation room. Fresh gloves must be used in this laboratory. Sample loaded PCR tubes are then transported to the PCR Amplification laboratory (dirty area) for placement in thermal cyclers and production of PCR amplified DNA. Amplified product is also analyzed in this laboratory during RT-PCR or by gel electrophoresis or fluorescence detection following completed thermal cycling. Following analysis PCR tubes or plates are disposed into bins for high temperature incineration or other disposal method. Care must be taken to ensure PCR product is contained in air tight containers or bags prior to removal from the laboratory. PCR tube racks that must return against the workflow must be decontaminated with sodium hypochlorite prior to washing and sterilization by autoclave [2]. Equipment and consumables used in the PCR Amplification laboratory must not travel to the NA Extraction and PCR set-up laboratory or to the Reagent Preparation room.

The ideal airflow is to have a positive or outflow of air from the super clean reagent preparation room and inward flowing air to laboratories used for specimen processing and PCR set up. Laboratories used for PCR amplification of target NA and analysis should have inward flowing air ideally at negative pressure in regard to surrounding rooms. Conditioned air must be single pass and not recycled as

PCR formed DNA would be dispersed through the building. Building air inlets and exhaust should be distant from each other.

1.3 Good Molecular Biology Laboratory Practices

The efficiency and sensitivity of PCR is determined by the purity, activity and robustness of reagents and the quality and number of target NA sequence. The specificity of PCR is determined by the degree of homology between the primers and probes used in the assay with other non-target NA sequences and in the proportion of assay samples that did not contain the target NA sequence but had been contaminated with target NA from other samples or from NA produced in previous PCR assays.

Practices that optimize sensitivity include;

1. Use of molecular biology grade reagents and consumables including; water, chemicals, enzymes and plastic consumables that have been certified not to contain DNase or RNase (if performing RNA amplification).
2. Collect separate dedicated specimens for PCR in an appropriate manner and transported in a condition and time period that has been proven to maintain the viability of target NA.
3. Process the specimen to a required level of purity determined by test validation. For some specimens and assay configurations processing may not be necessary other than the initial denaturation step of the PCR. For other specimens and assays multiple processing steps may be required to obtain NA in a form sufficiently pure to ensure consistent results.
4. Store samples under conditions that maximize sample integrity and longetivity. Archive purified DNA or RNA samples at –70°C for long-term storage.
5. Ensure all equipment is calibrated and operating correctly.
6. Always use the correct pipette for the volume to be dispensed and avoid trying to pipette volumes under 2 μl as dispensing errors increase with smaller volumes.
7. Optimize PCR assay reagent concentrations and thermal cycling conditions to make a robust assay that meets clinical requirements and is validated according to NPAAC Standard [3].
8. Use assay controls to monitor performance such as;

    a. Internal control that can be used to monitor the efficiency of sample processing and PCR amplification.
    b. Detection limit positive control dilution series that will demonstrate assay reproducibility.
    c. Inhibition control to detect samples that contain substances that prevent PCR amplification of target NA.

9. Pulse centrifuge PCR tubes prior to loading into the PCR instrument to ensure all reagents are mixed and in the bottom of the tube for optimum thermal cycling.

Practices that optimize specificity include;

1. Use published primers and probes from studies with the same or similar patient demographics. Validate the assay in accordance with diagnostic criteria [3].
2. When choosing gene sequence regions for primers and probes use a recognized program for design and ensure the target is unique for the agent and that all strains will be detected. The latter is accomplished by extensive sequence search of microorganisms and human DNA databases.
3. Include no DNA controls (NDC's) that contain all of the components of the PCR except template DNA at a frequency that can be used to determine possible contamination events and rates at both specimen processing and PCR set-up. Add specimen processing NDC's to PCR MM tubes prior to sample addition at PCR set-up to determine if contamination is occurring at processing. Add PCR set-up NDC's after each sample or sample set to determine if contamination is occurring at PCR set-up. If there is concern of contamination increase the number of NDC's to determine likely source.
4. Carryover contamination is aided by air flow. Specimens must be processed within a BSC to protect the operator from infectious disease agents. Following inactivation of agents at processing, transfers of NA to PCR tubes should be performed in PCR workstations with static air to limit potential for carryover. PCR product analysis should be performed in a static air flow and definitely not within a recirculation BSC as PCR amplified DNA segments pass easily through HEPA filters and the product will be dispersed throughout the laboratory.
5. Use only molecular biology grade plugged filter pipette tips or positive displacement pipettes with disposable tips for all liquid transfers to eliminate carryover.
6. All equipment should remain in the designated laboratory and must be cleaned and decontaminated prior to removal for maintenance. Pipettes should be bagged and returned to the designated laboratory.
7. Consumables must not return to clean laboratories against the workflow.
8. PCR tube racks that must return against the workflow are soaked in 2–10% sodium hypochlorite for 4 h to destroy any contaminating DNA prior to washing and sterilization by autoclave [2].
9. Pulse centrifuge all tubes before opening to pull liquid away from the lid and reduces the possibility of contaminating pipettes and gloves.
10. Put on a new pair of gloves when starting work in a designated laboratory and change gloves regularly when working.
11. Change your laboratory gown in each designated area and leave it there. Never return a gown from a dirty laboratory to a clean laboratory against the workflow.
12. Wash hands following removal of laboratory gown and prior to leaving each designated laboratory.
13. Store reagents in small aliquots to limit extent of a contamination incident. Always use new plastic ware during preparation and discard remnants of aliquots not totally used in the procedure.

14. Add all components of the PCR into a MM and aliquot into PCR tubes before addition of sample NA. PCR MM must be prepared in the Reagent Preparation room.
15. Take care during all pipette liquid transfers of samples to limit the potential for aerosol by avoiding bubble formation.
16. Keep tubes closed that are not in immediate use.
17. Add positive controls in amounts that will enable detection at the end of assay set-up following the addition of all samples.
18. Clean cabinets and laboratory surfaces regularly with 2–10% sodium hypochlorite to destroy NA. This is particularly important in laboratories analyzing PCR product.
19. Pipette tips used in the PCR process should be discarded in containers containing 1% sodium hypochlorite to disinfect any microorganisms and destroy remaining NA.
20. PCR tubes containing amplified DNA must remain capped and placed into sealed bags for disposal.
21. Great care must be taken for disposal of electrophoresis gels and buffer as they contain large amounts of amplified target DNA. Avoid creation of aerosols and bag gels for disposal. Absorb DNA stain from buffer with activated charcoal (if required) and carefully flush buffer down the sink.

PCR assays provide scientists with a sensitive and specific tool for diagnosis of infectious disease. Carryover contamination events occur in all microbiology laboratories and produce false positive results in neighbouring samples. With culture systems these events are usually limited in number of affected specimens and time frame. Laboratories performing PCR are also subject to this source of error. However, amplified DNA formed in PCR assays provides another source of contamination for subsequent assays that can be insidious, irregular, persist for long periods, provide confusion to patient diagnosis and reduce confidence in assay results. The implementation of PCR testing must be planned by experienced scientists, with appropriate laboratory areas designated and good molecular biology laboratory practices diligently adhered. The inclusion of controls to alert to the possibility of false negative and false positive results and to monitor the performance of the PCR assay will provide reassurance as to the quality and accuracy of results.

References

1. Burman WJ, Reves RR (2000) Review of false positive cultures for *Mycobacterium tuberculosis* and recommendations for avoiding unnecessary treatment. Clin Infect Dis 31:1390–1395
2. National Pathology Accreditation Advisory Council (2006) Laboratory Accreditation Standards and Guidelines for Nucleic Acid Detection and Analysis
3. National Pathology Accreditation Advisory Council (2007) Requirements for the Development and Use of In-House In Vitro Diagnostic Devices (IVDs)

Chapter 2
PCR Methodology

Ian Carter, Catriona Halliday, Theo P. Sloots, Todd M. Pryce, Ian D. Kay, Gerald B. Harnett, Glenys R. Chidlow, and Philip M. Giffard

Abstract PCR methodologies have become firmly entrenched in many clinical laboratories for the detection of a wide range of organisms, because they offer major advantages of improved sensitivity and rapidity over traditional diagnostic methods. However, many variables need to be considered in performing a reliable PCR assay, ranging from nucleic acid extraction, storage, composition of the PCR reaction mix used, to the dynamics of the amplification reaction. To control for these variables, there is an obvious need for standardised reagents and quality assurance programmes to obtain reproducible and clinically significant results. The diagnostic potential of the PCR technology has been greatly enhanced with the development of multiplex, real-time, and quantitative PCR methods, and these are now routinely performed in many diagnostic laboratories. More recently, PCR has been applied to bacterial typing, and a reasonable prediction is that in the near future, bacterial typing will be performed by either some variant of next-generation sequencing, or by HRM analysis of selected markers, depending on the amount of information required.

Keywords Extraction · Storage · DNA · RNA · Conventional PCR · Components of PCR · Amplification · Detection · Limitations · Real-time PCR · Probes · Instrumentation · Quantitative PCR · Internal controls · Multiplex PCR · Applications · Molecular typing

2.1 Nucleic Acid Extraction, Purification and Storage

Ian Carter

The application of PCR and other methods in molecular biology require the extraction of nucleic acid from biological samples and a number of approaches have been

---

I. Carter (✉)
SEALS Microbiology Division, Prince of Wales Hospital, Randwick, NSW 2031, Australia
e-mail: ian.carter@sesiahs.health.nsw.gov.au

M. Schuller et al. (eds.), *PCR for Clinical Microbiology*,
DOI 10.1007/978-90-481-9039-3_2,

devised for performing this extraction. Nucleic acids generally do not occur as free molecules but rather in bacteria, cells, virus particles, fungi, protozoa etc. as they are covered with cell membranes and walls which are composed of proteins, lipids and sugars. Nucleic acids themselves form complexes with histone and other proteins and to extract nucleic acids which are present in this manner, the cell membranes and walls covering them must be disrupted and the proteins of the complexes mentioned above denatured or degraded to thereby become soluble, so that the nucleic acids are freed and then extracted.

In the case of isolating a nucleic acid, the nucleic acid-protein complex needs to be denatured or degraded to free the desired nucleic acid from the complex so that it can be solubilised and extracted.

In recent years, a number of approaches have been developed for rapid extraction of nucleic acids from various materials, including blood, blood, serum, faeces, urine, tissue (including paraffin-embedded), cell cultures, plasmid DNA from bacterial lysates and genomic DNA and total RNA from blood, animal and human or plant tissues and cell cultures.

The methods not only provide an ease and convenience of processing, but they allow the processing of a high volume of samples [70]. Many websites exist to assist the scientist – one excellent website combining many links to protocols can be found at http://www.molecularstation.com/.

Nucleic acids have been conventionally extracted by one or a combination of the following methods:

- Extraction with phenol and phenol/chloroform mixtures for purification of DNA and RNA. Proteins and restriction enzymes are removed by phenol and chloroform in disrupting protein secondary structure causing proteins to denature and precipitate from solution. Although each of these solvents is capable of performing this function alone, the two materials together remove proteins from solution much more effectively. Nucleic acids are recovered in the liquid phase.
- Nucleic acids are released by means of strongly denaturing and reducing agents, including hydrolytic enzymes (e.g. protease, lysozyme, lyticase), from cells and tissues and subsequently extracted and purified with a mixture of chloroform and phenol. The nucleic acids are finally obtained from the aqueous phase by ethanol precipitation or narrowing down by means of dialysis [63] or capturing via silicon resins or magnetic particles.
- Proteinase K[1]/phenol method, in which a proteolytic enzyme such as proteinase K or a surfactant is added to disrupt the cell membrane or wall and the protein of

[1] Proteinase K is a commonly used protease which is commercially available as a lyophilized powder or in aqueous solutions or suspension (Sigma-Aldrich, St. Louis, MO). The typical activity of proteinase K preparations is about 30 Units/mg. The concentrations of proteinase K in the extraction composition is preferably at least about 25 units/ml or greater, at least about 50 units/ml or greater or at least about 100 units/ml or greater. Maximal concentrations are effected by solubility of the enzyme as well as the ability to sufficiently denature the enzyme at the end of extraction so

a complex of interest is degraded to free nucleic acids; then phenol/chloroform are added and the mixture is centrifuged to have the nucleic acids transferred into the aqueous phase; the aqueous phase is recovered by separation and ethanol, isopropanol or the like is added to the recovered aqueous phase, thereby rendering the nucleic acids insoluble [63].
- The AGPC method, in which a liquid mixture of guanidinium isothiocyanate and phenol is added to a sample of interest to disrupt the cell membrane and wall, so that the protein of the complex is denatured to become soluble; nucleic acids are then freed and chloroform is added to transfer the nucleic acids to the aqueous phase; the aqueous phase is recovered by separation and thereafter, ethanol, isopropanol or the like is added to the recovered aqueous phase, thereby rendering the nucleic acids insoluble [15]. This method often uses a proprietary formulation of this reagent called Trizol (TRIzol is a chemical solution used in RNA/DNA/protein extraction and is the brand name of the product from *Invitrogen*).
- The guanidinium method, in which guanidinium hydrochloride or guanidinium thiocyanate is added to a sample of interest to disrupt the cell membrane and wall, so that the protein of the complex is denatured to become soluble and to remove dissolved impurities; nucleic acids are then freed and ethanol or isopropanol is added to render the free nucleic acids insoluble. Elution of any bound nucleic acids from the supporting material is with water or low-salt buffers such as 10 mmol/l Tris or TE (10 mmol/l Tris, 1 mmol/l EDTA). An advantage of this method is that chaotropic salts ensure the irreversible denaturing and thus inactivation of nucleases. Essential disadvantages of the method are that the concentration of the chaotropic salts to some extent have to be strongly adjusted to the material used, and also the lysis of biological material, such as fungal or plant tissue, is sometimes only very inefficient (these may require initial protease inclusion to dissolve these thicker cell walls). If enzymes (proteinase, RNase) are used in the purification methods, the concentration of chaotropic agents must be reduced below values which otherwise bring about inactivation of nucleases [15, 63].
- The sodium iodide method, in which sodium iodide containing glycogen which has affinity for the nucleic acid to be extracted is added to a sample of interest, whereby the cell membrane and wall are disrupted and the protein of the complex is denatured, so that it becomes soluble; nucleic acids are then freed and isopropanol is added to render the free nucleic acids and glycogen insoluble [32].

Nucleic-acid-containing solutions can also be obtained by incubation of nucleic-acid-containing materials with lysis buffers, which contain either (i) chaotropic salts such as guadinine salts, (ii) alkaline compounds such as NaOH, (iii) neutral, anionic or cationic detergents such as sodium dodecyl sulphate (SDS), Triton

---

that there will be no decrease in DNA polymerase activity during the subsequent PCR amplification of the extracted nucleic acid.

X-100, TWEEN-20 or hexadecyl trimethyl ammonium bromide CTAB or (iv) enzymes such as proteinase K or lysozyme in bacteria, lyticase in yeasts, chitinase in fungi, or proteases in tissues. All approaches are to lyse cells and release the nucleic acid along and may be followed with extraction using phenol and/or chloroform [63].

These methods were superseded by the more rapid and easier method devised by Boom et al. [10]. The method is based on the lysing and nuclease-inactivating properties of the chaotropic agent guanidinium thiocyanate together with the nucleic acid-binding properties of silica particles or diatoms in the presence of this agent. By using size-fractionated silica particles, nucleic acids (covalently closed circular, relaxed circular and linear double-stranded DNA; single-stranded DNA; and rRNA) could be purified from 12 different specimens in less than 1 h and were recovered in the initial reaction vessel.

These "classical" methods are especially time-consuming (sometimes taking up to 48 h), require a considerable amount of equipment, and relatively large quantities of biological material and, in addition, involve a considerable health risk (amongst other things due to the use of chloroform and phenol).

The newer methods are commercially marketed in the form of easy to use extraction kits (some in broad total nucleic acid extraction format) see Table 2.1, and are based on the principle that nucleic acids bind to mineral supports in the presence of high ionic strength, especially chaotropic salts. Finely ground glass powder (e.g. Promega, MoBio), diatomaceous earth (Sigma), silica gels (Qiagen) or chemically modified materials such as silica carbide can also be used and have also proved successful as supporting materials. Ambion, Applied Biosystems, Epicentre Technologies, Invitrogen, MO BIO Laboratories, Inc., QIAGEN, Roche Diagnostics and Sigma Aldrich, are some of the larger Companies with manual extraction kits and methods and there are many other smaller companies vying for business. However many of the smaller companies are being bought out by the larger

Table 2.1 "Larger" companies with manual DNA and RNA extraction kits

Ambion	http://www.ambion.com/
Bioline	http://www.bioline.com
Epicentre Biotechnologies	http://www.epibio.com/main.asp
Invitec	http://www.invitek.eu/
Invitrogen	http://www.invitrogen.com/site/us/en/home/Products-and-Services/Applications/Nucleic-Acid-Purification-and-Analysis.html
Machery Nagel	http://www.mn-net.com/tabid/1293/Default.aspx
MO BIO Laboratories Inc	http://www.mobio.com/nucleic-acid-purification/
Promega	http://www.promega.com/applications/dna_rna/gdna.htm
Roche Highpure Kit	https://www.roche-applied-science.com/sis/napure/index.jsp
Qiagen range of kits	http://www1.qiagen.com/Products/GenomicDnaStabilizationPurification/ClinicalSamples.aspx
Sigma Aldrich	http://www.sigmaaldrich.com/life-science/molecular-biology/dna-and-rna-purification.html

companies and this will most likely continue over the next few years. Additionally many methods have been the subject of intellectual property and patent clams and this has also limited further development with some of the methodology.

Some of these commercial kits can be used on different robotic extraction platforms – which make these robotic systems more attractive as more rapid, better and cheaper kits become available. Refer to Table 2.2 for major systems available in Australia. Automated systems allow you to run from 1 to 96 samples at once with minimal hands on time. DNA and/or RNA can be extracted in time frames from around 10 min up to 2 h. Some of the larger systems are also an "Open System" which allows tailoring of ones own methods and/or commercial kits on the machine. Other systems are closed with no change in a specific procedure possible. The major suppliers are Roche, Qiagen, Abbott, Biomerieux, Thermo Scientific, Promega, Invitrogen, Ambion and Epicentre Biotechnologies with continual advances from most – what exists today may change tomorrow which makes specifics a little more challenging! However an automated system is not for every laboratory as they are not cheap to purchase and depending on the target and PCR methods employed, such "quality" nucleic acid may not even be necessary.

Methods based on silica do not involve the use of chaotropic salts. An advantage of such a commercially available method is that nucleic acids can be extracted even from materials with very small nucleic acid content with a universal protocol. A disadvantage is that the nucleic acid preparations may not meet high quality standards (including photometric measurement-absorption ratios of less than 1.70 at 260 nm to 280 nm) and this may increase the potential of inhibition within the PCR reaction. It is possible that PCR inhibitors (heme protein, anionic surfactants, cationic surfactants, non-ionic surfactants and zwitterionic surfactants) can sometimes interfere with subsequent PCR amplification. These inhibitors may therefore be present in the extraction solution or obtained as part of the extraction process from the biological sample or from some other source. The presence of PCR inhibitors in the extraction solution would result in little or no amplification of nucleic acids and this would be deemed to constitute absence of effective extraction from the sample. That is where the inclusion in the PCR of some form of internal control is of importance and why standards push for their inclusion.

Therefore the chosen method by which high-purity total DNA and total RNA (including tRNA, mRNA, rRNA, mitochondrial RNA and hnRNA [Heterogeneous nuclear RNA – a variety of RNAs found in the nucleus, including primary transcripts]) can be prepared, which is universally applicable with regard to the nucleic-acid-containing source material and is quick and simple to handle, will require evaluation to optimise methods for each application of extraction and PCR combination for the specific specimen type and method chosen.

It is therefore essential to start with what your target for PCR actually is. This is reflected in the basic premise that the PCR itself can only amplify DNA via the action of DNA polymerase enzymes. Many targets may be some form of the less stable RNA, be it messenger, transfer or viral RNA and it is this more labile feature of RNA that makes extraction, purification and handling of RNA a more stringent procedure. Ideally the process used should be quick, simple, reproducible and this has

Table 2.2 Automated systems for nucleic acid extraction

Company	Unit	No of samples	Company website
Abbott Molecular	*m*24sp	1–24	http://international.abbottmolecular.com/m2000_1602.aspx
	*m*2000sp	96	http://international.abbottmolecular.com/m2000_1602.aspx
Applied Biosystems	MagMAX™ Express-96	96	https://products.appliedbiosystems.com/ab/en/US/adirect/ab?cmd=catNavigate2&catID=605139&tab=DetailInfo
	ABI PRISM™ 6700	96	https://products.appliedbiosystems.com/ab/en/US/adirect/ab?cmd=catNavigate2&catID=600647&tab=DetailInfo
Biomerieux	NucliSENS easyMAG	24	http://www.biomerieux-usa.com/servlet/srt/bio/usa/dynPage?doc=USA_PRD_LST_G_PRD_USA_12
	NucliSENS miniMAG	12	http://www.biomerieux-usa.com/servlet/srt/bio/usa/dynPage?open=USA_PRD_LST&doc=USA_PRD_LST_G_PRD_USA_14&lang=en
Invitrogen	Iprep	12	http://www.invitrogen.com/site/us/en/home/Products-and-Services/Applications/Nucleic-Acid-Purification-and-Analysis/Automated-Nucleic-Acid-Purification/iPrep-Purification-Instrument.html
Promega	Maxwel l 16	16	http://www.promega.com/maxwell16/default.htm
Qiagen	QIASymphony SP	96	www1.qiagen.com/Products/QIAsymphonySP.aspx
	BioSprint 96	96	www1.qiagen.com/Products/Automation/BioSprint96.aspx
	EZY1 advanced XL	1–14	http://www1.qiagen.com/Products/EZ1AdvancedXL.aspx
	EZY1	1–6	http://www1.qiagen.com/Products/EZ1Advanced.aspx
	QIAcube	12	http://www1.qiagen.com/products/automation/qiacube.aspx?r=7249#Tabs=t5
	BioRobot MDx	24–96	http://www1.qiagen.com/Products/Automation/BioRobotMDxSystem.aspx
	BioRobot MDx DSP	96	http://www1.qiagen.com/Products/Automation/BioRobotMDxDSP.aspx
	BioRobot universal system	96	http://www1.qiagen.com/products/automation/BioRobotUniversalSystem.aspx
	Autopure LS	8 or 16	http://www1.qiagen.com/Products/Automation/AutopureLS.aspx
	QIAxtractor	8–96	http://www1.qiagen.com/Products/QIAxtractor.aspx
Roche Diagnostics	MagNA pure LC 2.0 System	32	https://www.roche-applied-science.com/sis/automated/index.jsp
	Magna Pure Compact	1–8	https://www.roche-applied-science.com/sis/automated/index.jsp
Thermo Scientific	Kingfisher	24	http://www.thermo.com/com/cda/product/detail/1,,11598,00.html
	Kingfisher flex	96	http://www.thermo.com/com/cda/product/detail/1,,10136240,00.html
	Kingfisher mL	15	http://www.thermo.com/com/cda/product/detail/1,,12616,00.html
Diagnostic Technology	Tiagen LabTurbo 36	1–36	http://www.labturbo.com/index.php?op=products&item=overview&itemLevel=lab36_compact
	Tiagen LabTurbo 96	12–96	http://www.labturbo.com/index.php?op=products&item=overview&itemLevel=lab96

been improved with the many different commercial methods available. Put simply – nucleic acids must be extracted in such a manner that they can be subsequently amplified by PCR.

Coupled to these advances are the increasingly available commercial robotic extraction systems which increase the throughput and lessen the boredom of multiple smaller scale extractions. These systems may use magnetic particle-based extraction (the majority of high through-put systems) or silica resin based extraction systems. These systems may extract up to 96 samples per run but some use more plastic disposables than others and some may be used coupled with a liquid handling system for higher throughput of PCR testing. This then also becomes a budgetary decision.

A look at the methods within this text will show that the variety of nucleic acid extraction methods is vast. Many are commercial, many are performed on robotic platforms and many may be a simple extraction procedure without the extensive lysis and washing procedures demanded for high quality nucleic acid.

Other simple procedures for nucleic acid extraction consist of lysing microorganism membranes by a combination of three alternative modes of lysis: chemical lysis using a detergent, mechanical lysis by agitation in the presence of beads (good for cell wall disruption), and heat shock lysis by repeated freezing and incubation at very high temperature (around 95–100°C). All are time consuming.

The simplest procedure able to be used in some assays is the "boil for 10 min method". This effectively lyses cells and may inactivate high risk microorganisms. It is still used in some laboratories and for specific assays fully evaluated for the use of such a procedure. However it is the potential existence of other factors within the sample itself which may inhibit the actual PCR method used. This may be cellular or histone proteins, lipids, cell wall components or heme proteins derived from blood which has been shown to be inhibitory to some PCR methods. This inhibition is most often seen with the extraction of DNA from faecal samples. The extraction of the more labile RNA from stools also requires methods more suitable to extraction, stability and avoidance of inhibition. Some commercial kits have been designed with inhibitor removal buffers or tablets for these specific applications.

Three main rapid extraction procedures are available in the form of ready to use solutions. These can eliminate the time-consuming and labor-intensive deproteinisation, organic extraction, dialysis, and alcohol precipitation protocols required in traditional DNA purification procedures. Additionally, in many instances, they can replace robotic extraction procedures and save valuable time:

1. Bio-Rad Laboratories has a 20 ml Chelex-based resin solution for PCR-ready DNA purification from blood, cultured cells, or bacteria. The procedure is rapid: incubate sample with InstaGene® matrix at 56°C for 15–30 min, then boil for 8 min. and microcentrifuge for 2 min to pellet the resin. DNA in the supernatant is ready for PCR. InstaGene® matrix is made with a specially formulated 6% w/v Chelex resin and it makes DNA sample preparation fast, easy, and cost-effective. The Chelex matrix binds to PCR inhibitors and adsorbs cell lysis

products rather than DNA, preventing DNA loss due to irreversible DNA binding and produces an improved substrate for PCR amplification.
http://www.bio-rad.com/prd/en/US/adirect/biorad?cmd=BRCatgProductDetail&productID=111001

2. Applied Biosystems has a 20 ml PrepMan® Ultra Preparation reagent and is applicable for a variety of different sample preparation applications. It has been used successfully to prepare DNA template from bacteria, yeast, filamentous fungi, both from a plate or from tissue smears, human cells (buccal swab), mammalian whole blood and from Gram-negative food-borne pathogens for use in PCR amplification reactions. Using a simple boil and spin protocol, PrepMan®Ultra Reagent efficiently inactivates PCR inhibitors and significantly reduces the need to repeat the template preparation step. PrepMan®Ultra Reagent is a novel formulation, developed entirely at Applied Biosystems, and is a homogeneous solution that does not contain Chelex or any other type of resin or matrix. It is based on ethyl glycol monobutyl ether with hydroxylated organoamine. The protocol is rapid and easy: resuspend the sample in 200 μl PrepMan® Ultra reagent or add the sample directly to 200 μl PrepMan® Ultra reagent, boil for 10 min, cool for 2 min, microcentrifuge for 2 min, transfer 5 μl of the supernatant to the assay.
https://products.appliedbiosystems.com/ab/en/US/adirect/ab?cmd=catNavigate2&catID=602362&tab=Overview
3. EPICENTRE Biotechnologies has a range of rapid extraction procedures depending on sample type and "target" of extraction – be it DNA or RNA. http://www.epibio.com/main.asp

Purified RNA should be stored at –20°C or –70°C in RNase-free water. One must ensure that when RNA is purified using a chosen kit or method that no degradation will occur upon storage. Purified DNA should be stored at –20°C or –70°C under slightly basic conditions (e.g., Tris-HCl, pH 8.0) because acidic conditions can cause hydrolysis of DNA. It is preferable to store diluted solutions of nucleic acids in aliquots and thaw them once only. It is also recommended to store aliquots in siliconised or low absorption tubes to avoid adsorption of nucleic acids to the tube walls, which would reduce the concentration of nucleic acids in solution.

New nucleic acid stabilisation technologies allow for the storage of DNA and RNA at room temperature in a cost-effective, environmentally friendly manner. Some of the currently available products include Biomatrica, GenVault and the Qiagen QIAsafe DNA Tubes and 96-well Plates. These innovative technologies provide room temperature storage which saves on refrigeration costs and enables easy transportation. Sample recovery is as easy as "just add water"!

A recent study [83] evaluated two novel products for room temperature DNA storage: Biomatrica's DNA SampleMatrix technology and GenVault's GenTegra DNA technology. The study compares the integrity and quality of DNA stored using these products against DNA stored in a freezer by performing downstream testing with short range PCR, long range PCR, DNA sequencing, and SNP microarrays. In addition, the investigators tested Biomatrica's RNAstable product for its ability to preserve RNA at room temperature for use in a quantitative reverse transcription PCR assay.

2.2 Conventional PCR

Catriona Halliday

2.2.1 Introduction

The discovery of the polymerase chain reaction (PCR) in 1985 revolutionised the diagnosis of infectious diseases in clinical laboratories by allowing rapid, sensitive and specific detection and identification of pathogens directly from clinical specimens, without the need for culture. PCR-based assays enable the amplification of a few target molecules (theoretically a single cell) to detectable levels, from both viable and non-viable cells. These applications are gradually complementing or replacing culture-based, biochemical and immunological assays in routine diagnostic laboratories [84].

2.2.2 Components of PCR

A conventional PCR reaction mix consists of target DNA, two primers, heat-stable DNA polymerase, deoxynucleotide triphosphates (dNTPs including dATP, dCTP, dGTP and dTTP), and a buffer usually containing Mg^{2+}. Primers are short (20–30 base pairs) oligonucleotides of known sequence that are complementary to the two 3′-ends of the target DNA [54]. The specificity of the primers determines the accuracy of the PCR assay as poor quality nucleic acids or non-target background DNA can influence the specific annealing of the primers. This can result in non-specific amplification and possible misinterpretation of results [84].

When the target nucleic acid is RNA, reverse transcriptase is included in the PCR reaction to convert RNA to cDNA. This procedure is known as reverse transcription PCR (RT-PCR).

A nested PCR assay is a type of conventional PCR that uses two pairs of primers in two separate, successive reactions. The initial reaction amplifies a target region of DNA with an outer primer pair. The resulting PCR product is used as template

C. Halliday (✉)
Centre for Infectious Diseases and Microbiology, Westmead Hospital, Westmead, NSW 2145, Australia
e-mail: Catriona.Halliday@swahs.health.nsw.gov.au

DNA for the second reaction, which employs a second set of primers that are located internally to those used in the first reaction. These assays have better sensitivity and specificity than single amplification assays and are useful for pathogen detection in clinical specimens. However, they are more prone to contamination from carryover of PCR product from the first reaction to the second [52].

2.2.3 PCR Amplification and Product Detection

PCR amplification is automated and performed on thermocyclers programmed to heat and cool to different temperatures for varying lengths of time.

Each PCR cycle involves three steps: (i) denaturation, (ii) annealing, and (iii) extension, and the cycles are repeated 20–40 times. The steps for RT-PCR are essentially the same, once the RNA has been transcribed to cDNA at 40–50°C. During denaturation, the DNA template is heated (94–96°C) to separate the two DNA strands. The temperature is then cooled (50–65°C) during the annealing step to allow the specific primers to hybridize to the 3′ ends of the separated DNA strands. The annealing temperature is dependent on the length and composition of the primers. Finally, during extension (72°C), the heat stable *Taq* DNA polymerase catalyses the elongation of the primers by incorporating the complimentary dNTPs that bind to the target DNA. The extended primers form two new strands of target DNA for the next PCR cycle. In theory, the amount of target DNA should double after each PCR cycle [54].

"Hot-start" PCR was developed to improve PCR amplification and specificity by reducing non-specific amplification during PCR set-up. The Hot-start *Taq* DNA polymerase enzyme is inactive at ambient temperature, preventing the extension of non-specifically annealed primers or the formation of primer dimer. The functional activity of the enzyme is restored during incubation at 95°C for 5–10 min [17].

Upon completion of a conventional PCR assay, the amplified products are usually analysed by agarose gel electrophoresis using DNA-binding fluorescent dyes (e.g. ethidium bromide) under UV illumination and fragment length as an indicator for identification [52].

2.2.4 Conventional PCR in the Diagnostic Laboratory

PCR-based assays are now accepted as the standard method for detecting many viruses and bacteria in diagnostic microbiology laboratories, however, their use is lagging for diagnosis of fungi and parasites due to the absence of commercial kits with quality controls. The development of real-time PCR technology and automated DNA extraction systems is expected to improve the reliability of conventional assays [11].

Generally, conventional PCR-based assays are only able to detect a single parameter, which can limit their scope unless a particular pathogen is suspected. Broad-range PCR assays, which target universal regions such as 16S-23S rRNA and heat-shock proteins, have been developed to allow simultaneous testing for more than one organism or to screen clinical specimens for pathogens. If the PCR assay yields an amplicon, the aetiologic agent must be identified by DNA sequencing [52]. Multiplex PCR assays, which incorporate multiple sets of primers in a single reaction to simultaneously detect numerous pathogens have also been developed, but they are better suited to real-time PCR technologies.

2.2.5 Limitations of Conventional PCR

The considerable increase in analytical sensitivity and specificity of PCR-based assays compared to conventional diagnostic tests are their major advantage, however, there are also limitations. Each individual PCR assay requires careful optimisation of reagents (Mg^{2+} and primers) and amplification conditions. Primer design is extremely important for effective PCR amplification as cross-reaction with non-target DNA can result in non-specific products. The cost of performing molecular tests is high in comparison to traditional diagnostic tests and re-imbursement is often low, particularly for assays developed "in-house" [54]. Additionally, laboratories performing these assays need to invest considerable costs in dedicated "DNA-free" laboratory space and equipment. This is essential to minimise contamination of subsequent specimens by PCR amplicons that can lead to false positive results. This must be monitored by the inclusion of non-template or water controls in every PCR. The development of closed-tube, real-time PCR technologies and melting curve analysis has greatly reduced the risk of contamination and false positive results. DNA degradation or PCR inhibitors in clinical specimens can lead to false negative results and this must be monitored by the inclusion of internal positive controls.

2.2.6 Summary

There is no doubt that the development of PCR has revolutionised the diagnosis of infectious diseases in routine microbiology laboratories. However, many first generation conventional PCR assays are being replaced by real-time PCR platforms which offer increased sensitivity, specificity and rapidity, reduced contamination and greater potential for automation [52]. Despite their limitations, conventional PCR assays will continue to have a role in smaller, regional diagnostic laboratories that perhaps cannot afford the higher cost of reagents and instrumentation needed for real-time PCR assays.

2.3 Real-Time PCR

Theo P. Sloots

2.3.1 Introduction

Since its introduction, real-time PCR has made a major contribution to the diagnosis of infectious disease in most clinical laboratories. Its success has been due to the development of novel chemistries and instrumentation enabling detection of PCR products on a real-time basis within a closed system over a range of cycles. Also, real-time PCR instruments cycle the temperatures more rapidly than conventional thermocyclers, and, because of the increased sensitivity of the fluorescent detection system, offer a much broader dynamic range compared to conventional PCR. The inbuilt detection system offers product confirmation and quantification, with an electronic result output which lends itself to high through-put electronic reporting.

2.3.2 Real-Time PCR Technology

Generally, real-time PCR chemistries consist of fluorescent probes which release a fluorescent signal that increases in direct proportion to the amount of PCR product produced in the reaction [41, 42]. The higher the starting copy number of the nucleic acid target, the sooner a significant increase in fluorescence is observed.

A number of different fluorescent chemistries are now widely used, including non-specific DNA intercalating dyes, dye-primer systems and target-specific oligoprobes. Each system has its own unique characteristics, but the strategy for each is similar; they must link a change in fluorescence to amplification of DNA.

The non-specific DNA intercalating dyes exhibit little or no fluorescence when in solution, but emit a strong fluorescent signal upon binding to double-stranded DNA. Oligoprobes depend on Fluorescence Resonance Energy Transfer (FRET) to generate the fluorescence signal via the coupling of a fluorogenic donor molecule and a quencher or reporter molecule to the same or different oligonucleotide substrates.

Non-specific dyes such as SYBR® green [65], YO-PRO 1 [31], SYTO9 [51] and more recently BOXTO (TATAA Biocenter, Sweden) are relatively inexpensive and do not require additional oligoprobe design. Also they are not affected by mutations in target sequence which may impair the binding of specific probes thereby influencing the final result [87].

T.P. Sloots (✉)
Queensland Paediatric Infectious Diseases Laboratory, Sir Albert Sakzewski Virus Research Centre, Queensland Children's Medical Research Institute, Children's Health Service District, Brisbane, QLD 4029, Australia
e-mail: t.sloots@uq.edu.au

SYBR® green is the most widely used chemistry, and provides the simplest and most economical format for detecting and quantifying PCR products.

It is present in the reaction mix at the start, and binds to the minor groove of double stranded DNA, emitting 1,000-fold greater fluorescence than when it is free in solution. Thus, as a amplification product accumulates, fluorescence increases. The advantages of SYBR® green are that it is inexpensive, easy to use, and quite sensitive. The disadvantage is that it will bind to any double-stranded DNA in the reaction, including non-specific reaction products such as primer-dimer, which may give false-positive results or an overestimation of the target concentration. To help assess specificity, the dissociation curve of the amplified product can be analysed to determine the melting point. If two or more melting peaks are evident, it suggests that more than one amplified sequence was obtained, and the amplification was not specific for a single DNA target. For single PCR product reactions with well designed primers, SYBR® green can work extremely well, with spurious non-specific background only showing up in very late cycles. Therefore, in practice, the non-specific reporters are most suitable for highly optimised PCR assays.

There are many different dye-primer based signaling systems for real-time PCR, ranging from simple light upon extension (LUX) primers to the more complex scorpion primers [67]. The template specificity of the dye-primer system is the same as for the intercalating DNA dyes except for the scorpion primer, where the signal generated by the primer is dependent on a complementary match with sequence located within the PCR amplification product.

The use of oligoprobes to detect amplification product adds a further level of specificity to the reaction by immediate confirmation of the target sequence. There are a variety of oligoprobe-based assays in use today, including (i) hydrolysis (TaqMan®) probes [26], (ii) minor groove binding (MGB) probes [1], (iii) molecular beacons [49], (iv) hybridisation probes [91], (v) scorpion primer/probes, (vi) locked nucleic acid (LNA) probes [43], and (vii) peptide nucleic acid (PNA) light-up probes, and combination thereof. Each of these has the capacity to use multiple reporter dyes with multiple quenchers for efficient FRET pairs.

Hydrolysis or TaqMan® probes (also called 5′-nuclease probes because the 5′-exonuclease activity of DNA polymerase cleaves the probe) were among the first to be used in real-time PCR and are arguably the most widely used fluorescent probe format. They are sequence-specific oligonucleotides that carry a fluorescent dye at the 5′ base, and a quenching dye on the 3′ base, and are designed to anneal to a complementary sequence on the amplification product. Whilst the probe is intact, the quencher and reporter are in close proximity, and the quencher absorbs the signal from the reporter through FRET. During amplification the 5′-nuclease activity of the DNA polymerase hydrolyses the probe [27], separating the fluorescent reporter dye and the quencher, allowing the reporter's energy to be released as a fluorescent signal. The level of fluorescence increases in each cycle proportional to the rate of probe cleavage, and is indicative of a positive reaction. Examples of common quencher fluorophores include TAMRA, DABCYL, and BHQ, whereas many reporter dyes are available (e.g., FAM, VIC, NED, etc.). Hydrolysis probes have greater specificity because only sequence-specific amplification is measured.

Minor groove binding (MGB) probes are a modification of the hydrolysis oligoprobe chemistry [1]. This system uses a reporter dye at the 5′ terminus and a non-fluorescent quencher (NFQ) at the 3′ end. In addition, the 3′ end also carries a MGB molecule which further stabilises the oligoprobe-target duplex by folding into the minor groove of the double stranded DNA [1]. Unhybridised, the MGB probe assumes a random coil configuration which results in quenching of the fluorescent signal. On specific hybridization with the target, the molecule becomes linear before being cleaved by the DNA polymerase resulting in the emission of fluorescence. The advantage of MGB probes is that they may be very short (12–17 nt), and are ideal for targets with limited consensus sequence.

Like TaqMan probes, molecular beacons also contain a fluorescent dye at the 5′ end and a quencher molecule at the 3′ end and use FRET to detect a fluorescent signal. However, molecular beacons remain intact during the amplification reaction, and rehybridise to the target sequence for signal measurement during every cycle of the PCR. When free in the reaction mix, molecular beacons assume a stem-loop configuration, with the fluorescent and quencher molecules in close proximity, thereby preventing the probe from fluorescing. When hybridised to a target, the fluorescent dye and quencher are separated, quenching through FRET does not occur, and a fluorescent signal is released upon excitation with an appropriate light source. Molecular beacons, like TaqMan probes, can be used for multiplex assays by using spectrally separated fluorescent and quencher molecules on separate probes, one each for the target under investigation.

With scorpion probes, sequence-specific priming and PCR product detection is achieved using a single oligonucleotide. Like beacons, the molecule contains a fluorophore at the 5′ end and a quencher at the 3′ end. In the unhybridised state it maintains a stem-loop configuration bringing the two dyes in close proximity, and the fluorescence emitted by the fluorophore is absorbed by the quencher. The 3′ portion of the stem also contains sequence that is complementary to the extension product of the primer. This sequence is linked to the 5′ end of a specific primer via a non-amplifiable monomer. After extension of the scorpion primer, the specific probe sequence binds to its complement within the extended amplification product thus opening up the hairpin loop. This separates the fluorescent reporter and quencher molecules emitting a signal of the appropriate wavelength

Hybridisation probes (or HybProbes) are commonly used with the LightCycler instrument (Roche Diagnostics, Switzerland), and consist of two oligoprobes. One, the donor probe, is labeled with fluorescent dye at the 3′ end and the second, the acceptor probe, is labeled at the 5′ end with a reporter dye which absorbs resonance energy from the donor probe. Fluorescence by the acceptor probe will only occur through FRET when both the donor probe and the acceptor probe have annealed to the amplification product in close proximity to each other. Increasing fluorescence is a measure of amplification product formation. Unlike TaqMan® probes the process is non-destructive and reversible.

An added advantage of this system is the ability to perform melting curve analysis to confirm the identity of the amplification product. Melting curve analysis provides an extra element of specificity to the PCR, because sequence variation in probe target sites will result in a shift of melting temperature. This may also act as an important quality control feature to confirm the correct identity of the amplification product, and provides a simple and elegant method to genotype mutations, including single base mutations [7, 39].

2.3.3 Real-Time PCR in the Diagnostic Laboratory

Review of the current literature shows that real-time PCR has been widely applied in clinical laboratories for the detection of bacterial, viral and fungal pathogens. As a result, DNA and RNA are now widely accepted as important and universal diagnostic targets. The biggest impact of real-time PCR has been in the rapid diagnosis of life-threatening diseases such as meningococcal disease, SARS, avian influenza (H5N1) and herpes simplex encephalitis [86]. However, real-time PCR diagnostics generally, also offers significant improvements over more traditional methods for the detection of a wide range of organisms, particularly organisms that may not be isolated by culture or those that require extended isolation processes. A comparison with conventional culture-based methods has convincingly demonstrated greater sensitivity of the molecular assays.

2.3.4 Instrumentation for Real-Time PCR

Real-time PCR technology requires appropriate instrumentation such as a thermal cycler with optics for the collection of fluorescence excitation and emission and a computer with appropriate data acquisition and analysis software. Because fluorescent chemistries require both a specific input of energy for excitation and a detection of a particular emission wavelength, the instrumentation must be able to do both simultaneously and at the desired wavelengths. Three basic ways are used to supply the excitation energy for fluorophores: by (i) lamp, (ii) light-emitting diode (LED), or (iii) laser. Instruments that utilise lamps (tungsten halogen or quartz tungsten halogen) generally also include filters to restrict the emitted light to specific excitation wavelengths. An example of instruments using lamps include the Applied Biosystem ABI 7500 and Stratagene Mx4000; LED systems include the Roche LightCycler and Qiagen Rotor-Gene, whilst the ABI Prism 7700 and 7900HT use a laser for excitation. Detectors to collect emission energies include charge-coupled device cameras, photomultiplier tubes, or other types of photodetectors. Only the desired wavelength(s) are collected by use of narrow wavelength filters or channels. Usually, multiple discrete wavelengths can be measured at once, which allows for multiplexing of assays measuring different targets.

Another important feature for real-time PCR is the ability of the thermocycler to change temperatures rapidly, and to maintain a consistent temperature among all sample wells. Differences in temperature across the block could lead to different PCR amplification efficiencies and varying results. Consistent heating is achieved by using a heating block (Peltier based or resistive), heated air, or a combination of the two. Solid heating blocks generally change temperature much more slowly than heated air, resulting in longer cycling times.

Data generated during the real-time PCR requires appropriate data-acquisition and analysis software. Generally, PCR data are presented by graphical output of assay results including amplification and dissociation (melting point) curves. The amplification curve gives data regarding the kinetics of amplification of the target sequence, whereas the dissociation curve reveals the characteristics of the final amplified product.

In Australia, real-time PCR instruments are available from several manufacturers and differ in configuration and sample capacity as well as overall sensitivity, and may have platform-specific differences in how the software processes data. The price of real-time PCR instruments varies widely, currently about $65,000–$150,000, but is well within purchasing capacity of diagnostic facilities that have the need for high throughput quantitative or qualitative analysis.

2.3.5 Considerations in the Use of Real-Time PCR

The generation of accurate quantitative PCR results is dependent on the strict control and standardisation of many assay parameters. The main sources of error include the nucleic acid extraction process and the presence of PCR inhibitors in clinical specimens. The best way to control for these factors is by using a robust internal control strategy such as previously suggested, in which a quantification standard of known copy number is incorporated into each sample, and carried through sample extraction, reverse transcription amplification, hybridisation and detection [67].

Also, sequence variation in the primer or probe binding sites may lead to false-negative results or otherwise can have a significant impact on quantitative PCR results. As few as two mismatches at the 3′ end of a single primer may affect the efficiency of the PCR reaction. This is particularly important in quantitative PCR and can result in underestimating the true microbial load by up to 3 logs [87]. The overall impact of this on the final result is dependent on the reaction conditions used, the composition of the primers, the annealing temperature and master mix composition.

Similarly, sequence variation may impact upon fluorescent signal. In a real-time PCR assay for respiratory syncytial virus (RSV) using a MGB probe, notable differences were observed in assay results due to sequence divergence between the MGB probe and the target sequence. The amplitude of some linear amplification curves was greatly reduced and in some cases false-negative results were obtained. The authors of this study concluded that MGB probes should be used with caution if mutation in the target sequence is common such as occurs in RNA viruses [88]. Therefore, careful and extensive optimisation of the real-time PCR conditions must be performed to obtain meaningful results, and to ensure that the efficiency of the reaction does not vary due to sequence variation or minor differences between samples.

2.3.6 Summary

Real-time PCR methodologies for the detection of a wide range of organisms are firmly entrenched in many clinical laboratories and offer major advantages of improved sensitivity and rapidity over traditional methods. However, as the use of real-time PCR assays evolved, there has been an obvious need for standardised reagents and quality assurance programmes in order to obtain reproducible and clinically significant results. Also, we need to take heed of the inherent limitations associated with the targeted nature of PCR. These are often difficult to control, particularly in virology, where the heterogeneous nature of the viral genome may lead to significant difficulties in assay design and may impact on assay performance. However, an awareness of these issues will ultimately result in a better understanding of this new technology, and enable us to fully explore its potential as a modern diagnostic tool.

2.4 Quantitative PCR

Todd M. Pryce and Ian D. Kay

2.4.1 Introduction

Quantitative PCR (Q-PCR) is routinely performed by many microbiology laboratories with real-time PCR (RT-PCR) capabilities. RT-PCR requires instrumentation capable of PCR product (amplicon) detection in a cycle-by-cycle, or in a "real-time" fashion using fluorescent chemistry.

PCR assays are either qualitative or quantitative. Both assays detect nucleic acids targets (DNA or RNA) in a given sample; the target is amplified by the PCR process and detected by measurement of a fluorescent signal at every PCR cycle until completion of the assay. However, qualitative RT-PCR assays are used for simple detection of nucleic acids targets (DNA or RNA) in a sample; if a specific target is amplified and detected then the sample is considered PCR-positive for the

T.M. Pryce (✉)
Department of Microbiology and Infectious Diseases, PathWest Laboratory Medicine WA, Royal Perth Hospital, Wellington Street Campus, Box X2213 GPO, Perth, WA, 6847, Australia
e-mail: Todd.Pryce@health.wa.gov.au

target, or if the target has not been amplified then the sample is considered PCR-negative. Hence, qualitative PCR assays are designed to give either positive or negative results only. These assays are commonly in diagnostics for detection or exclusion of pathogens. In contrast, Q-PCR assays are designed to determine the concentration or copies of a detected target present in the original sample. These assays are commonly used for monitoring response to therapy. This is achieved in Q-PCR using nucleic acid standards and exploiting the predictable kinetics of the PCR reaction.

2.4.2 PCR Kinetics and Q-PCR

The kinetics of a PCR reaction plotted graphically has a distinctive shape, with three distinct phases; (a) the background, or early phase; (b) the exponential growth phase; (c) the plateau phase (Fig. 2.1). In the early phase the oligonucleotides hybridise to the target sequence and the PCR reaction has commenced, but at an undetectable level. In the exponential growth phase the target is amplified in an exponential fashion and the fluorescence levels become detectable above the background. In the plateau phase the reactants have been consumed or have deteriorated and the PCR

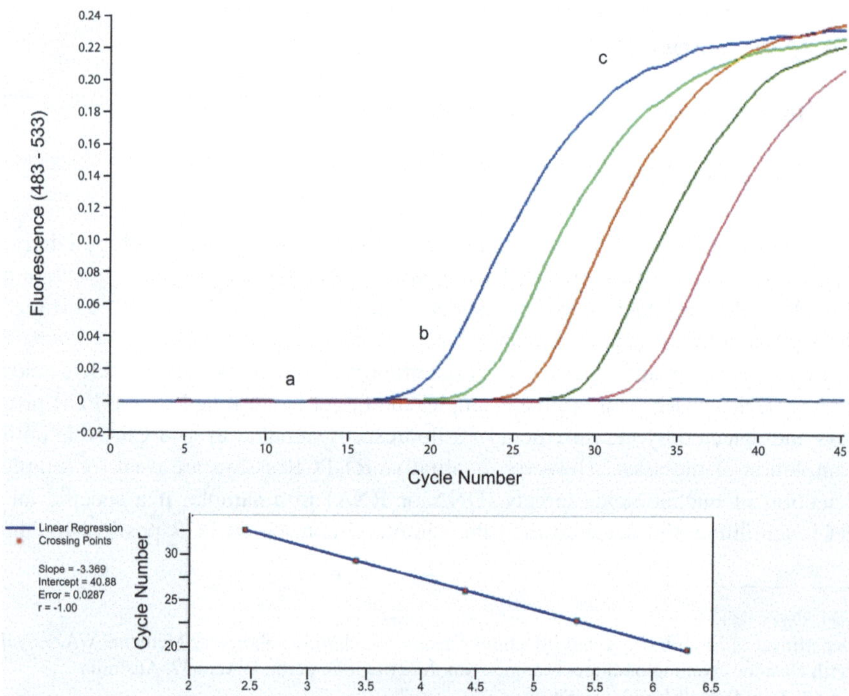

Fig. 2.1 Quantitative PCR standards and standard curve

reaction is no longer operating efficiently. The exponential phase is the important stage of Q-PCR. During this phase accurate quantification of the target DNA is possible.

To clarify, the exponential phase can be mathematically explained using the equation:

$$\mathrm{NC} = \mathrm{No} \times (E + 1)^C$$

where C is the number of cycles, E is amplification efficiency (also expressed as $\%\mathrm{E} = E \times 100\%$), NC is the number of amplicon molecules, and No is the initial number of target molecules. In simple terms, each cycle produces an increase in *NC* in proportion to amplification efficiency. Hence, 100% efficiency produces a doubling in the number of amplicon molecules. Additionally, the quantity of NC present after any specific number of cycles is dependent on No. Rearrangement of the equation provides the mathematical relationship upon which Q-PCR is based, however in reality the amplification efficiency is less than 100%. The PCR efficiency can be calculated from the slope of the curve. Hence, once a standard curve has been established, unknown samples can be amplified by the same process and compared to the standards to determine the target concentration in the original sample. These standards can be either be external or internal to the assay.

2.4.3 Absolute Quantification with External Standards

The most common and easiest way to produce quantitative results is to create a standard curve using external standards. External standards are usually 10-fold serial dilutions of the target. The standards are usually plasmids but can be whole organisms or nucleic acid. These standards are amplified and detected using the same assay conditions as for samples. The PCR cycle at which product fluorescence intensity rises or is higher than the background is called the crossing point (*Cp*). At this point the exponential PCR phase begins. Theoretically, the rate of amplification is maximal with PCR products doubling every cycle; hence quantification is performed at this stage. Following completion of the assay the *Cp* for each standard is determined and standard curve is prepared using instrument software. A typical standard curve is a plot of the cycle number *Cp* at the (*Y*-axis) versus the log of initial template amount (*X*-axis), derived from an assay based on serial dilutions. The standard curve is a least square fit line drawn through all dilutions (Fig. 2.1; bottom). Using several standards which cover the expected clinical range a line of best fit can be determined using the *Cp* values. It is then a case of determining the unknown samples *Cp* and reading the resulting concentration. Absolute quantification with external standards are useful but they do not control for changes in amplification efficiencies which may occur and vary from sample to sample. Hence, if PCR inhibition occurs due to inhibitors in the patient sample, this degree of inhibition will not be represented in the standard curve, therefore a lower quantitative value may be produced. However, for samples demonstrating negligible inhibition, absolute quantification with external standards is relatively reliable and easy to set up.

2.4.4 Absolute Quantification Using Internal Controls

Absolute quantification using internal controls is an advanced Q-PCR method that controls for nucleic acid extraction and PCR inhibition. This approach involves the use of exogenous control DNA. In this approach, a homologous DNA fragment of known concentration is engineered which has the same primer binding regions as the target, but with different probe-binding sequence and fluorescent marker. This DNA fragment can be added to an individual sample prior to nucleic acid extraction and PCR. The fragment is then co-extracted and amplified along with the target. Two probes with different fluorescent labels are used; one to detect the target and the other to detect the internal control. The advantage of this system is the amplification efficiency of the reaction will be the same for the target as well as the control. So if there is an inhibitor in the sample it will affect the target and the control equally and the result should be more accurate. The disadvantage of this system is the need to generate a second PCR fragment that may compete with the detection of the target if not carefully designed.

2.4.5 Detection Formats

Real-time PCR instruments currently available use fluorescent dyes to generate a signal. The most commonly used dyes are the following:

2.4.5.1 Intercalating Dyes

SYBR Green: see Section 2.3.2 for description of action. Disadvantages of using SYBR Green for Q-PCR are that it binds to all double stranded DNA (from cellular DNA in the sample) or non specific DNA products generated during the reaction; and high concentrations of SYBR Green can cause reduced amplification efficiencies due to toxicity to Taq Polymerase. The advantage is SYBR Green is simple to use and does not require the use of probes or multi-colour instrumentation.

2.4.5.2 Sequence-Specific Fluorescent Dyes

(a) Fluorescent Resonance Energy Transfer (FRET): see Section 2.3.2 for description of action. The use of FRET probes has an advantage over hydrolysis probes as they can be used to determine the Melting Temperature (*Tm*) of a particular PCR fragment. This *Tm* is the temperature at which 50% of the probe has dissociated from the target. The *Tm* varies according to the base sequence of the target. For example, a target rich in G/C bases will have a higher *Tm* than a target with high A/T content. This is due to the stronger hydrogen bonds that join C and G residues. This makes FRET probes particularly useful for rapid genotyping. Using *Tm*, the degree of specificity can measured and a specific target can be confirmed. Hence FRET is a sequence-dependant method of PCR product detection. However, FRET probes can be difficult to design as the head-to-tail design may span 40 or more conserved nucleotide bases.

(b) 5′–3′ Hydrolysis Probes: these assays are typically called TaqMan assays, see Section 1.2.3 for description of action. Unlike FRET probes, hydrolysis probes cannot be used for *Tm* analysis, however they are easier to design as a span of only 18–25 nucleotide bases of the target is required.

2.4.6 Clinical Use

The clinical use of Q-PCR is broad, however Q-PCR is most commonly used to determine the concentration or load of a pathogen in a clinical sample. For example, quantitative PCR can be used for determining the baseline level of a particular infecting virus e.g. hepatitis C virus (HCV) in plasma. If patients are on therapy (e.g. Interferon), the HCV viral load can be measured to track the response to treatment. A falling titre indicates successful treatment whilst an increasing or static level may indicate treatment failure.

In other cases, organisms may be present in the host at low levels and may not cause significant disease. Qualitative PCR may be positive in these situations. However, by determining the organism load by quantitative PCR, a break-point level can be determined which indicates the organism is now at a titre which may cause disease. For example, many people have had previous exposure to cytomegalovirus (CMV). The virus may be present in the latent stage and qualitatively positive in white cells. Detection of active CMV viraemia, or reactivation, is important in immunocompromised patients. A rising CMV viral load can indicate reactivation which may lead to CMV disease. Response to therapy can also be measured.

2.4.7 Conclusion

There are other types and subtypes of molecular-based quantitative methods, however Q-PCR using either SYBR Green, FRET, or hydolysis probes are the more commonly used approaches. Most commercial Q-PCR methods use internal control standards and are automated, thereby increasing reliability and reproducibility and at the same time controlling for PCR inhibition.

2.5 Multiplex PCR in Diagnostic Microbiology

Gerald B. Harnett and Glenys R. Chidlow

Abstract Multiplex PCR assays have many attractive features in terms of economy and offer a practical means to provide molecular-based assays for the increasing

G.B. Harnett (✉)
Department of Microbiology and Infectious Diseases, PathWest Laboratory Medicine, QEII Medical Centre, Nedlands, WA 6009, Australia
e-mail: Gerry.Harnett@health.wa.gov.au

range of known infectious agents. Apart from the rapidly expanding list of newly discovered viruses, PCRs for bacterial virulence factors and antibiotic resistance genes are likely to be in increasing demand. There are currently several methods which provide multiplex PCR capability and it is uncertain which will become the preferred technology

2.5.1 Introduction

Conventional PCR assays generally detect a single target nucleic acid sequence using a set of oligonucleotide primers with or without a probe to confirm the identity of the amplified PCR product DNA. Multiplex PCR assays detect multiple targets within a single amplification reaction using corresponding multiple primer pairs. The use of multiplex assays in diagnostic microbiology has many obvious attractions. In theory, the multiplexing of assays should enable many infectious agent targets to be detected with economies in labour, consumables and with the requirement for only a small volume of sample. In fact, since the sample extraction is usually the most expensive component of a PCR assay, the reduction in the sample extract volume required to test for multiple targets probably provides the most attractive feature of multiplex assays. Once a PCR is at peak efficiency the only conceivable method to increase its sensitivity is to increase the effective amount of sample introduced into the assay. For multiple targets then, this often means performing multiple extractions which is expensive of time and consumables and may be limited in some cases by the volume of sample available.

Another expensive PCR reagent is the reverse transcriptase (RT) enzyme needed for RNA targets.This is several times more expensive than DNA polymerase and many respiratory and gastrointestinal disease agents are RNA viruses, the list of which has grown considerably in recent times. To test for a comprehensive range of these agents is impractical for most laboratories if single assays are to be used. Another facet of potential multiplex use is in providing coverage for sequence variations in target agents.

As sequencing capability has become widespread in laboratories, and numerous entries made into public sequence databases, it is becoming evident that sequence variation in target agents is a major factor limiting the long-term reliability of molecular-based assays. This is especially true for many RNA viruses which evolve very rapidly. Multiplex assays then, have the potential to mitigate this problem by targeting multiple gene sequences of an agent, thus reducing the chances of random variations compromising the performance of an assay. In fact, it is probably true that in future this principle will need to be incorporated into many PCR assays for infectious agents to ensure their reliability. These comments apply particularly to real-time assays. Although having considerable advantages compared with PCRs having gel-based amplicon detection, real-time assays are more susceptible to sequence variation failure especially if the variation occurs in the probe target region.

2.5.2 Multiplex Technologies

There are several alternative multiplex PCR technologies now available. The oldest of these uses primers for different targets which produce amplicons of different sizes distinguishable by gel-electrophoresis [61]. This method can be used as a single PCR or nested for extra sensitivity. It has the advantage that it is less susceptible to sequence variation than real-time methods but is more labour intensive and relies on subjective interpretation of gel electrophoresis results. A recent development is the use of capillary electrophoresis for this purpose and there are several commercial alternative equipment options available which afford very accurate measurement of amplicon size. Real-time technologies are now in general use and can be multiplexed using fluorphore-labelled probes which are detected at different wavelengths. However, in spite of real-time thermocyclers having up to six wavelength options most workers find that it is difficult to devise multiplexes which perform satisfactorily with more than three different probe labels in an assay. It is of course possible to perform a multiplex assay with large numbers of primer sets and to then pass the product to real-time triplex assays. This introduces risk of laboratory cross-contamination, a known hazard of nested PCRs, but this can be reduced by limiting the initial multiplex PCR to 10–20 cycles, a principle used by Stanley and Szewezuk in their Tandem assays [69]. Luminex bead technology has the potential to detect up to 100 targets in an assay but practical considerations probably limit this to about 30. This method consists of performing a generic PCR or one with multiple primer sets in which the primers carry a biotin label. Products from the multiplex are hybridised with Luminex beads carrying probes specific for the multiple targets. The beads can be uniquely identified as the 100 bead types available each contain different proportions of fluorescent material and each probe is thus identified [64]. A fluorescent-labelled avidin is applied and washing steps remove background fluorescent material before the beads are processed in a flow-cytometer. A novel multiplex approach has been suggested recently which involves the use of Mass Tag PCR [19]. Numerous targets can be detected with DNA microarray technologies but the sensitivity of these is limited by the amount of target material available unless generic amplification is first performed [4]. Array technology is also expensive but will become useful for the detection of virulence factors and antibiotic resistance genes in bacterial cultures where target quantity is unlimited and when these genes sequences have been more fully identified.

2.5.3 Practical Applications

An important consideration in the design of a multiplex PCR assay is whether only one of the agents included is usually present or whether there is a likelihood of multiple infections in the samples to be tested. In the latter case, the question of competition for reagents within the PCR, between the multiple agents becomes an issue. In general, reagent concentration needs to be increased in multiplex assays to allow for full efficiency. It is important to determine whether a large copy number of one agent will inhibit the reactivity of a small copy number of another [81]. For example in a cerebrospinal fluid (CSF) sample one would only expect infection with one flavivirus causing encephalitis but in genital samples multiple HPV virus infections are commonplace and may be mixtures of high and low risk types for the causation of cervical carcinoma in widely varying copy numbers. In our laboratory we are endeavouring to put in place some of the principles described in this article in the in-house PCR assays that we use. Our initial simple use of multiplexing is in providing an inbuilt control of sample extraction efficiency and removal of RT and/or PCR inhibitors, in a semi-quantitative fashion. Standardized amounts of equine herpesvirus (EHV) and/or MS-2 RNA coliphage are incorporated in extraction lysis buffers and assays for these agents are carried out as multiplexed systems with different probe labels than for target agents. This ensures that unsatisfactory extraction or incomplete removal of inhibitors is detected and also checks that thermocycling has been performed properly. We have designed an assay with three probe labels for HSV-1, HSV-2 and varicella virus which has three probe label types and allows for duplicate tests for the viruses and incudes an EHV control. Another assay modifies the principle of tandem PCR suggested by Stanley and Szewezuk [69]. In this assay for *Chamydia trachomatis* and *Neisseria gonorrhoeae,* an initial PCR multiplex containing six primer sets is cycled for 20 cycles only in a conventional thermocycler. Diluted products are then transferred to real-time triplex assays for processing in a real-time thermocycler. This multiplex reduces competition for reagents if multiple infection is present and includes three targets for *C. trachomatis*, two for *N. gonorrhoeae* and an EHV control. Although in their early stages we have a number of other multiplex assays in development and are confident that this approach, using whichever multiplex technology becomes dominant, will provide the only realistic approach to give wide coverage of infectious agents.

2.6 Molecular Subtyping

Philip M. Giffard

2.6.1 Introduction

The fundamental taxonomic unit for bacteria remains the species. In higher organisms, a species is effectively defined as a collection of organisms without significant barriers to gene flow. This rule cannot be consistently applied to bacteria because

P.M. Giffard (✉)
Menzies School of Health Research, Institute of Advanced Studies, Charles Darwin University, Casuarina, NT, Australia
e-mail: Phil.Giffard@menzies.edu.au

of the considerable variation in propensities for horizontal gene exchange (HGT) among different bacterial taxa. Therefore, bacterial species are currently defined using a pragmatic mix of phylogenetic, gene flow and phenotypic data, with the weights given to these criteria, and the level of diversity within a species varying from case to case. It is however universal that a bacterial species is not composed of a single clone. In other words, bacterial species contain diversity and can be divided into finer taxonomic units. These are generally named and recognised using less formal conventions than are used for the standard Latin scientific nomenclature. An exception to this is the category of subspecies, which is used in conventional formal taxonomy. However, in the great majority of cases, a taxonomic unit that is a subset of a species is known simply as a "type", or a derivative of this word such as "serotype", genotype or "subtype". In the last 30 years, a plethora of bacterial typing technologies and methods have emerged, and bacterial typing has become commonplace in many laboratories.

Many of the more recent methods incorporate the PCR. Deciding what typing method to use is not straightforward. A major reason for this is that all methods have a different performance. A question frequently asked is "Does typing method (X) resolve all strains?". This question is not meaningful, in part because there is no accepted definition of the term "strain". Greater than 0.1% of bacterial cell divisions yield a point mutation somewhere in the genome, and other classes of genetic change occur at a much higher frequency than this. Accordingly, there is no such thing as a bacterial population of any significant size in which all the cells are genetically identical, and "clones" or lineages cannot be regarded as immutable entities. In essence, what typing methods do is indicate whether or not the most recent common ancestor of two or more isolates post-dates a particular time in the past. The higher the resolution of the typing method, the more recent that time point.

In general, the absolute value of that time point is not clearly understood. Of course, increasing sophistication of typing methods and associated data analysis and exchange facilities allows degrees of genetic relatedness to be determined, and the relationship between the typing method output and the actual population structure of the relevant species to be elucidated. Detailed descriptions of the complexities of understanding and comparing the performances of typing methods have been published by van Belkum et al. and Faria et al. [22, 80].

The field of bacterial typing is an enormous one, so this essay is of necessity a brief outline. Likewise the references provided are a very small fraction of the relevant literature. They are designed to assist the reader to understand the state of the art and also to find colleagues with expertise in these methods. Therefore, there is some bias to recent publications from Australian researchers.

2.6.2 Why Bacterial Typing Is Performed

Several bacterial typing methods provide sufficient resolution that only isolates with extremely recent common ancestries will have the same type. Such methods can be used to test hypotheses of direct epidemiological linkage. This is relevant in investigating e.g. outbreaks of food-borne disease, break-downs of infection control in health care facilities, the long distance dissemination of dangerous clones, and biological attacks. While such typing methods often do not directly indicate clinical properties, the association between types and such properties may have been previously determined. Therefore, the clinically relevant properties of an isolate such as the resistance and virulence phenotype can sometimes be inferred from the type by inductive reasoning. Typing methods that provide results that indicate the evolutionary history of the isolate can be used to determine and monitor the diversities and population structures of bacterial species. This is frequently performed as a largely academic exercise. However, the implications can be far from trivial. Mapping the patterns of dissemination onto phylogeny can provide profound insight into the history of a bacterial species. The practical applications of this include investigation of the impacts of new or changed vaccination protocols on circulating serotypes, and the time scales and the likely vectors of important dissemination events of problematic clones of bacterial pathogens.

2.6.3 Typing Methods Based on Phenotype

In general, early typing methods involved examination of the phenotype of the cell, while more recent typing methods involve genetic analysis. Probably the most widely used phenotypic approach is serotyping, which determines the reactivity of the cells with a standard bank of antisera. This is an indirect means of probing variability of cell surface molecules, principally carbohydrates and proteins. Serotyping is still seen as a valuable approach, and this is due in large part to simplicity of execution. Also, in species such as *Streptococcus pneumoniae*, for which there are vaccines, close correlation between serotype and vaccine susceptibility can be very useful when determining the impact of vaccination programs [18, 40].

Another widely used phenotype-based method with a long history is phage typing. This has been particularly applied to *Salmonella* [2] and involves the determination of the susceptibilities of the cells to a standard panel of phages. It is now known that the principal basis for variation in phage susceptibility is variation in the prophage content in the genome. In general, lysogens of a phage are immune to infection by the same phage [36]. Phage typing requires some specialist skills, and has generally been confined to reference labs, in particular with regard to *Salmonella*. Phage typing is now largely superseded by more direct genetic analyses.

Direct chemical analysis of bacteria has long shown promise as a bacterial identification and typing approach. There has been a recent resurgence of this with the development of dedicated and robust mass spectrometry devices. This approach is primarily used for identification to the species level, but it does have some ability to resolve within species [66].

2.6.4 Typing Methods Based on Electrophoresis of DNA Fragments Derived from the Whole Genome

Numerous nucleic acid-based bacterial typing methods have been developed in the last two decades. Many of these are based on the conversion of the genome into a series of DNA fragments of varying size that can be resolved by electrophoresis so as to generate a genetic fingerprint. A distinctive feature of this family of methods is that there is nothing inherent in the techniques that allows inference as to just what genetic change causes a change in the fingerprint. In other words, the fragments in the gel are anonymous, although they can usually be identified if the complete genome sequence is known.

A genome can be converted into fragments by either cleavage or synthesis. Probably the most conceptually straightforward method is cleavage of the genome with a very infrequently cutting restriction enzyme, followed by resolution of the resulting very large fragments by variable field agarose electrophoresis. This technique makes use of a hexagonal electrode that allows rapid alteration of the angle of the voltage gradient. For reasons that are not fully understood, this results in the resolution of DNA fragments much larger than can be resolved by conventional agarose electrophoresis. The early versions of this method made use of a field that periodically reversed its direction. This was termed pulsed-field electrophoresis (PFGE). A variant of this approach that makes use of a hexagonal electrode and consequent multiple directions for the electric field has become commonplace. This is correctly known as "(contour) clamped homogenous electric fields" (CHEF) gel electrophoresis, although the PFGE abbreviation remains ubiquitous [57]. Variation in PFGE fingerprints likely arise from large genome rearrangements, and gene gain or loss events that change the size of restriction fragments. Point mutations that create or destroy restriction sites may also occur, but these are likely to be very rare given the very small percentage of the genome that the restriction sites comprise. PFGE provides very high resolving power, and there is a convention that even if there is up to three band differences between the fingerprints of different isolates, very recent epidemiological linkage between these isolates cannot be ruled out [76]. It should be noted that this convention is not fully accepted, and numerous variants have been proposed. There is sufficient detail in the data that relative evolutionary distances can be estimated and phylogenetic trees constructed [92]. PFGE remains an extensively used method. For instance, it is arguably the gold standard for monitoring *Staphylococcus aureus* dissemination in Australia [55].

Methods that use DNA synthesis to convert a genome into fragments for electrophoresis are essentially all based on the PCR. The simplest approach is termed random amplification of polymorphic DNA (RAPD) analysis [58]. This makes use of one or two (generally one) 10mer PCR primers of random sequence. PCR products will form whenever by chance two primers molecules anneal to the template in opposite orientations with their 3′ends facing towards each other, and at a distance that is not too large for PCR product synthesis. The resulting fragments are quite small and can easily be resolved by conventional agarose electrophoresis. A persistent problem with RAPD analysis is poor reproducibility, particularly between laboratories. This is probably due to the weak annealing between the very short 10mer primers and their targets which may not even be 100% complementary with the primer, and consequent extreme sensitivity to reaction conditions.

A variation of the RAPD approach that is designed to have better reproducibility is based upon sequences that are repeated throughout bacterial genomes. These are known variously as "REP", "ERIC", and "BOX" sequences [5, 12, 74, 89], depending upon the sequence and the bacterial species, and there is considerable strain to strain variation regarding just where these repeats are. PCR-based typing methods have been developed that are very similar to RAPD methods, but make use of primers that target these repeat sequences. These methods are less prone to reproducibility problems, probably because of stronger annealing between the primers and their genuine targets.

Amplified fragment length polymorphism (AFLP) analysis combines both genome cleavage, and the synthesis of PCR products [23, 38]. The principal behind this method is the cleavage of the genome with two restriction enzymes so as to generate a large number of fragments, and then the amplification by PCR of a subset of these fragments. The number of fragments in the subset is optimised so as to provide a good compromise being large enough to provide high discriminatory power and small enough to for clear band separation. The fragments are amplified by first ligating adaptors to the sticky ends, and then performing PCR using primers that target the adaptor sequences. The selectivity of the amplification is obtained by using primers with 3′ extensions that will therefore only anneal to adaptors that by chance are adjacent to a particular base(s) in the amplified fragments, and also by labelling only one primer, so that fragments derived purely from the other primer are not visualised. In general, denaturing thin polyacrylamide slab gels (i.e. DNA sequencing gels) or capillary electrophoresis are used to separate the amplified fragments. AFLP analysis is highly reproducible and discriminatory. Like PFGE, there is sufficient detail in the fingerprints to allow estimation of relative evolutionary distances [46].

2.6.5 Typing Methods Based upon Known Polymorphic Genes or Sites

The second broad class of DNA-based bacterial typing methods is based on the interrogation of genes that are known to be variable, or specific variable sites within such genes. In general, with these methods it is possible to directly infer the genetic changes that give rise to different types, although there is some variation from method to method with the precision that this can be achieved.

The earliest described examples involve restriction fragment length polymorphism (RFLP) analysis of highly variable loci. This can be done either by cleaving the entire genome with a restriction enzyme and then visualising the fragments of interest by Southern hybridisation, or by amplifying the locus of interest by PCR, and then carrying out the restriction digestion after that. Most methods involve analysis of only one locus. Ribotyping is an automatable Southern hybridization–based method that targets the genes that encode ribosomal RNA, as well as spacer regions [9, 13]. A well known example of a PCR-based method is *flaA* RFLP analysis in *Campylobacter jejuni* and *Campylobacter coli* [53]. In recent years RFLP-based methods have largely been superseded. This is because they are inherently multi-step to perform, and not all sequence changes will change the location of the restriction sites, thus limiting resolving power. In addition, with these methods, direct inference of the precise genetic changes that lead to changes in banding patterns is not possible.

Probably the most conceptually straightforward approach to bacterial typing is sequencing one or more genes or gene fragments. There are several examples of single locus-based sequence typing methods [6, 20, 25, 45, 47]. Many of these target genes that are hypervariable, often because of immune selection directed against the surface-located gene product. These methods are often very informative and efficient. However, they are limited in that horizontal gene transfer (HGT) can mean that a single gene is an inadequate marker for the evolutionary position of an entire genome. Therefore, single locus sequence typing can be very effective for testing hypotheses of epidemiological linkage but less effective for studies of population structure. The response to this was the development just over a decade ago of multilocus sequence typing (MLST) [44]. This involves sequencing of standardized fragments of approximately 450 bp from multiple (almost always seven) housekeeping genes. Housekeeping genes encode cytoplasmic enzymes involved in core metabolism or other fundamental cellular processes. They therefore do not evolve by positive selection, so sequence differences accumulate in a clock-like fashion and do not reflect differing selective pressures in different strains or lineages. A very important aspect of MLST is on-line databases of variants at the loci (alleles), and the alleles that are found together in isolates (sequence types). These web sites also contain information regarding the isolates, and a variety of analytical tools. They are a really powerful resource for studying bacterial population structures. There are now MLST schemes for essentially all the major bacterial pathogens, and the associated web sites can be reached via http://www.mlst.net/ and http://pubmlst.org/. MLST is still primarily a research tool, as carrying out seven sequence determinations remains a time and cost challenge for routine high throughput applications. Also, because slowly evolving housekeeping genes are targeted, the resolution can be insufficient for testing hypotheses of very recent epidemiological linkage.

An interesting development in very recent years is the appearance of methods in which the MLST loci are analysed by mass spectrometry rather than sequence analysis [21, 28]. In these cases, it is base composition rather than the sequence that is obtained. By definition this has less information, but this does not seem to significantly degrade the performance of the MLST.

Another approach derived from MLST has been the use of bioinformatic methods to derive resolution optimised sets of single nucleotide polymorphisms (SNPs) from MLST databases, or similar compendia of known sequence variation [29, 60, 72]. The descriptor of resolution can be the power to identify particular sequence types

(STs) or groups of STs, or alternatively the power to discriminate all STs from all STs. In the latter case, this is assessed by calculating the Simpsons Index of Diversity which in this context is the probability that any two STs selected at random will be discriminated if the SNPs are interrogated. Published SNP-based bacterial typing methods in general make use of allele specific PCR, or competitive hybridization of Taqman probes, in real-time PCR devices [8, 60]. This typing format is attractive because real-time PCR is inherently single-step closed-tube. However, there are many methods for interrogating SNPs and in principal any could be applied to SNP-based bacterial typing.

The housekeeping genes used in MLST evolve slowly, and this limits MLST resolution. A direct approach to circumventing this is to use loci that evolve more rapidly. Microsatellite loci in multicellular organisms are composed of, or contain, sequence repeats. The repeating units are very short; 1–∼6 nucleotides. They evolve rapidly due to slipped strand mis-pairing during DNA replication. It is now known that repeat-containing loci are not hard to find in bacteria, and, as expected, they also evolve rapidly and so have high informative power for typing. In bacteria, they are known as variable number tandem repeat (VNTR) loci. There have been many recent publications concerning VNTR-based bacterial typing methods, and this is a small sample [30, 46, 56, 62, 68, 79, 82]. The repeating units in bacterial VNTR loci are more variable in length than microsatellites in higher organisms, but they appear to be usefully polymorphic whatever the repeat length. Variation in loci with long repeat lengths is probably due to homologous recombination rather than slipped strand mis-pairing. VNTR based tying methods can be single locus or multi-locus. Recent trends are towards multilocus methods, and these are known generically as "multilocus VNTR analysis" (MLVA). VNTR loci can be interrogated by DNA sequencing, or, as is commonly done with microsatellite analysis, by length determination. Length-based MLVA methods are inherently efficient because the loci can be amplified in a multiplex PCR reaction using primers labelled with different fluorophores, and the products then resolved by capillary electrophoresis. However, with complex repeat loci, in which not all the repeating units are identical, sequence differences can be missed when only the length is determined.

MLVA methods are very promising, but have one inherent limitation in that VNTR loci can evolve in a reversible fashion i.e. repeat units can be lost then gained again or *vice versa*. Thus the same allelic states can result from different evolutionary histories i.e. homoplasy can occur. This can on occasion confound MLVA analysis. A strategy to circumvent this is to use a combination of markers that include both VNTRs and more slowly evolving markers that are less prone to homoplasy, such as SNPs that define phylogentic lineages. This strategy has been termed phylogenetic hierarchical analyses using nucleic acids (PHRANA) [33, 34]. The rationale is that the slowly evolving markers divide the species into lineages in which the average evolutionary distance between strains is sufficiently low that the probability of homoplasy in the VNTR markers is greatly reduced. In reality, there are innumerable possible combinations of rapidly and slowly evolving markers, and they will all provide a particular compromise between resolving power, lack of homoplasy, and ease and cost of execution. One emerging example is the use of sequences of multiple rapidly evolving surface antigen encoding genes to type *Neisseria gonhorrhoeae* [85].

A completely different approach to bacterial typing is to make use of variations in gene content, rather than gene sequence. One of the major unexpected findings to emerge from the explosion of bacterial genome sequencing in the last decade has been the extent of variation in gene content between different isolates in the same species. The term "pan-genome" has been coined to describe the gene content of a species [77], and a pan-genome can contain many more genes than any cell. It therefore follows that typing on the basis of gene presence or absence can have considerable resolving power. This approach is sometimes termed binary typing, because the informative genes exhibit binary variation i.e. they exist in two states: present or not present. The fact that many determinants of virulence and resistance to antimicrobials are carried on mobile elements that are inherently likely to exhibit binary variation increases the potential informative power of this approach. Micro-array technology has been used to study genome-wide binary variation. Comparative genome hybridization (CGH) arrays are used for this task [75, 93]. They are generally equipped with probes deduced from all the open reading frames in multiple genome sequences within a species. Analysis by CGH array is too laborious to be regarded as a bacterial typing method, although smaller arrays with probes for e.g. putative virulence factors are increasingly being applied to the analysis of large numbers of isolates [50].

Binary typing is also used to analyse individual hypervariable loci. This is usually done by conventional multiplex PCR, or real-time PCR. The typing of the SCC*mec* mobile element that defines MRSA is a good example of this [16]. SCC*mec* exhibits particularly complex binary variation. In general it is typed using markers or combinations of markers that are diagnostic for known types. However, an alternative approach to marker selection involving maximisation of Simpsons Index of Diversity has been reported [71]. This approach has also been applied to the derivation of resolution-optimised sets of binary markers from CGH data [59].

Essentially any method that detects the presence of specific genes can be applied to binary gene-based bacterial typing. As mentioned above, large scale analysis of binary variation is carried out using micro-arrays. Many extant methods for smaller scale analyses make extensive use of multiplex PCR and conventional agarose gel electrophoresis. Real-time PCR is increasingly being used. Interestingly, a nylon membrane-based reverse line blot technique that is essentially a precursor of array technology has been shown to remain effective and competitive as a medium-high throughput method [35].

The ultimate genetic fingerprint is a complete genome sequence. The emergence of next generation sequencing devices in the last 5 years has resulted in a large reduction in the cost of genome sequencing [3]. However the cost of this is still much higher than the cost of other typing methods, in terms of consumables, instrumentation and time, so this approach is not currently viable as a routine high-throughput typing method.

2.6.6 Projections of Future Developments

The explosive increase in comparative genome information will continue, and this will lead to increasingly detailed and sophisticated understanding of bacterial population structures. This in turn will make it increasingly difficult to justify using methods that do not clearly place an isolate within a population structure defined by comparative genomics. This probably means that typing methods that interrogate known polymorphic sites or loci will remain in the ascendancy, while those that generate anonymous banding patterns may lose favour. There will probably continue to be a range of techniques. It is easy to envisage a cut-down version of next-generation sequencing that derives sequences from random genome fragments. It could be that in an environment where there are thousands of complete genomes to serve as on-line comparators, sequencing random genes will be just as effective for rapidly obtaining a high resolution fingerprint as sequencing targeted genes. This approach could perhaps be applied directly to mixed cultures or clinical samples.

One method not discussed above is high resolution melting (HRM) analysis. This is rapidly emerging as a very robust and effective approach to resolving sequencing variants [90]. The basis of its most common embodiment is the accurate monitoring of the reduction in fluorescence, as DNA stained with a double strand specific fluorescent dye melts in response to a controlled temperature increase. The attraction of HRM is that the amplification plus the HRM analysis constitutes a homogeneous, single step and closed tube procedure. Its potential for miniaturization is excellent because of a lack of requirement for moving parts and/or microfluidics. HRM analysis can be added to pre-existing PCR-based primary diagnosis or binary typing methods so as to provide additional information at essentially no cost. The consumables costs of stand-alone HRM analyse are typically less than \$1.00, because the requirements are only unlabelled primers and a generic PCR master mix. HRM analysis has been shown to be able to discriminate multiple sequence variants [14, 24, 37, 48, 73], and to yield data that can be compared between laboratories [78], which raises the possibility of on-line comparison of HRM curves, in a manner analogous to sequence comparison using Genbank or MLST sites.

In summary, a reasonable prediction is that in the near future, bacterial typing will be performed by either some variant of next-generation sequencing, or by HRM analysis of selected markers, depending on the amount of information required. Analysis of mixed/clinical samples will become commonplace, and typing will be combined with diagnosis. Typing data will be interpreted with reference to massive amounts of comparative genomic data, and this will greatly facilitate the monitoring of dissemination at all scales of time and space, and the inference of clinically relevant properties.

References

1. Afonina IA, Reed MW, Lusby E et al (2002) Minor groove binder-conjugated DNA probes for quantitative DNA detection by hybridization-triggered fluorescence. Biotechniques 32:940–944, 946–949
2. Anderson ES, Ward LR, Saxe MJ et al (1977) Bacteriophage-typing designations of *Salmonella typhimurium*. J Hyg (Lond) 78:297–300

3. Ansorge WJ (2009) Next-generation DNA sequencing techniques. N Biotechnol 25:195–203
4. Anthony RM, Brown TJ, French GL (2000) Rapid diagnosis of bacteremia by universal amplification of 23S ribosomal DNA followed by hybridization to an oligonucleotide array. J Clin Microbiol 38:781–788
5. Baldy-Chudzik K (2001) Rep-PCR – a variant to RAPD or an independent technique of bacteria genotyping? A comparison of the typing properties of rep-PCR with other recognised methods of genotyping of microorganisms. Acta Microbiol Pol 50:189–204
6. Beall B, Facklam R, Thompson T (1996) Sequencing emm-specific PCR products for routine and accurate typing of group A *Streptococci*. J Clin Microbiol 34:953–958
7. Bernard PS, Lay MJ, Wittwer CT (1998) Integrated amplification and detection of the C677T point mutation in the methylenetetrahydrofolate reductase gene by fluorescence resonance energy transfer and probe melting curves. Anal Biochem 255:101–107
8. Best EL, Fox AJ, Owen RJ et al (2007) Specific detection of *Campylobacter jejuni* from faeces using single nucleotide polymorphisms. Epidemiol Infect 135:839–846
9. Bingen EH, Denamur E, Elion J (1994) Use of ribotyping in epidemiological surveillance of nosocomial outbreaks. Clin Microbiol Rev 7:311–327
10. Boom et al (1990) Rapid and simple method for purification of nucleic acids. J Clin Microbiol 28:495–503
11. Bretagne S, Costa J-M (2006) Towards a nucleic acid-based diagnosis in clinical parasitology and mycology. Clin Chim Acta 363:221–228
12. Brusetti L, Malkhazova I, Gtari M et al (2008) Fluorescent-BOX-PCR for resolving bacterial genetic diversity, endemism and biogeography. BMC Microbiol 8:220
13. Castanheira M, Bell JM, Turnidge JD et al (2009) Carbapenem resistance among *Pseudomonas aeruginosa* strains from India: evidence for nationwide endemicity of multiple metallo-beta-lactamase clones (VIM-2, -5, -6, and -11 and the newly characterized VIM-18). Antimicrob Agents Chemother 53:1225–1227
14. Chan WF, Maharjan RP, Reeves PR et al (2009) Rapid and accurate typing of *Bordetella pertussis* targeting genes encoding acellular vaccine antigens using real time PCR and high resolution melt analysis. J Microbiol Methods 77:326–329
15. Chomczynski P, Sacchi N (1987) Single-step method of RNA isolation by acid guanidinium thiocyanate-phenol-chloroform extraction. Meth Anal Biochem 162:156–159
16. Chongtrakool P, Ito T, Ma XX et al (2006) Staphylococcal cassette chromosome mec (SCCmec) typing of methicillin-resistant *Staphylococcus aureus* strains isolated in 11 Asian countries: a proposal for a new nomenclature for SCCmec elements. Antimicrob Agents Chemother 50:1001–1012
17. Csako G (2006) Present and future of rapid and/or high-throughput methods for nucleic acid testing. Clin Chim Acta 363:6–31
18. Dagan R (2009) Impact of pneumococcal conjugate vaccine on infections caused by antibiotic-resistant *Streptococcus pneumoniae*. Clin Microbiol Infect 15 Suppl 3: 16–20
19. Dominguez SR, Briese T, Palacios G, Hui J, Villari J, Kapoor V, Tokarz R, Glodé MP, Anderson MS, Robinson CC, Holmes KV, Lipkin WI (2008) Multiplex mass tag-PCR for respiratory pathogens in pediatric nasopharyngeal washes negative by conventional diagnostic testing shows a high prevalence of viruses belonging to a newly recognised rhinovirus clade. J Clin Virol 43:219–222
20. Dyet KH, Martin DR (2006) Clonal analysis of the serogroup B meningococci causing New Zealand's epidemic. Epidemiol Infect 134:377–383
21. Ecker DJ, Massire C, Blyn BL et al (2009) Molecular genotyping of microbes by multilocus PCR and mass spectrometry: a new tool for hospital infection control and public health surveillance. Methods Mol Biol 551:71–87
22. Faria NA, Carrico JA, Oliveira DC et al (2008) Analysis of typing methods for epidemiological surveillance of both methicillin-resistant and methicillin-susceptible *Staphylococcus aureus* strains. J Clin Microbiol 46:136–144

23. Fry NK, Savelkoul PH, Visca P (2009) Amplified fragment-length polymorphism analysis. Methods Mol Biol 551:89–104
24. Ghorashi SA, Noormohammadi AH, Markham PF (2010) Differentiation of *Mycoplasma gallisepticum* strains using PCR and high resolution melting curve analysis. Microbiology 156:1019–1029
25. Goering RV, Morrison D, Al-Doori Z et al (2008) Usefulness of mec-associated direct repeat unit (dru) typing in the epidemiological analysis of highly clonal methicillin-resistant *Staphylococcus aureus* in Scotland. Clin Microbiol Infect 14:964–969
26. Heid CA, Stevens J, Livak KJ et al (1996). Real time quantitative PCR. Genome Res 6:986–994
27. Holland PM, Abramson RD, Watson R (1991) Detection of specific polymerase chain reaction product by utilizing the 5'–3' exonuclease activity of *Thermus aquaticus* DNA polymerase. Proc Natl Acad Sci U S A 88:7276–7280
28. Honisch C, Chen Y, Mortimer C et al (2007) Automated comparative sequence analysis by base-specific cleavage and mass spectrometry for nucleic acid-based microbial typing. Proc Natl Acad Sci U S A 104:10649–10654
29. Huygens F, Inman-Bamber J, Nimmo GR et al (2006) *Staphylococcus aureus* genotyping using novel real-time PCR formats. J Clin Microbiol 44:3712–3719
30. Hyytia-Trees E, Lafon P, Vauterin P et al (2009) Multilaboratory validation study of standardized multiple-locus variable-number tandem repeat analysis protocol for shiga toxin-producing *Escherichia coli* O157: a novel approach to normalize fragment size data between capillary electrophoresis platforms. Foodborne Pathog Dis 23:112–117
31. Ishiguro T, Saitoh J, Yawata H et al Y (1995) Homogeneous quantitative assay of hepatitis C virus RNA by polymerase chain reaction in the presence of a fluorescent intercalater. Anal Biochem 229:207–213
32. Ishizawa M, Kobaashi Y, Miyama T, et al (1991) Simple procedure for DNA isolation from Human Serum. Nucleic Acids Res 19:5792
33. Keim P, Pearson T, Okinaka R (2008) Microbial forensics: DNA fingerprinting of Bacillus anthracis (anthrax). Anal Chem 80:4791–4799
34. Keim P, Van Ert MN, Pearson T et al (2004) Anthrax molecular epidemiology and forensics: using the appropriate marker for different evolutionary scales. Infect Genet Evol 4:205–213
35. Kong F, Gilbert GL (2006) Multiplex PCR-based reverse line blot hybridization assay (mPCR/RLB) – a practical epidemiological and diagnostic tool. Nat Protoc 1:2668–2680
36. Kropinski AM, Sulakvelidze A, Konczy P et al (2007) Salmonella phages and prophages – genomics and practical aspects. Methods Mol Biol 394:133–175
37. Kugelman G, Tapsall JW, Goire N et al (2009) Simple, rapid, and inexpensive detection of *Neisseria gonorrhoeae* resistance mechanisms using heat-denatured isolates and SYBR green-based real-time PCR. Antimicrob Agents Chemother 53:4211–4216
38. Lan R, Reeves PR (2007) Amplified fragment length polymorphism analysis of *Salmonella enterica*. Methods Mol Biol 394:119–132
39. Lay MJ, Wittwer CT (1997) Real-time fluorescence genotyping of factor V Leiden during rapid-cycle PCR. Clin Chem 43:2262–2267
40. Leach AJ, Morris PS, McCallum GB et al (2009) Emerging pneumococcal carriage serotypes in a high-risk population receiving universal 7-valent pneumococcal conjugate vaccine and 23-valent polysaccharide vaccine since 2001. BMC Infect Dis 9:121
41. Lee LG, Connell CR, Bloch W (1993) Allelic discrimination by nick-translation PCR with fluorogenic probes. Nucleic Acids Res 21: 3761–3766
42. Livak KJ, Flood SJ, Marmaro J et al (1995) Oligonucleotides with fluorescent dyes at opposite ends provide a quenched probe system useful for detecting PCR product and nucleic acid hybridization. PCR Methods Appl 4:357–362
43. Mackay IM, Mackay JF, Nissen MD et al (2007) Real-time PCR: history and fluorescent chemistries. In: Mackay IM (ed) Real-Time PCR in Microbiology: From Diagnosis to Characterisation, 1st edn. Caister Academic Press, UK

44. Maiden MC, Bygraves JA, Feil E et al (1998) Multilocus sequence typing: a portable approach to the identification of clones within populations of pathogenic microorganisms. Proc Natl Acad Sci U S A 95:3140–3145
45. Meinersmann RJ, Helsel LO, Fields PI et al (1997) Discrimination of *Campylobacter jejuni* isolates by fla gene sequencing. J Clin Microbiol 35:2810–2814
46. Melles DC, Schouls L, Francois P et al (2009) High-throughput typing of *Staphylococcus aureus* by amplified fragment length polymorphism (AFLP) or multi-locus variable number of tandem repeat analysis (MLVA) reveals consistent strain relatedness. Eur J Clin Microbiol Infect Dis 28:39–45
47. Mellmann A, Weniger T, Berssenbrugge C et al (2008) Characterization of clonal relatedness among the natural population of *Staphylococcus aureus* strains by using spa sequence typing and the BURP (based upon repeat patterns) algorithm. J Clin Microbiol 46:2805–2808
48. Merchant-Patel S, Blackall PJ, Templeton J et al (2010) *Campylobacter jejuni* and *Campylobacter coli* genotyping by high resolution melting analysis of a flaA fragment. Appl Environ Microbiol 76:493–499
49. Mhlanga MM, Malmberg L (2001) Using molecular beacons to detect single-nucleotide polymorphisms with real-time PCR. Methods 25:463–471
50. Monecke S, Berger-Bachi B, Coombs G et al (2007) Comparative genomics and DNA array-based genotyping of pandemic *Staphylococcus aureus* strains encoding Panton-Valentine leukocidin. Clin Microbiol Infect 13:236–249
51. Monis PT, Giglio S, Saint CP (2005) Comparison of SYTO9 and SYBR Green I for real-time polymerase chain reaction and investigation of the effect of dye concentration on amplification and DNA melting curve analysis. Anal Biochem 340:24–34
52. Mothershed EA, Whitney AM (2006) Nucleic acid-based methods for the detection of bacterial pathogens: present and future considerations for the clinical laboratory. Clin Chim Acta 363:206–220
53. Nachamkin I, Bohachick K, Patton CM (1993) Flagellin gene typing of *Campylobacter jejuni* by restriction fragment length polymorphism analysis. J Clin Microbiol 31:1531–1536
54. Nadder TS, Langley MR (2001) The new millennium laboratory: molecular diagnostics goes clinical. Clin Lab Sci 14:252–259
55. Nimmo GR, Coombs GW (2008) Community-associated methicillin-resistant *Staphylococcus aureus* (MRSA) in Australia. Int J Antimicrob Agents 31:401–410
56. Octavia S, Lan R (2009) Multiple-locus variable-number tandem-repeat analysis of *Salmonella enterica* serovar Typhi. J Clin Microbiol 47:2369–2376
57. Peters TM (2009) Pulsed-field gel electrophoresis for molecular epidemiology of food pathogens. Methods Mol Biol 551:59–70
58. Power EG (1996) RAPD typing in microbiology – a technical review. J Hosp Infect 34: 247–265
59. Price EP, Huygens F, Giffard PM (2006) Fingerprinting of *Campylobacter jejuni* by using resolution-optimized binary gene targets derived from comparative genome hybridization studies. Appl Environ Microbiol 72:7793–7803
60. Price EP, Thiruvenkataswamy V, Mickan L et al (2006) Genotyping of *Campylobacter jejuni* using seven single-nucleotide polymorphisms in combination with flaA short variable region sequencing. J Med Microbiol 55:1061–1070
61. Rodas C, Iniguez V, Qadri F, Wiklund G, Svennerholm A-M, Sjöling Å (2009) Development of multiplex PCR assays for detection of enterotoxigenic *Escherichia coli* colonization factors and toxins. J Clin Microbiol 47:1218–1220
62. Ross IL, Parkinson IH, Heuzenroeder MW (2009) The use of MAPLT and MLVA analyses of phenotypically closely related isolates of *Salmonella enterica* serovar Typhimurium. Int J Med Microbiol 299:37–41
63. Sambrook J, Russell DW (2001) Molecular Cloning – A Laboratory Manual. Cold Spring Harbor Laboratory Press, Cold Spring Harbor, NY
64. Schmitt M, Bravo IG, Snijders PJF, Gissmann L, Pawlita M, Waterboer T (2006) Bead-based multiplex genotyping of human papillomaviruses. J Clin Microbiol 44:504–512

65. Schneeberger C, Speiser P, Kury F et al (1995) Quantitative detection of reverse transcriptase-PCR products by means of a novel and sensitive DNA stain. PCR Methods Appl 4: 234–238
66. Seng P, Drancourt M, Gouriet F et al (2009) Ongoing revolution in bacteriology: routine identification of bacteria by matrix-assisted laser desorption ionization time-of-flight mass spectrometry. Clin Infect Dis 49:543–551
67. Shipley GL (2006) An introduction to real-time PCR. In: Dorak MT (ed) Real-Time PCR, 1st edn. Taylor & Francis, UK
68. Slack A, Symonds M, Dohnt M et al (2006) An improved multiple-locus variable number of tandem repeats analysis for *Leptospira interrogans* serovar Australis: a comparison with fluorescent amplified fragment length polymorphism analysis and its use to redefine the molecular epidemiology of this serovar in Queensland, Australia. J Med Microbiol 55:1549–1557
69. Stanley K, Szewezuk E (2005) Multiplexed tandem PCR: gene profiling from small amounts of RNA using SYBR Green detection. Nucleic Acids Res 33:e180
70. Steiner JJ, Poklemba CJ, Fjellstrom RG et al (1995) A rapid one-tube genomic DNA extraction process for PCR and RAPD analyses. Nucleic Acids Res 23:2569–2570
71. Stephens AJ, Huygens F, Giffard PM (2007) Systematic derivation of marker sets for *Staphylococcal* cassette chromosome mec typing. Antimicrob Agents Chemother 51:2954–2964
72. Stephens AJ, Huygens F, Inman-Bamber J et al (2006) Methicillin-resistant *Staphylococcus aureus* genotyping using a small set of polymorphisms. J Med Microbiol 55:43–51
73. Stephens AJ, Inman-Bamber J, Giffard PM et al (2008) High-resolution melting analysis of the spa repeat region of *Staphylococcus aureus*. Clin Chem 54:432–436
74. Syrmis MW, O'Carroll MR, Sloots TP et al (2004) Rapid genotyping of *Pseudomonas aeruginosa* isolates harboured by adult and paediatric patients with cystic fibrosis using repetitive-element-based PCR assays. J Med Microbiol 53:1089–1096
75. Taboada EN, Mackinnon JM, Luebbert CC et al (2008) Comparative genomic assessment of multi-locus sequence typing: rapid accumulation of genomic heterogeneity among clonal isolates of *Campylobacter jejuni*. BMC Evol Biol 8:229
76. Tenover FC, Arbeit RD, Goering RV et al (1995) Interpreting chromosomal DNA restriction patterns produced by pulsed-field gel electrophoresis: criteria for bacterial strain typing. J Clin Microbiol 33:2233–2239
77. Tettelin H, Riley D, Cattuto C et al (2008) Comparative genomics: the bacterial pan-genome. Curr Opin Microbiol 11:472–477
78. Tong SY, Lilliebridge RA, Holt DC et al (2009) High-resolution melting analysis of the spa locus reveals significant diversity within sequence type 93 methicillin-resistant *Staphylococcus aureus* from northern Australia. Clin Microbiol Infect 15:1126–1131
79. U'Ren JM, Schupp JM, Pearson T et al (2007) Tandem repeat regions within the *Burkholderia pseudomallei* genome and their application for high resolution genotyping. BMC Microbiol 7:23
80. van Belkum A, Tassios PT, Dijkshoorn L et al (2007) Guidelines for the validation and application of typing methods for use in bacterial epidemiology. Clin Microbiol Infect 13 Suppl 3:1–46
81. van Doorn L-J, Moljin A, Kleter B, Quint W, Colau B (2006) Highly effective detection of human papillomavirus 16 and 18 DNA by a testing algorithm combining broad-spectrum and type-specific PCR. J Clin Microbiol 44:3292–3298
82. Vu-Thien H, Corbineau G, Hormigos K et al (2007) Multiple-locus variable-number tandem-repeat analysis for longitudinal survey of sources of *Pseudomonas aeruginosa* infection in cystic fibrosis patients. J Clin Microbiol 45:3175–3183
83. Wan E, Akana M, Pons J et al (2010) Green technologies for room temperature nucleic acid storage. Curr Issues Mol Biol 12:135–142
84. Weile J, Knabbe C (2009) Current applications and future trends of molecular diagnostics in clinical bacteriology. Anal Bioanal Chem 394:731–742

85. Whiley DM, Goire N, Ray ES et al (2010) *Neisseria gonorrhoeae* multi-antigen sequence typing using non-cultured clinical specimens. Sex Transm Infect 86:51–55
86. Whiley DM, Mackay IM, Syrmis MW et al (2004) Detection and differentiation of herpes simplex virus types 1 and 2 by a duplex LightCycler PCR that incorporates an internal control PCR reaction. J Clin Virol 30:32–38
87. Whiley DM, Sloots TP (2005) Sequence variation in primer targets affects the accuracy of viral quantitative PCR. J Clin Virol 34:104–107
88. Whiley DM, Sloots TP (2006) Sequence variation can affect the performance of minor groove binder TaqMan probes in viral diagnostic assays. J Clin Virol 35:81–83
89. Wilson MK, Lane AB, Law BF et al (2009) Analysis of the pan genome of *Campylobacter jejuni* isolates recovered from poultry by pulsed-field gel electrophoresis, multilocus sequence typing (MLST), and repetitive sequence polymerase chain reaction (rep-PCR) reveals different discriminatory capabilities. Microb Ecol 58:843–855
90. Wittwer CT (2009) High-resolution DNA melting analysis: advancements and limitations. Hum Mutat 30:857–859
91. Wittwer CT, Herrmann MG, Moss AA et al (1997) Continuous fluorescence monitoring of rapid cycle DNA amplification. Biotechniques 22:130–131, 134–138
92. Xi M, Zheng J, Zhao S et al (2008) An enhanced discriminatory pulsed-field gel electrophoresis scheme for subtyping *Salmonella* serotypes Heidelberg, Kentucky, SaintPaul, and Hadar. J Food Prot 71:2067–2072
93. Zhang L, Foxman B, Drake DR et al (2009) Comparative whole-genome analysis of *Streptococcus mutans* isolates within and among individuals of different caries status. Oral Microbiol Immunol 24:197–203

Chapter 3
Laboratory Accreditation Standards

R. Wood and G. Barber

Abstract The role of a medical molecular microbiology laboratory is to consistently produce reliable and meaningful clinical data, essential for both health assessment and disease control. This role defines the standard for quality. Quality is the result of planned and monitored activity. It requires the setting of goals and the monitoring and analysis of processes and procedures to ensure the required outcome is reliably and repeatedly attained. This includes the need to consider how quality is attained when the laboratory produces its assays "in-house". The importance of activities to ensure ongoing quality must be acknowledged and supported appropriately from the highest levels of management. For Australian medical molecular microbiology laboratories, the decision to achieve and maintain quality is not optional, it is mandated.

Keywords Quality · Legislation · Accreditation · Standards · Regulation · IVD · Risk management · Validation · Verification · Quality management · Quality assurance · Quality control · Security · OGTR · Compliance

3.1 Quality in the Medical Molecular Microbiology Laboratory

R. Wood

3.1.1 Part 1: Overview

3.1.1.1 Quality and the Role of the Laboratory

The role of a medical molecular microbiology laboratory is to consistently produce reliable and meaningful clinical data, essential for both health assessment and

---

R. Wood (✉)
MASM, Consultant to the Department of Essential Health Technologies,
World Health Organisation, Switzerland
e-mail: woodro4@gmail.com

M. Schuller et al. (eds.), *PCR for Clinical Microbiology*,
DOI 10.1007/978-90-481-9039-3_3,

disease control. This role defines the standard for quality. Quality in the laboratory is referred to as good laboratory practice (GLP) and is achieved through the successful interpretation and ongoing implementation of recognized standards and practices. The International Organisation for Standardization/International Electrotechnical Commission (ISO/IEC) Guide 2:1996 "Standardization and related activities – General vocabulary" defines a standard as: "A document established by consensus and approved by a recognized body that provides for common and repeated use, rules, guidelines or characteristics for activities or their results, aimed at the achievement of the optimum degree of order in a given context". Standards include such guidance documents as standards, codes, specifications, handbooks and guidelines. A standard sets a minimum benchmark for quality.

3.1.1.2 Legislating for Quality

In Australian medical laboratories, attainment of quality is acknowledged through attainment of relevant accreditation. During accreditation, a laboratory has its processes, procedures and outputs investigated to ascertain whether it has successfully interpreted and implemented relevant standards. Accreditation is often a voluntarily activity, however for Australian medical laboratories, it is a legislated requirement.

Accreditation provides the assurance of quality laboratory practice for Australian governments, and as such serves as the mechanism that enables a laboratory to receive government reimbursement for testing activities. The National Association of Testing Authorities/Royal College of Pathologists of Australasia (NATA/ RCPA) is the accreditation body for these purposes appointed by the Commonwealth, auditing compliance to standards and guidelines identified and mandated through the Commonwealth's National Pathology Accreditation Advisory Council (NPAAC) process. Consequent to the requirement of accreditation for reimbursement, most medical laboratories in Australia operate using GLP.

The role of NPAAC includes the requirement to ensure the mandated standards and guidelines are relevant and current. However, the creation of relevant standards and the legal processes for mandating the use of a laboratory standard will always be retrospective to the introduction of its associated test or testing process. The onus lies therefore on the laboratory to ensure that quality practices occur, i.e. standards are identified and are voluntarily implemented, even in the absence of a mandated procedure or principle.

3.1.1.3 Quality and Standards

There are two types of standards; those relating to principles for management, and technical standards. The recognition of international standards, when available, ensures that practices can be benchmarked world wide. Most scientists are cognisant of the ISO/IEC standard series. Other standards organisations contribute significantly to setting international benchmarks through the publication of highly technical standards. Familiarisation with the standards from these organisations is highly recommended. These include, but are not limited to, standards developed and recognized within Australia (Standards Australia), the United States of America (USA) such as Clinical and Laboratory Standards Institute (CLSI), European (European Committee for Standardisation or CEN) and British standards (BSI British Standards). A compendium of international and national standards and guidance relevant to a medical molecular microbiology laboratory has been compiled in an annex at the back of this chapter.

An additional source of information that will in parts, assist the laboratory in achieving quality in-house "production", is regulatory guidance prepared for industry by the USA Food and Drug Administration (FDA). Certain FDA guidance documents are also listed in the Annex.

3.1.1.4 Quality and Regulation

National regulation exists in some countries to ensure that country has access to safe, effective and high quality assays. Regulatory requirements often mandate specific activities to assure good manufacturing practice (GMP), that is, to assure the quality of a product by ensuring key attributes in the development and manufacturing stages. Implementation of relevant standards is a common regulatory requirement.

Many diagnostic assays manufactured commercially and supplied in Australia will have fulfilled the requirements of Australian, North American or European regulatory systems. This will usually indicate that the assays have been manufactured using a quality systems standard from the ISO/IEC series. Additionally assay design, validation and other aspects will often have been undertaken by the manufacturer with direct reference to standards from those series mentioned earlier.

The nature of most molecular microbiology laboratories is that although an increasing number of commercial assays are available, many of the tests performed by the laboratory are manufactured in-house, that is, by the laboratory. An in-house test is defined by NPAAC as "in-house in vitro diagnostic device" or, abbreviated, in-house IVD. It should be noted that a number of "kits" provided commercially are labelled "for research use only". For the laboratory to use these kits for clinical purposes, it must undertake steps to ensure that the clinical and analytical uses of the assay have been sufficiently validated. As such, research use only kits also fulfil the NPAAC definition of an in-house IVD if they are used for clinical purposes. If a laboratory changes a commercial assay in any manner, for example, to implement the use of a different sample type than that identified by the manufacturer, then this change also complies with the definition of an in-house IVD. The laboratory is responsible for validating the use of this sample type.

GMP is different to GLP. The goal of GLP is to ensure the reliability and relevance of results. Obviously the quality of the tests used is a major contributor to GLP. Standards for GLP will not necessarily address issues required to ensure GMP. When a laboratory undertakes to develop an assay, it should be cognisant of this difference, and consider the means by which GMP will be achieved for each batch of that assay to be developed. The concept of being a manufacturer is not necessarily at the forefront of a scientist's mind when "working up" a new in-house IVD.

The management standard identified as representing best practice to achieve GMP for a laboratory assay is ISO 13485:2005 "Medical devices – quality management systems – Requirements for regulatory purposes". Most major regulators throughout the world have adopted this standard or its principles in legislation.

NPAAC identified that elements of this ISO standard are already incorporated in the many standards mandated for laboratory accreditation, but recognised that not all requirements would be met through existing mandated standards. Consequently, NPAAC produced the standard "Requirements for the development and use of in-house in vitro diagnostic devices (IVDs) (2007 Edition)" to specifically address manufacturing quality with respect to in-house IVDs. Attention should be paid to each aspect of this standard to ensure ongoing GMP.

Australian laboratories need to remain aware of regulatory changes being introduced in Australia by the Therapeutic Goods Administration (TGA) that are relevant to both commercial and in-house IVDs. This new legislation will describe the requirements for ensuring quality, safety, and performance of an IVD. All laboratories producing in-house assays will be required to fulfill relevant regulatory requirements. Involvement of the regulator (TGA) in assessment of these aspects of an in-house IVD will depend of the risk class of the assay. For low-risk assays, NATA will perform this assessment function on behalf of the regulator. Laboratories should refer to the TGA website (www.tga.gov) for further information.

3.1.2 Part 2: Specific Aspects Relating to Quality

The above information has been written to provide an overview of the goals to be attained to ensure quality, and some well recognised tools to achieve these goals. To describe the requirements from even the general principle based standards would fill several chapters, yet at the risk of exclusion, three aspects stand out as worthy of further attention.

3.1.2.1 Risk Management

Risk management in the manufacture of an in-house IVD is addressed in the NPAAC standard "Requirements for the development and use of in-house in vitro diagnostic devices (IVDs) (2007 Edition)". Additionally, ISO/TS 22367:2008 "Medical laboratories – Reduction of error through risk management and continual improvement" addresses risk in the laboratory. More detail on risk management in the manufacture of IVDs can be found in ISO 14971:2007 "Medical devices – Application of risk management to medical devices". A separate annex in this standard addresses issues unique to the manufacture of IVDs.

Proper implementation of risk management in the production of an in-house assay will assist in identifying critical aspects associated with the manufacture of a quality device including the level of validation required for the assay, and the optimal controls that should be implemented.

3.1.2.2 Validation and Verification

The need to identify and validate both the clinical and analytical utility of the assay is of importance in the development of any assay to be used diagnostically. The concise identification of the clinical and analytical utilities should direct all other studies, including but not limited to those for sample type, testing populations, and interfering substances.

Understanding the difference between validation and verification is important to justify the degree of testing required to fulfill these requirements. According to ISO 9000:2001 "Quality management systems – Requirements", validation is "confirmation, through the provision of objective evidence, that requirements for a specific intended use or application have been fulfilled". This is subtly but significantly different from verification, which according to ISO 9000:2001 is "confirmation, through the provision of objective evidence, that specified requirements have been fulfilled".

To provide an example of the difference, one may consider the activities needed to provide sufficient clinical evidence for a gene to methicillin resistant *Staphylococcus aureus*. The clinical utility and limitations for use of the mecA gene are well documented, and therefore if the test was for the detection of this gene, a reference to the relevant literature and testing of a limited number of well defined clinical isolates may provide sufficient objective evidence. This is a verification process. If a new, previously unidentified gene is being utilised in the assay to detect additional drug resistance, the clinical utility of this novel gene in the diagnosis of these resistant organisms would require rigorous testing. In other words, the clinical utility of the detection of the new gene must be validated. Until there is sufficient evidence, results of this assay would require an accompanied comment to the referring practitioner indicating that the assay still only has "research use only" status, or similar information conveying the absence of sufficient proof of a clinical association. The NPAAC standard "Requirements for the development and use of in-house in vitro diagnostic devices (IVDs) (2007 Edition)" states that the following wording must be used in these circumstances: "The test used has not yet been validated to the current NPAAC standards and results should be interpreted accordingly".

3.1.2.3 Quality Assurance

To know quality is assured, the laboratory is responsible for monitoring key performance indicators and reacting according to the findings. A well recognised monitoring method is participation in relevant external quality assessment schemes (EQAS). It is important that the EQA scheme is recognised by NATA for accreditation purposes. In Australia, two national EQAS providers are available and both are responsive, where justified, in creating programs for markers not previously offered. An EQAS should offer not only a regular program using samples that will challenge the testing system, but also offer an educational component.

Laboratories with EQAS results outside those of the consensus must examine their procedures and provide justification if satisfied their results are correct, albeit different. All of these activities must be well documented and reviewed by laboratory management. During the accreditation process, an audit of the laboratory's successful participation in EQAS will be performed for all tests included in the scope of accreditation.

Use of Quality Control (QC) procedures should be instigated during the routine use of an assay. QC samples used to monitor a test run should be chosen carefully, so that they are sensitive indicators to significant variations possible for each assay. In choosing QC samples for an assay, consider those aspects of the assay that are critical to producing a reliable result. More than one QC sample per run may be warranted. Obvious examples include the incorporation of amplification controls, contamination controls, limit of detection controls or controls near a critical clinical decision point in quantitative assays.

However, QC procedures have broader quality applications. QC procedures should be implemented at all critical points in the development and manufacture of an in-house assay. For example, a relevant QC procedure is likely to include acceptability testing for new lots/batches of reagents. Acceptance limits must be set and actions defined if the limits are not met. These activities "build in" quality rather than function as monitors of a quality outcome. Both types of QC activity should be considered essential if the assay is manufactured in-house.

3.1.2.4 Conclusion

Quality is the result of planned and monitored activity. It requires the setting of goals and the thorough analysis of processes and procedures to ensure the required outcome is reliably and repeatedly attained. The importance of activities to ensure ongoing quality must be acknowledged and supported appropriately from the highest levels of management. For Australian medical molecular microbiology laboratories, the decision to achieve and maintain quality is not optional, it is mandated.

3.1.3 Annex

3.1.3.1 Compendium of Standards and Guidelines for Medical Molecular Microbiological Laboratories

ISO Standards

ISO 15189:2007

Medical laboratories – Particular requirements for quality and competence

ISO 15190:2003

Medical laboratories – Requirements for safety

ISO 15193:2009

In vitro diagnostic medical devices – Measurement of quantities in samples of biological origin – Requirements for content and presentation of reference measurement procedures

ISO 15194:2009
In vitro diagnostic medical devices – Measurement of quantities in samples of biological origin – Requirements for certified reference materials and the content of supporting documentation
ISO 15195:2003
Laboratory medicine – Requirements for reference measurement laboratories
ISO 15198:2004
Clinical laboratory medicine – In vitro diagnostic medical devices – Validation of user quality control procedures by the manufacturer
ISO 17511:2003
In vitro diagnostic medical devices – Measurement of quantities in biological samples – Metrological traceability of values assigned to calibrators and control materials
ISO/TS 22367:2008
Medical laboratories – Reduction of error through risk management and continual improvement
ISO/TS 22367:2008/Cor 1:2009
ISO/TR 22869:2005
Medical laboratories – Guidance on laboratory implementation of
ISO 15189: 2003
ISO 22870:2006
Point-of-care testing (POCT) – Requirements for quality and competence
ISO 13485:2003
Medical devices – Quality management systems – Requirements for regulatory purposes
ISO/TR 14969:2004
Medical devices – Quality management systems – Guidance on the application of ISO 13485: 2003
ISO 14971:2007
Medical devices – Application of risk management to medical devices
(Please note: This standard contains an annex specifically referring to the risk management of IVDs)
ISO/TR 16142:2006
Medical devices – Guidance on the selection of standards in support of recognized essential principles of safety and performance of medical devices

3.1.3.2 European Committee for Standardization

(CEN) European Standard (EN)
CEN EN 13612:2002
Performance evaluation of in-vitro diagnostic medical devices
CEN EN 13640:2002
Stability testing of in-vitro diagnostic reagents
CEN EN 13641:2002

Elimination or reduction of risk of infection related to in-vitro diagnostic reagents

CEN EN 13975:2003

Sampling procedures used for acceptance testing of in vitro diagnostic medical devices – Statistical aspects

CEN EN 13641:2002

Elimination or reduction of risk of infection related to in vitro diagnostic reagents

CEN EN14136:2004

Use of external quality assessment schemes in the assessment of the performance of in vitro diagnostic examination procedures

3.1.3.3 Clinical and Laboratory Standards Institute (CLSI) Standards

Evaluation of Precision Performance of Quantitative Measurement Methods; Approved Guideline – Second Edition

Evaluation of the Linearity of Quantitative Measurement Procedures: A Statistical Approach; Approved Guideline

Method Comparison and Bias Estimation Using Patient Samples; Approved Guideline – Second Edition

Preliminary Evaluation of Quantitative Clinical Laboratory Measurement Procedures; Approved Guideline – Third Edition

User Protocol for Evaluation of Qualitative Test Performance; Approved Guideline – Second Edition

Laboratory Statistics – Standard Deviation; A Report

Evaluation of Matrix Effects; Approved Guideline – Second Edition

User Verification of Performance for Precision and Trueness; Approved Guideline – Second Edition

Protocols for Determination of Limits of Detection and Limits of Quantitation; Approved Guideline

Risk Management Techniques to Identify and Control Laboratory Error Sources; Proposed Guideline

A Framework for NCCLS Evaluation Protocols; A Report

Estimation of Total Analytical Error for Clinical Laboratory Methods; Approved Guideline

Evaluation of Stability of In Vitro Diagnostic Method Products; Proposed Guideline

Assessment of the Clinical Accuracy of Laboratory Tests Using Receiver Operating Characteristic (ROC) Plots; Approved Guideline

Molecular Diagnostic Methods for Infectious Diseases; Approved Guideline – Second Edition

Using Proficiency Testing to Improve the Clinical Laboratory; Approved Guideline – Second Edition

Assessment of Laboratory Tests When Proficiency Testing is Not Available; Approved Guideline

United States of America Food and Drug Administration Guidance (FDA)
Class II Special Controls Guidance Document: Nucleic Acid Amplification Assay for the Detection of Enterovirus RNA
Establishing Performance Characteristics of In Vitro Diagnostic Devices for Detection or Detection and Differentiation of Influenza Viruses (Draft)
In Vitro Diagnostic Devices to Detect Influenza A Viruses: Labeling and Regulatory Path – Guidance for Industry and FDA Staff
Guidance for Industry and FDA Staff – Class II Special Controls Guidance Document: Reagents for Detection of Specific Novel Influenza A Viruses
Guidance for Industry and FDA Staff – Class II Special Controls Guidance Document: RNA Preanalytical Systems (RNA Collection, Stabilization and Purification Systems for RT-PCR used in Molecular Diagnostic Testing)
Guidance for Industry: Class II Special Controls Guidance Document: In Vitro HIV Drug Resistance Genotype Assay
Guidance for Industry: In the Manufacture and Clinical Evaluation of In Vitro Tests to Detect Nucleic Acid Sequences of Human Immunodeficiency Viruses Types 1 and 2

3.2 Security Sensitive Biological Agents (SSBA) and Office of the Gene Technology Regulator (OGTR) Standards

G. Barber

3.2.1 Security Sensitive Biological Agents (SSBA) Regulatory Scheme

The Security Sensitive Biological Agents (SSBA) Regulatory Scheme is established by the *National Health Security Act 2007* (NHS Act) which is supported by the *National Health Security Regulations 2008* and the SSBA Standards. The SSBA Regulatory Scheme aims to improve the control of biological agents of security concern and to build on Australia's obligations under the Biological and Toxins Weapons Convention and UN Security Council Resolution 1540.

The basis of the SSBA Regulatory Scheme is the List of SSBAs that was established by the Minister for Health and Ageing in November 2008. Over 200 biological agents were assessed for inclusion on the List. Specific criteria was used to divide the List into two Tiers: Tier 1 agents pose the highest security risk and have been regulated since 31 January 2009, while Tier 2 agents pose a moderate security risk and have been regulated from January 2010.

G. Barber (✉)

The SSBA Standards set out the minimum security requirements for risk assessment, personnel security, physical security, Information Technology security, inactivation of agents and transport. The SSBA Regulatory Scheme requires entities and facilities that handle SSBAs to register with the Department of Health and Ageing and submit information about the administration of their SSBA holdings including the location and the nature of their activities with the SSBAs.

Compliance with the SSBA Regulatory Scheme is monitored by inspections carried out by Office of the Gene Technology Regulator inspectors.

For further enquiries contact the SSBA Regulatory Scheme at: Email: ssba@health.gov.au

Postal Address:
Laboratory Capacity and Regulation Section
Department of Health and Ageing
GPO Box 9848, MDP 140
Canberra ACT 2601
Web: http://www.health.gov.au/ssba

3.2.2 Regulation of Gene Technology

The *Gene Technology Act 2000* (the Act), which came into force on 21 June 2001, introduces a national scheme for the regulation of genetically modified organisms (GMO) in Australia, in order to protect the health and safety of Australians and the Australian environment by identifying risks posed by or as a result of gene technology, and by managing those risks through regulating certain dealings with genetically modified organisms

The Office of the Gene Technology Regulator (OGTR) has been established within the Australian Government Department of Health and Ageing to provide administrative support to the Gene Technology Regulator in the performance of his functions under the *Gene Technology Act 2000*.

The Act prohibits any person or organisation dealing with a GMO unless they have the appropriate licence or in the case of certain classes of dealings after appropriate notification has been made to the Regulator.

Under the Act, the Gene Technology Regulator may issue technical and procedural guidelines in relation to GMOs (under section 27 of the Act), in relation to certification of facilities to specified containment levels (section 90) and in relation to accreditation of organisations (section 98). The Act, the Gene Technology Regulations 2001 or instruments issued by the Regulator can require compliance with these guidelines in conducting dealings with GMOs or in obtaining and maintaining certification or accreditation.

Inspectors from the OGTR conduct regular inspections (including unannounced spot checks) to monitoring compliance with the Act.

Any person or organisation that is dealing, or is considering dealing, with a GMO should contact the Office of the Gene Technology Regulator to ensure they are meeting the requirements of the legislation.

E-mail: ogtr@health.gov.au
Free call: 1800 181 030
Fax: 02 6271 4202
Mail: Office of the Gene Technology Regulator (MDP 54)
GPO Box 9848
Canberra ACT 2601

Part II
Clinical Aspects and Diagnosis of Infectious Disease

Chapter 4
Meningitis and Encephalitis

David L. Gordon

Abstract Until the advent of a vaccine, the most common cause of bacterial meningitis in Australia used to be *Haemophilus influenzae type b* (Hib). Since a vaccine was introduced, the two most common bacterial causes of meningitis in children are *Neisseria meningitidis* and *Streptococcus pneumoniae*. Viral meningitis is usually much less severe than bacterial meningitis, except in cases where the virus has caused encephalitis as well as meningitis. Viral causes of meningitis or encephalitis include varicella zoster, influenza, mumps, measles and until recently, polio. Many of these infections can be prevented with immunization, and following extensive preventive programmes for these viruses, enteroviruses have now become the most common cause of viral meningitis. Fungal infections of the central nervous system may also occur. The most common form of fungal meningitis is caused by the fungus *Cryptococcus neoformans*. Cryptococcal meningitis is common in AIDS patients. Although treatable, fungal meningitis often recurs in nearly half of affected persons. Arboviruses are the primary cause of acute encephalitis whilst herpes viruses are the other major cause of encephalitis in the developed world. Despite advances in antiviral therapy over the past two decades, herpes simplex encephalitis (HSE) remains a serious illness with significant risks of morbidity and death. The accurate diagnosis of encephalitis and meningitis due to infectious agents has been difficult using traditional methods. The introduction of molecular diagnostic assays has greatly facilitated their diagnosis, but clinical (CSF) samples need to be collected early in the infection process to maximize the likelihood of pathogen detection.

Keywords Meningitis · Encephalitis · Central nervous system infections

---

D.L. Gordon (✉)
Department of Microbiology and Infectious Diseases, Flinders University,
Bedford Park, SA 5042, Australia
e-mail: d.gordon@flinders.edu.au

M. Schuller et al. (eds.), *PCR for Clinical Microbiology*,
DOI 10.1007/978-90-481-9039-3_4,

Infectious diseases of the central nervous system (CNS) vary widely in severity and are amongst the most devastating of all infectious diseases, often with dramatic clinical presentations, potentially rapid progression and persisting long-term neurologic sequelae.

The diagnosis of CNS infections is complex, and difficulties are compounded by: (i) the wide spectrum of potential pathogens (including bacteria, viruses, fungi and parasites); (ii) effects of previously administered antimicrobials; (iii) often limited CSF availability; and (iv) the occurrence of post-infectious and non-infectious conditions that are clinically indistinguishable. In a significant proportion of cases, especially for encephalitis, a specific etiologic agent may never be identified. Improvements in molecular diagnostic tests are of particular significance in the diagnosis of CNS infections because determination of the specific etiology often has important implications for therapy, prognosis and public health considerations.

Patients with meningitis have typically been considered to present with fever, neck stiffness and altered mental state. In adults with bacterial meningitis however, only two-thirds have this classic triad, although all patients have at least one of these symptoms [2]. Children, especially neonates, may have more subtle presentations. Attia et al. [1] evaluated the clinical findings in acute meningitis and emphasised the limitations of classic symptoms and signs for diagnosis. Presentation can be related to the pathophysiology of meningitis: fever and rash (systemic infection); headache and neck stiffness (meningeal irritation); focal signs of seizures (cerebral vasculitis); and headache, altered mental state and cranial nerve palsies (increased intra-cranial pressure and cerebral oedema). The contribution of the host inflammatory response to neuronal injury and outcome of bacterial meningitis is now well-recognised and has led to attempts to improve sequelae by modulation of this by co-administration of corticosteroids.

Encephalitis is an inflammatory process of the brain, with associated altered mental state, confusion, delirium, cognitive and behavioural changes, focal neurological signs or seizures. Often symptoms of meningitis overlap with those seen in encephalitis, resulting in a 'meningoencephalitis' syndrome.

Encephalitis is most frequently associated with viral infections, particularly herpes simplex virus (HSV), enteroviruses and arboviruses (Japanese encephalitis, Murray Valley encephalitis and West Nile virus), but may occur with bacterial, fungal, protozoal or helminthic infection [6]. Pathologically, acute viral encephalitis is associated with cell death from lysis, necrosis or induction of apoptosis, and perivascular inflammation and glial reaction, predominantly in the gray matter. Encephalitis may also result from a presumed immune-mediated 'post-infectious' response as in acute disseminated encephalomyelitis (ADEM). Clinically this is difficult to distinguish from direct viral infection unless a clear precipitant such as recent measles or mumps infection has occurred; the pathological process is however primarily focused on white matter with prominent perivenular inflammation and demyelination. Abnormalities on MRI are frequently present [5].

The approach to diagnosing meningitis and encephalitis involves studies outside the CNS, examination of CSF and rarely brain, and exclusion of non-infectious conditions such as systemic lupus erythematosis, sarcoidosis, seizures, migraine,

Table 4.1 Common causes of meningitis and encephalitis[a]

	Meningitis	Encephalitis
Bacterial	*Streptococcus pneumoniae* *Neisseria meningitidis* *Haemophilus influenzae* *Listeria monocytogenes*[b] *Escherichia coli*[c] *Streptococcus agalactiae*[c] *Mycobacterium tuberculosis* Staphylococci[d] Gram negative bacilli[d]	*Listeria monocytogenes*[b]
Viral	HSV Enteroviruses	HSV, Enteroviruses, Arboviruses, HIV, VZV, CMV EBV, HHV6, mumps, adenoviruses Rabies
Fungal	*Cryptococcus neoformans*	
Parasitic	*Acanthomoeba* sp. *Naegleria fowleri*	*Toxoplasma gondii* Malaria

[a]This is not a complete list, selected and more common causes shown
[b]Neonates, elderly, immunosuppressed
[c]Neonates
[d]Post-neurosurgery, head injury

and toxic and metabolic causes which present as an encephalopathy (cerebral dysfunction without evidence of inflammation). Initial diagnostic studies should focus on detection of the most common pathogens (Table 4.1), with consideration of important epidemiological risk factors such as age, immunosuppression, travel and dietary history, occupation and animal exposure. Blood cultures should be performed in all patients prior to antimicrobial administration. Isolation or detection of organisms at other sites (e.g. enteroviruses in pharyngeal swabs) provides presumptive, although not definitive evidence, of the pathogen. Serological studies for HIV, syphilis, arboviruses, *Mycoplasma pneumoniae* and Rickettsiae may be useful in selected cases, especially for encephalitis presentations. A position statement on the diagnosis and management of encephalitis was recently published by the Infectious Diseases Society of America, and provides further detailed information [6].

CSF should be obtained in all cases, unless contraindicated. The concern regarding risks of cerebral herniation following lumbar puncture (LP) is greatly exaggerated and it is unclear if this association truly exists. An LP can be performed without prior radiology if the patient does not have new-onset seizures, an immunocompromised state, papilloedema, focal neurological signs or moderate-severe impairment of consciousness [3]. If LP is delayed, blood cultures should be collected, and antimicrobials administered immediately. Brief antimicrobial use prior to LP will not affect the Gram stain, cell count, glucose or protein results [4], but may reduce the yield from culture, especially for *Neisseria meningitidis*. Molecular techniques may be particularly useful in this regard. Initial CSF analysis

Table 4.2 Typical CSF findings in bacterial vs viral meningitis[a]

	Bacterial	Viral
WBC count (per μl)	1,000–5,000	<300
Predominant cells	Neutrophils	Lymphocytes[b]
Glucose	Reduced	Normal
Protein	Elevated	Normal
Gram stain	Positive (50–90%)	Negative

[a]Adapted from Fitch et al. [3]
[b]Neutrophils may be predominant early
Note: These are general findings and cannot be used to exclude bacterial meningitis

includes Gram stain, cell count, protein and glucose and may give some indication of a bacterial versus viral etiology (Table 4.2), as can determination of serum C-reactive protein and procalcitonin. However, none of these tests, alone or in combination, can be used clinically to exclude a bacterial etiology. Following routine bacterial culture, subsequent testing of CSF including mycobacterial and fungal culture, antigen detection, viral studies and PCR frequently needs prioritization, as the volume of CSF received is limited. Testing needs to be dictated by the cell count, biochemistry, and most importantly by discussing the clinical presentation and progress with the clinician. Conversely, a normal leukocyte count and protein level is almost always associated with negative PCR results for common viruses and there is little value in performing the test.

The diagnosis and management of meningitis and encephalitis remains a major challenge for laboratories and clinicians alike. The introduction of molecular techniques has revolutionized the diagnosis of conditions such as HSV encephalitis and enterovirus meningitis and has had important management implications. However substantial progress remains to be made in determining the etiologic agent in many instances of CNS infection.

References

1. Attia J, Hatala R, Cook DJ et al (1999) Does this adult patient have acute meningitis? JAMA 281:175–181
2. Durand ML, Calderwood SB, Weber DJ et al (1993) Acute bacterial meningitis in adults – a review of 493 episodes. New Eng J Med 328:21–28
3. Fitch MT, van de Beek D (2007) Emergency diagnosis and treatment of adult meningitis. Lancet Infect Dis 7:191–200
4. Proulx N, Fréchette D, Toye B et al (2005) Delays in the administration of antibiotics are associated with mortality from adult acute bacterial meningitis. Quart J Med 98:291–298
5. Sonneville R, Klein I, de Broucker T et al (2009) Post-infectious encephalitis in adults: diagnosis and management. J Infect 58:321–328
6. Tunkel AR, Glaser CA, Bloch KC et al (2008) The management of encephalitis: clinical practice guidelines by the Infectious Diseases Society of America. Clin Infect Dis 47:303–326

Chapter 5
Respiratory Infections

Michael D. Nissen, Stephen B. Lambert, David M. Whiley, and Theo P. Sloots

Abstract Until recently, conventional culture techniques and immunofluorecence assays were considered the gold standard for the detection of respiratory viruses, even though results are mostly available too late or lacked specificity and sensitivity. These methods are now widely replaced with appropriate DNA- and RNA-based amplification techniques, in particular real time PCR amplification, for the detection of an extended number of agents responsible for acute respiratory infections. Real-time PCR offers rapid results, efficiencies in work flow and a reduced risk of false positive results due to contamination. As a result, better patient management or reduction of unnecessary antibiotic administration will be possible leading to enhanced efficiencies in health care. In applying molecular methods to diagnostic use, the laboratory can optimise its diagnostic strategy by applying a combination of real-time amplification tests for respiratory viruses and the non-viral respiratory bacterial pathogens. However this must be done within a context of resource availability, technical expertise available and clinical utility. It seems certain that molecular microbiology will continue to develop, leading to further applications in diagnostic technology, thereby improving our understanding of disease processes and enhancing our knowledge of the pathogens responsible.

Keywords Respiratory infection · Diagnosis · Virus · Bacteria · Fungi · Clinical · Pathogenesis · Epidemiology · Emerging viruses

5.1 Introduction

Acute respiratory infections (ARIs) continue to be the leading cause of acute illnesses worldwide and remain the most important cause of mortality in infants and young children.

T.P. Sloots (✉)
Queensland Paediatric Infectious Diseases Laboratory, Sir Albert Sakzewski Virus Research Centre, Queensland Children's Medical Research Institute, Children's Health Service District, Brisbane, QLD 4029, Australia
e-mail: t.sloots@uq.edu.au

M. Schuller et al. (eds.), *PCR for Clinical Microbiology*,
DOI 10.1007/978-90-481-9039-3_5,

They account for about two million deaths each year [22] and rank first among causes of disability-adjusted life-years (DALYs) lost in developing countries (94.6 million, 6.3% of total [43]). The populations most at risk for developing a fatal respiratory disease are the very young, the elderly, and the immunocompromised.

While upper respiratory infections (URIs) are very frequent but seldom life-threatening, lower respiratory infections (LRIs) are responsible for more severe illnesses such as influenza, pneumonia, tuberculosis, and bronchiolitis that are the leading contributors to ARIs' mortality (Fig. 5.1). Pneumonia, with a global burden of 5,000 childhood deaths every day, is a tangible threat that needs to be dealt with accordingly.

The incidence of ARIs in children aged younger than 5 years is estimated to be 0.29 and 0.05 episodes per child-year in developing and industrialized countries, respectively, which translates into 151 million and 5 million new episodes each year, respectively [28]. Most cases occur in India (43 million), China (21 million), Pakistan (10 million), Bangladesh, Indonesia and Nigeria (56 million each). Pneumonia is responsible for about 21% of all deaths in children aged younger than 5 years, leading to estimates that of every 1,000 children born alive, 12–20 die from pneumonia before their fifth birthday [43].

The main aetiological agents responsible for ARIs in children include *Streptococcus pneumoniae, Haemophilus influenzae* type b (Hib), *Staphylococcus*

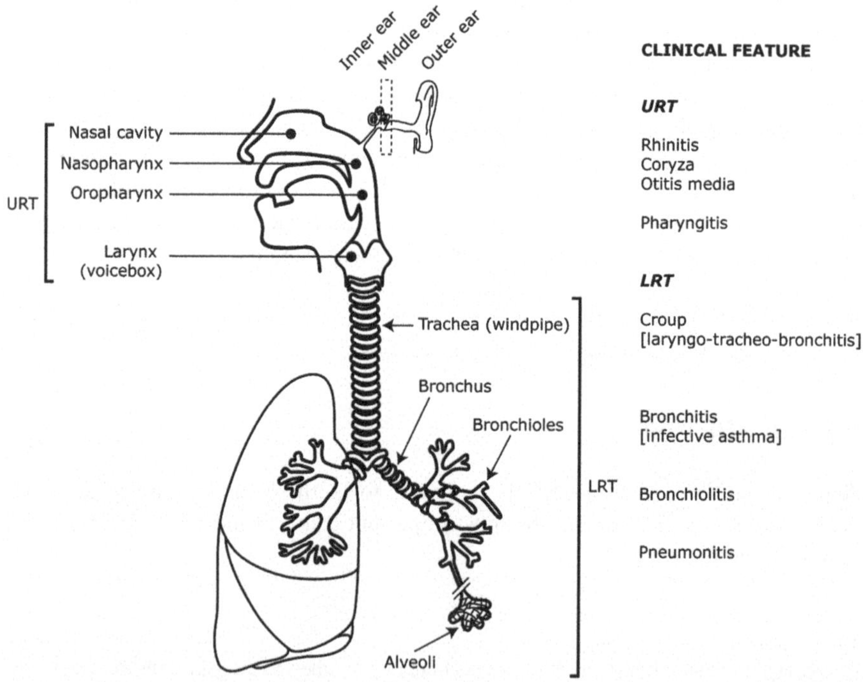

Fig. 5.1 A schematic representation of the human respiratory tract. The upper and lower respiratory tract and components of the ear are indicated together with other major respiratory sites

aureus and other bacterial species, respiratory syncytial virus (RSV), measles virus, human parainfluenza viruses type 1, 2, and 3 (PIV-1, PIV-2 and PIV-3), influenza virus (INF) and varicella virus.

5.2 Clinical Aspects and Epidemiology

5.2.1 Clinical Presentation

A wide variety of well known and newly identified agents cause respiratory illness and disease in humans. It is not possible to differentiate with certainty the aetiological agent in an infection based on clinical symptoms alone. In pre-school aged children, for example, illnesses due to respiratory viruses have seasonal variations and differences in the presence of fever and other symptoms, the likelihood of household transmission, and the impact they have in terms of medical visits and disruption to family life. But none of these features in isolation or combination is sufficiently specific to link illness to a pathogen with certainty.

Diagnosis in individual mild illnesses may not alter management, but can prevent unnecessary hospitalisation, antibiotic therapy, or further invasive investigation. Laboratory confirmation of the cause of infections has been made more sensitive and rapid through the use of PCR technology. PCR has taught us that the constellation and severity of symptoms can cluster with particular infectious agents. For example, recent findings from the New Vaccine Surveillance Network in the United States show that despite respiratory syncytial virus (RSV), parainfluenza viruses (PIVs), and human coronaviruses (HCoVs) all being common in early childhood; RSV and PIVs are more common causes of hospital admission with acute febrile and respiratory illness than HCoVs [33, 41]. Despite such clustering, in individual illnesses it can be said that even viruses more typically associated with severe childhood illness can cause milder symptoms, modified by immune or possibly genetic factors, and that severe disease, whilst more commonly caused by a small group of well-studied viruses, can result from infection due to any virus. Viruses that are typically considered to cause infrequent or mild disease, such as influenza C virus and PIV-4, may cause more significant illness in vulnerable populations, including the young and immunocompromised.

5.2.2 Pathogenesis (Transmission, Incubation, Site of Infection)

Influenza, the most studied of respiratory viruses, provides broad insights for other common myxovirus and paramyxovirus respiratory agents. A review of healthy adult human volunteer studies showed that viral shedding increased sharply between 0.5 and 1 day after influenza virus challenge, peaking on day two; shedding can be detected 24 to 28 h before clinical onset, and has a mean duration of 4.8 days; two-thirds of subjects had symptomatic infection, and total symptom scores peaked on day three [11]. The natural history of infection may differ in the elderly and children. For example, pre-symptomatic influenza virus shedding has been seen for 6 days before clinical onset and mean duration of virus isolation from hospitalised children not receiving an antiviral was 6.8 days [29].

Respiratory viruses can be transmitted through a number of modes: direct contact and fomites, large droplet, and airborne small particles. The importance of each of these modes depends on the virus in question, the site of infection, and the environment. For example, the eyes and nose appear to be much more important routes of infection for RSV than the mouth.

5.2.3 Epidemiology (Frequency, Seasonality, Age Groups)

Modern molecular methods have resulted in the identification of previously unknown viruses from specimens collected from the respiratory tract. Testing for new viruses along with known viruses, including rhinoviruses [6] by PCR, is filling the diagnostic void in respiratory illness and infection, and has improved our understanding of the epidemiology of such illnesses. In all but tropical climates there are a group of respiratory viruses that occur more frequently in the non-summer months, often peaking in winter; these viruses include influenza viruses, RSV, human metapneumovirus (HMPV), PIVs, and HCoVs (Fig. 5.2). Human rhinoviruses (HRVs) are the most commonly identified group of viruses in both community-managed and more severe respiratory illness in children and older age-groups, having a year round presence but being more common in the spring and autumn months [6]. Whilst it is clear rhinoviruses are a major pathogenic group, there is still uncertainty about the predictive value of a positive molecular test for a picornavirus, particularly from children. In tropical settings, influenza and other respiratory viruses can have a high background year-long presence [38].

Respiratory viruses circulate freely in all populations, but moderate to severe illness tends to disproportionately affect certain groups. Infections due to common viruses that result in disease severe enough to warrant laboratory testing, notification, or hospitalisation occur in the young, the very old, or both, such as with RSV and influenza [9, 14].

5.3 Commonly Recognised Respiratory Agents

5.3.1 Respiratory Disease Due to Viruses

In spite of the inclusion of the live attenuated measles vaccine in the Expanded Program of Immunization (EPI), measles virus was still responsible in 2002 for some 213,000 deaths worldwide, essentially due to insufficient vaccine coverage [27] . The situation has fortunately been substantially improved lately, but the leading cause of serious respiratory illness in young children is respiratory syncytial virus (RSV), the agent of infantile bronchiolitis, which is associated with substantial morbidity and mortality [21]. Parainfluenza viruses (PIV-1, PIV-2 and PIV-3), especially PIV-3, are second in incidence immediately after RSV. All children by the age of 2 years have had at least one episode of PIV and/or RSV illness. In addition, both viruses can cause severe disease in the elderly, especially in patients with a chronic respiratory or cardiac condition [18]. Although the disease burden due to these pathogens has not been accurately quantified in developing countries, extrapolation from known figures in industrialised countries, such as 125,000 reported cases of RSV per year in the USA, leads to the impressive global estimates of 64 million cases and 160,000 deaths per year from RSV infection worldwide. RSV was identified in 15–40% of pneumonia or bronchiolitis cases admitted to hospital in developing countries, followed by influenza viruses, parainfluenza viruses, human metapneumovirus and adenovirus [40]. The elderly also are at risk for severe RSV disease, and 14,000–60,000 RSV-related hospitalisations of the elderly are reported to occur annually in the USA [14].

Human metapneumovirus, a member of the *Paramyxoviridae*, is a recognised cause of a large fraction of severe ARIs in infant, elderly and immunocompromised populations [15]. Other viruses that cause respiratory infections are coronaviruses, adenoviruses and rhinoviruses. Recently discovered coronaviruses HCoV-HKU1 and HCoV-NL63 are significant pathogens that contribute to the hospitalisation of children for ARI [24, 35]. Among other members of the *Coronaviridae* are human coronaviruses HCoV-229E and HCoV-OC43, agents of the common cold. Another recently identified coronavirus is that of the severe acute respiratory syndrome (SARS), SARS-CoV, which emerged in southern China in late 2002 and spread in the spring of 2003 to some 30 countries within Asia, Europe and North America.

In the elderly, influenza-related pneumonia remains a leading cause of infectious disease-related deaths. The threat of an avian influenza pandemic has been looming ever since the emergence in 1997 in Hong Kong of the H5N1 avian influenza virus,

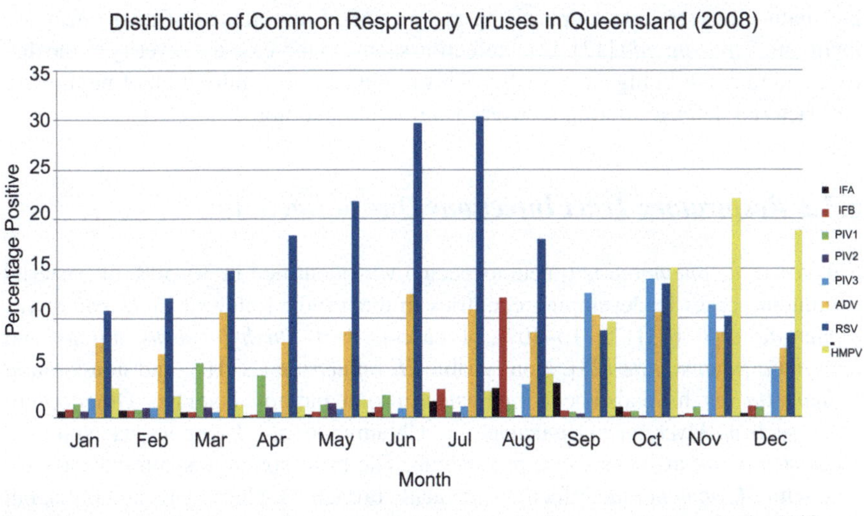

Fig. 5.2 Seasonal distribution of major respiratory viruses in Queensland, Australia for 2008

especially since the reappearance of human cases in 2003–2004. The new H5N1 variant is highly pathogenic for poultry and wild birds and can lethally infect cats and humans. At this time, however, it still is not possible to predict which virus is going to eventually cause a pandemic and when it is going to happen, but the preparation of pandemic influenza vaccines is being actively pursued, generating broad new knowledge on how to improve seasonal influenza vaccine immunogenicity. This has proven to be of the utmost importance with the recent emergence of influenza virus A H1N1 (2009) ("Human Swine Influenza") in Mexico and the USA in 2009.

Regarding influenza virus, the average global burden of inter-pandemic influenza may be on the order of 1 billion cases per year, leading to 300,000–500,000 deaths worldwide. However, the substantial reduction in ARI mortality observed in developing countries that have implemented simple case management, including provision of antibiotics to children with ARI, suggests that bacterial pneumonia contributes to a large proportion of deaths in these populations. Available data suggest that dual infections with viral and bacterial pathogens may be quite common, as seen by the fact that, in the industrialized world, epidemics of RSV and/or influenza coincide with epidemics of *S.pneumoniae* year after year [32]. While influenza virus is the most commonly met pathogen in this context, other respiratory viruses, including RSV, measles virus, parainfluenza viruses, or adenoviruses may also predispose to secondary bacterial infections. Several different bacterial species may be implicated, including *H. influenzae*, *Staphylococcus aureus, Streptococcus pyogenes, Mycoplasma pneumoniae*, and, most importantly of all, *S. pneumoniae* [25]. Half or more of the flu-associated mortality in the 1918–1919 Spanish Flu epidemic is believed to have resulted from pneumococcal super-infections.

The same is true for developing countries. As an example, the observation was made in South Africa that children vaccinated with the 7-valent conjugate pneumococcal vaccine showed 31% reduction in virus-associated pneumonias requiring hospitalisation, strongly emphasising the presumed importance of dual infections involving *S. pneumonia* [13]. Dual infection seems to increase the severity of the disease and to result in higher mortality. This might be due to inhibition of pulmonary antibacterial defenses during recovery from viral infectiions.

5.3.2 Respiratory Tract Infections Due to Bacteria

Streptococcus pneumoniae (pneumococcus) was identified in 30–50% of bacterial pneumonia cases in developing countries in the 1990s, followed by *Haemophilus influenzae* type b (Hib; 10–30% of cases), then *Staphylococcus aureus* and *Klebsiella pneumoniae* [28]. Non-typable *H. influenzae* (NTHI), and non-typhoid *Salmonella spp*. have also been implicated in some but not all studies. Other organisms, such as *Mycoplasma pneumoniae,* Chlamydia spp., Pseudomonas spp. and *Escherichia coli* also can cause pneumonia. The most common syndromes associated with *M. pneumoniae* infections are acute bronchitis, pharyngitis and otitis, but 10% of infected children develop pneumonia [39].

The introduction of Hib conjugate vaccines has resulted in a truly remarkable decline in Hib disease where the vaccine has been introduced. However, the vaccine is not yet routinely made available to a majority of children worldwide. As a result, 400,000 deaths are still estimated to occur from Hib disease each year [12]. In view of their safety and remarkable efficacy, the WHO has recommended the global implementation of the Hib conjugate vaccines.

S. pneumoniae is estimated to cause more than one-third of the 2 million deaths due to ARIs, especially in developing countries where the bacterium is one of the most important bacterial pathogens of infancy and early childhood [45]. Virtually every child in the world is colonised with one or more strains of pneumococcus and becomes a nasopharyngeal carrier during their first few years of life. Many children will go on to develop otitis media, and a few will eventually develop invasive pneumococcal disease including bacteraemic pneumonia and/or meningitis. The introduction of the conjugate pneumococcal vaccine in routine infant immunization should have a major impact on pneumonia in children less than 5 years of age worldwide, as already documented in the USA [26].

Tuberculosis (TB) continues to be a leading cause of deaths worldwide, with an estimated one third of humanity infected and about 1.7 million deaths each year, a global toll of 4,650 lives daily. The emergence of *Mycobacterium tuberculosis* (Mtb) strains carrying drug-resistance mutations against first-line dugs (MDR-TB) and, more recently, against both first- and second-line drugs (XDR-TB), shows that it will most probably be impossible to contain the TB pandemic with drugs alone. More than one hundred new TB vaccine candidates have been tested in animal models and some have moved into clinical trials. Testing such a wide variety of vaccine types using different strategies will obviously require time and a lot of coordination, especially as surrogate markers of protection still remain mostly unknown at this time.

Finally, it should be emphasized that nosocomial or hospital-acquired pneumonia is a major public health problem: pneumonia is the second most common type of all nosocomial infections, with an associated case fatality rate of 20–50%.

5.3.3 Respiratory Infections Due to Fungi

Infectious fungal respiratory diseases can be divided into those that occur opportunistically in immunosuppressed patients and those that occur in generally healthy individuals. Fungi which affect immunosuppressed individuals are frequently *Pneumocystis jiroveci*, and species of *Aspergillus* and *Candida* as well as *Cryptococcus neoformans* [5], while organisms such as *Histoplasma capsulatum, Coccidioides immitis* and *Blastomyces dermatitidis* are frequent pathogens in healthy individuals in certain endemic regions.

The causative fungi of respiratory infections vary with the population selected and the geographical region. The majority of fungal infections in lung-transplant recipients involve *Aspergillus* spp., followed by *Candida, Pneumocystis, Cryptococcus*, geographically restricted agents and newly emerging fungal pathogens. *Aspergillus* infection remains the main fungal complication in lung-transplantation recipients. There are various fungal agents responsible for pulmonary fungal infection in patients with haematological malignancies, but *Aspergillus* spp. and other molds such as zygomycetes or *Fusarium* spp. represent the most frequently isolated microorganisms [19]. Less commonly, pneumonia could be due to *Candida* spp., *Cryptococcus* spp. or *Pneumocystis jirovecii* [17, 20]. Although invasive aspergillosis is an uncommon complication of haematopoietic stem cell transplants (HSCT) and solid organ transplants (SOT), it continues to be associated with poor outcomes [5]. Invasive pulmonary aspergillosis is rare in patients with chronic obstructive lung disease and is commonly associated with high doses of corticosteroids and multiple broad-spectrum antibiotics [19].

Candida infection is predominant in patients with non-haematologic malignant tumours and in non-lung SOT recipients. *Candida albicans* and *Candida parapsilosis* were the predominant isolates of pulmonary candidiasis in ventilated preterm infants with a birth weight of less than 1,250 g; the incidence rate of pulmonary candidiasis during the first month of life was 8.6% (20/233 cases) [17]. In Intensive Care Units, *C. albicans* was also the most frequently isolated fungal species in all sites (68.9%). Isolation of fungi allowed a diagnosis of fungal infection in 121 patients (7.7%) [47].

Cryptococcus infection occurred in both immunosuppressed and immunocompetent individuals. Pulmonary cryptococcosis is usually the primary site of a disseminated or central-nervous-system cryptococcal infection that may be fatal [46]. Traditionally, capsule-deficient *Cryptococcus neoformans* was considered to have low virulence. However, a recent study showed that the presentations and outcomes did not differ significantly between patients with proven pulmonary cryptococcosis caused by capsule-deficient *Cr. neoformans* and six patients with pulmonary cryptococcosis caused by capsule-intact *Cr. Neoformans* [46].

The predisposing factors for fungal respiratory infections are increasing with the emergence of new immunosuppressive treatment. The increase of fungal associated respiratory tract infections may predominantly be attributed to the development of invasive diagnostic tools and the use of new methods for the identification of isolates, such as molecular techniques.

5.4 New Viruses Associated with the Respiratory Tract

Recent advances in molecular biology have greatly improved the detection of viral respiratory pathogens. Yet, even with the most sensitive molecular techniques, only 40–60% of infections are consistently diagnosed. This suggests that additional respiratory viruses are likely to exist. In fact, since 2001, seven previously undescribed viruses have been identified by analysis of clinical specimens from the human respiratory tract (Table 5.1).

These new viral agents were detected by novel molecular methods such as virus discovery based on cDNA-AFLP (amplified fragment length polymorphism) (VIDISCA), pan-viral DNA microarrays and high throughput sequencing [4]. More

Table 5.1 Newly recognized viruses in the human respiratory tract: a summary of distribution, clinical association and methods of discovery

Virus	Patient Group	Prevalence	Clinical signs	Method of discovery	Reference
HMPV	Children and the elderly	3–25%	Bronchiolitis, pneumonia, bronchitis, rhinorrhoea, cough, sore throat	Virus isolation, electron-microscopy, and random PCR	[34]
SARS CoV	All ages	Sporadic	Pneumonia	Virus isolation, electron-microscopy, and consensus coronavirus PCR	[23]
NL63 HKU1	Children and the elderly	1–10%	Bronchiolitis, pneumonia, rhinorrhoea, fever, cough, wheezing	Virus isolation, VIDISCA, consensus coronavirus PCR	[36, 44]
HBoV	Children	1–11%	Bronchiolitis, pneumonia, acute otitis media, asthma	Random PCR	[1]
KIV WUV	Children	1–7%	Bronchiolitis, pneumonia, cough	Random PCR	[2, 16]

broadly, the advent of these new technologies has greatly stimulated efforts to identify novel viruses in the respiratory tract and in other human disease states.

Of the viruses discovered over the last 7 years, human metapneumovirus (HMPV) and the newly emerging human coronaviruses (HCoV) are considered causative agents of respiratory disease. However, to date, SARS coronavirus has been restricted geographically and has only been associated with limited and sporadic outbreaks. Recently, human bocavirus (HBoV) and the new human polyomaviruses KIV, WUV and Merkel Cell polyomavirus (MCV) were detected in respiratory secretions, and although an association with the respiratory tract has been postulated, it still remains to be proven [2, 8, 16].

5.4.1 Human Metapneumovirus

HMPV infection is associated with a broad spectrum of clinical signs in patients of all age groups, and is the cause of upper and lower respiratory tract infection in infants and young children [30]. It is second only to RSV as a significant cause of bronchiolitis in early childhood and children are most likely to be hospitalised with severe disease. Studies have linked HMPV with acute otitis media and asthma exacerbations in children and with exacerbations of both asthma and chronic obstructive pulmonary disease (COPD) in adults. However, severe disease may occur in all patients with underlying medical conditions such as cardiopulmonary disease, the elderly and immunocompromised subjects. In these subjects the virus can cause prolonged and serious infections, particularly severe lung disease including fatality [10] (Fig. 5.3).

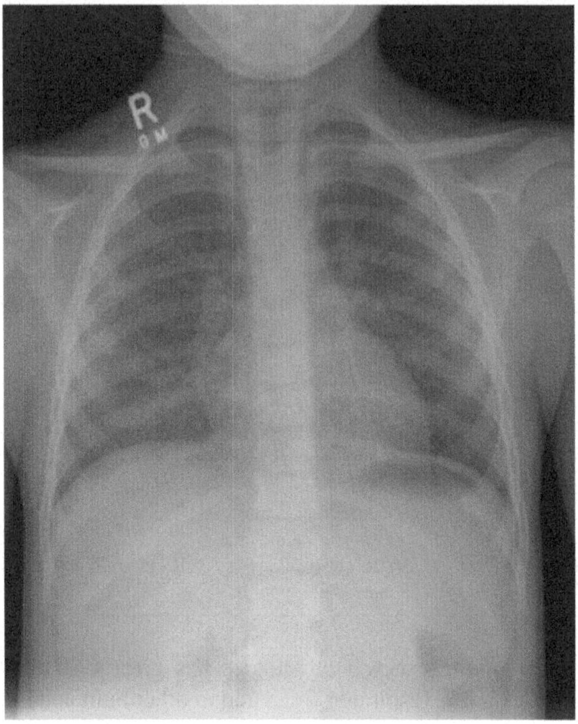

Fig. 5.3 Example of representative radiography showing features of RSV infection

5.4.2 SARS Coronavirus

The first reports of SARS coronavirus infection were published in 2003, and the causative agent was subsequently characterised as a novel human coronavirus [23]. The SARS epidemic was halted by a highly effective global public health response coordinated by the World Health Organization, and there is no further evidence that SARS CoV is currently circulating in humans. However, the SARS outbreak focused renewed attention on coronaviruses generally, resulting in the discovery of two more new human coronaviruses, NL63 and HKU1.

5.4.3 Human Coronaviruses NL63 and HKU1

HCoV-NL63 was first detected in 2004 in a child from The Netherlands with bronchiolitis, shown to be a cause of severe lower respiratory tract infection (LRTI) in young children, and an agent of laryngotracheitis (croup) [37].

HCoV-HKU1 was detected in 2005 in an adult with chronic pulmonary disease in Hong Kong [44] and was subsequently shown to be globally distributed. HKU1 infection presents with common respiratory symptoms as well as a more severe clinical presentation including bronchiolitis and pneumonia [24].

In healthy adults HCoV NL63 and HKU1 infections are generally not life threatening. This suggests that these coronaviruses, like 229E and OC43, only cause more-severe disease in young children, elderly persons, and the immunocompromised. They may be detected in 1 to 10% of patients with acute respiratory tract infections, and co-detection of these viruses with other respiratory viruses is common [37].

5.4.4 Human Bocavirus

Human bocavirus was first described in 2005 with a prevalence of 3.1% in Swedish children with LRTI [1] and subsequently in Australia with 5.2% prevalence in children with ARTI during winter [31]. Although a positive association of HBoV with ARTI was suggested, the results remain inconclusive, because a high prevalence of other respiratory viruses was found in HBoV-positive patients. To confirm the role of HBoV as a respiratory pathogen, more extensive studies including matched control populations need to be performed. One such study using control subjects [3] proposed HBoV as a cause of acute wheezing in children.

HBoV has been frequently detected in immunosuppressed adults but only rarely in immunocompetent adult subjects. However, it is uncertain if the presence of HBoV in these subjects is the result of re-infection, viral persistence or reactivation.

5.4.5 Novel Human Polyomaviruses KIV, WUV and MCV

Recently, three new human polyomaviruses, KIV, WUV and MCV were detected in specimens of patients with ARTI [2, 8, 16]. Allander et al. [2] reported a prevalence of 1% for KIV in nasopharyngeal aspirates (NPA) collected from a Swedish population and Gaynor et al. [16] showed a prevalence of 3% and 0.6% for WUV in respiratory samples from Australia and the USA respectively. More recently, although originally found in Merkel cell carcinoma, MCV was also found in 5.9% of NPA collected from Australian subjects with ARTI [8]. Since these first reports, KIV, WUV and MCV have been detected in a number of geographic locations, suggesting a global presence for these viruses.

One striking feature of early findings concerning KIV and WUV is their high rate of co-detection with other respiratory viruses. A co-detection rate of 74% has been observed for KIV and rates ranging from 68% to 79% for WUV [7]. So, even though an aetiological role in childhood respiratory disease has been proposed, it is difficult to assess a pathogenic role for these viruses unless observations are compared with those for matched control populations.

Further studies will need to be completed before the role of KIV and WUV as respiratory pathogens can be confirmed, and it remains possible that these viruses are not involved in respiratory disease, but that their presence in the respiratory tract simply reflects their mode of transmission.

5.5 Laboratory Diagnosis

Cheryl Bletchly

5.5.1 Respiratory Specimens and Transport

Laboratory diagnosis of respiratory virus infections requires specimens containing cells from the respiratory tract collected early in the clinical illness. The most appropriate specimens are NPAs and bronchoalveolar lavage (BAL). Where the generation of aerosols may pose an infection risk to collection personnel, nose and throat swabs are a viable alternative. Recent studies however, have shown that swabs are not as effective as NPAs for the detection of adenovirus and respiratory syncytial virus. Limited data is available suggesting that flocked swabs are superior to traditional swabs for detecting respiratory virus infections.

Collection swabs should be dry and not contain bacterial transport medium as these often contain substances that are inhibitory to PCR reactions.

In the case of influenza virus detection, the CDC recommends that Viral Transport Medium (VTM) is added to swabs to assist in virus preservation. If VTM is to be added to dry swabs, they should be vortexed to release cells in to the medium and transferred to a sterile vial for transportation to the laboratory. VTM cannot be added to swab receptacles as they will invariably leak in transit and pose an infection risk. All respiratory specimen types should be transported to the laboratory at 4 °C. Some NPAs and sputum samples received in the laboratory are very mucoid. These may either be diluted in VTM or digested with sputasol to facilitate the extraction process prior to molecular analysis.

It is preferable to use an extraction method that will extract both RNA and DNA with equal efficiency as additional testing for agents such as *Bordetella pertussis* and *Mycoplasma pneumoniae* are often requested if a viral cause cannot be found. Lung tissues are often collected at autopsy for either culture or molecular viral studies.

The specimen requirements for both molecular and culture based respiratory virus detection methods are similar and therefore laboratories can choose the detection method that suits their role.

All respiratory specimens must be handled with appropriate personal protection equipment as specified for PC2 laboratories and opened and aliquotted only in a Class II Biological Safety Cabinet. Specimens should be stored at 4°C until they are processed.

C. Bletchly (✉)
Molecular Diagnostic Unit, Microbiology Division, Pathology Queensland Central Laboratory, Brisbane, QLD 4029, Australia
e-mail: Cheryl_Bletchly@health.qld.gov.au

5.5.2 Traditional Methods

Traditional laboratory respiratory virus diagnosis involved virus culture with, or without immunofluorescence staining with specific antibodies. Culture is highly sensitive if the appropriate cell lines are utilised and once a virus is isolated it can be further characterised and amplified if required. Viral culture provides the added advantage that it will only detect infectious virus and not persistent nucleic acid. It has the disadvantages however of being expensive due to the requirement to maintain cells lines and tissue culture media and the requirement for expertise in sterile technique and interpretation of cytopathic effect (CPE) and fluorescence staining.

Direct Immunofluorescence Assay (DFA) of respiratory tract cells provides rapid results but does involve a series of manual manipulations and washes along with interpretation by highly skilled personnel (Fig. 5.4).

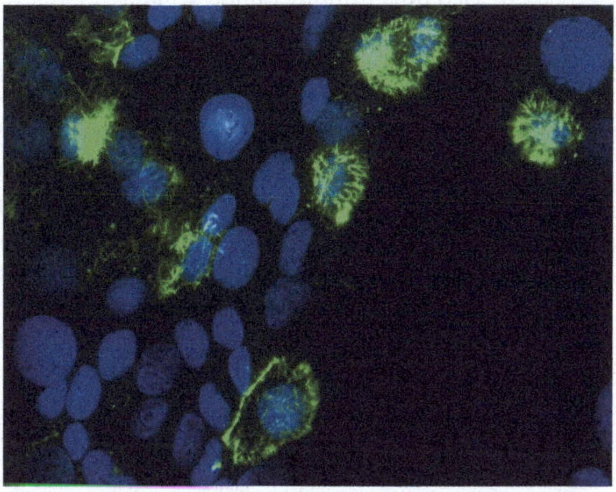

Fig. 5.4 Fluorescent staining pattern of LLC-MK2 cell infected with HMPV (02-001), stained with DAPI and mouse monoclonal antibody to the HMPV N protein (photo courtesy of Dr. F.M. Preston)

5.5.3 Criteria for Test Selection

The laboratory's primary role will most often dictate their method of choice for respiratory virus detection. Diagnostic laboratories require rapid result turn-around and will most likely opt for molecular detection which can be readily adapted to high throughput batching and multiplexing. Most molecular assays can be reported within 24 h of sample receipt whereas culture methods will take from 1 to 14 days and often much longer for confirmation.

Direct antigen tests are available commercially for some respiratory viruses but lack sensitivity (approximately 70%) although the positive predictive value is high. The expense and false negative rate of direct antigen tests need to be balanced with their rapidity and ease of use. Negative direct antigen results should be confirmed by a more sensitive assay such as PCR.

Public health laboratories will most likely employ viral culture to enable viral isolate characterisation and the ability to detect unsuspected viruses.

5.5.4 Commercial Assays

Despite the popularity of molecular detection methods for respiratory viruses very few well validated commercial assays are available for the wide range of respiratory pathogens that are of clinical significance. Most laboratories utilise "in-house" methods in combination with a rigorous quality assurance programmes.

5.6 PCR Detection of Respiratory Viruses

Cell culture and direct immunofluorescent assay (DFA) staining using monoclonal antibodies were previously the most commonly used laboratory techniques for detecting respiratory viruses. Although still used widely in the United States, these traditional techniques have gradually been superseded by highly sensitive and rapid reverse transcriptase polymerase chain reaction (RT-PCR) assays, with most laboratories in Australia now using RT-PCR methods. Additional advantages of RT-PCR detection of respiratory viruses include results that are not significantly affected by a loss of viral viability during specimen transport or storage, and that RT-PCR does not require the presence of intact, infected cells within the specimen. The PCR revolution has also been further stimulated through improvements in the technology, including the advent of real-time PCR and multiplex PCR methods, both of which reduce staff hands-on time, decrease result turnaround-times, increase through-put and are more user friendly compared to conventional PCR techniques.

Standard quality control practices should be implemented when testing for respiratory viruses by PCR, including the use of a suitable positive control, negative control, extraction control and inhibition control. *Sequence-related* issues (discussed in more detail in Chapter 8) are also very relevant to respiratory viruses. Consideration needs to be given to the type of probe used when designing real-time PCR methods for respiratory viruses. The main issue is that respiratory viruses, particularly the RNA viruses, show considerable genetic variation. For this reason, it is often difficult to identify a sufficiently large and conserved region to accommodate two hybridisation probes and so the single-probe TaqMan format is more commonly utilised for respiratory virus detection. On the other hand, we have also observed problems using the smaller minor-groove binder (MGB) TaqMan probes. The issue stems from MGB TaqMan probes being more susceptible to single nucleotide polymorphisms than standard TaqMan probes. For example, we have found that some RSV strains provided poor fluorescent signal, barely above the background negative signal, in an RSV MGB assay [42].

References

1. Allander T, Tammi MT, Eriksson M et al (2005) Cloning of a human parvovirus by molecular screening of respiratory tract samples. Proc Natl Acad Sci U S A 102:12891–12896
2. Allander T, Andreasson K, Gupta S et al (2007) Identification of a third human polyomavirus. J Virol 81:4130–4137
3. Allander T, Jartti T, Gupta S et al (2007b) Human bocavirus and acute wheezing in children. Clin Infect Dis 44:904–910
4. Ambrose HE, Clewley JP (2006) Virus discovery by sequence-independent genome amplification. Rev Med Virol 16:365–383
5. Antachopoulos C, Walsh TJ, Roilides E (2007) Fungal infections in primary immunodeficiencies. Eur J Pediatr 166:1099–1117
6. Arden KE, McErlean P, Nissen MD et al (2006) Frequent detection of human rhinoviruses, paramyxoviruses, coronaviruses, and bocavirus during acute respiratory tract infections. J Med Virol 78:1232–1240
7. Bialasiewicz S, Whiley DM, Lambert SB et al (2008) Presence of the newly discovered human polyomaviruses KI and WU in Australian patients with acute respiratory tract infection. J Clin Virol 41:63–68
8. Bialasiewicz S, Whiley DM, Lambert SB et al (2009) Merkel cell polyomavirus in respiratory specimens: a possible route of transmission? Emerg Infect Dis 15:492–494
9. Brotherton J, Wang H, Schaffer A et al (2005) Vaccine preventable diseases and vaccination coverage in Australia, 2003 to 2005. Commun Dis Intell 31 Suppl:S1–S152
10. Cane PA, van den Hoogen BG, Chakrabarti S et al (2003) Human metapneumovirus in a haematopoietic stem cell transplant recipient with fatal lower respiratory tract disease. Bone Marrow Transplant 31:309–310
11. Carrat F, Vergu E, Ferguson NM et al (2008) Time lines of infection and disease in human influenza: a review of volunteer challenge studies. Am J Epidemiol 167:775–785
12. Chandran A, Watt JP, Santosham M (2008) *Haemophilus influenzae* vaccines. In: Plotkin SA, Orenstein WA, Offit PA (eds) Vaccines, 5th edn. Saunders-Elsevier, UK
13. Cutts FT, Zaman SM, Enwere G et al (2005) Efficacy of nine-valent pneumococcal conjugate vaccine against pneumonia and invasive pneumococcal disease in The Gambia: randomised, double-blind, placebo-controlled trial. Lancet 365:1139–1146
14. Falsey AR, Hennessey PA, Formica MA et al (2005) Respiratory syncytial virus infection in elderly and high-risk adults. N Engl J Med 352:1749–1759
15. Foulongne V, Guyon G, Rodiere M (2006) Human metapneumovirus infection in young children hospitalized with respiratory tract disease. Pediatr Infect Dis J 25:354–359
16. Gaynor AM, Nissen MD, Whiley DM et al (2007) Identification of a novel polyomavirus from patients with acute respiratory tract infections. PLoS Pathog 3:595–604
17. Gerberding KM, Eisenhut CC, Engle WA et al (1989) Congenital candida pneumonia and sepsis: a case report and review of the literature. J Perinatol 9:159–161
18. Han LL, Alexander JP, Anderson LJ (1999) Respiratory syncytial virus pneumonia among the elderly: an assessment of disease burden. J Infect Dis 179:25–30.
19. Hope WW (2009) Invasion of the alveolar-capillary barrier by *Aspergillus spp.*: therapeutic and diagnostic implications for immunocompromised patients with invasive pulmonary aspergillosis. Med Mycol 47 Suppl 1:S291–S298
20. Huston SM, Mody CH (2009) Cryptococcosis: an emerging respiratory mycosis. Clin Chest Med 30:253–264
21. Karron RA (2008) Respiratory synsytial virus and parainfluenza virus vaccines. In: Plotkin SA, Orenstein WA, Offit PA (eds) Vaccines, 5th edn. Saunders-Elsevier, UK.
22. Kieny MP, Girard MP (2005) Human vaccine research and development: an over-view. Vaccine 23:5705–5707

23. Ksiazek TG, Erdman D, Goldsmith CS et al (2003) A novel coronavirus associated with severe acute respiratory syndrome. N Engl J Med 348:1953–1966
24. Lau SK, Woo PC, Yip CC et al (2006) Coronavirus HKU1 and other coronavirus infections in Hong Kong. J Clin Microbiol 44:2063–2071
25. McCullers JA (2006) Insights into the interaction between influenza virus and pneumococcus. Clin Microbiol Rev 19:571–582
26. Metlay JP, Fishman NO, Joffe M et al (2006) Impact of pediatric vaccination with pneumococcal conjugate vaccine on the risk of bacteremic pneumococcal pneumonia in adults. Vaccine 24:468–475
27. (2004) Progress in reducing global measles deaths: 1999–2002. Wkly Epidemiol Rec 79:20–21
28. Rudan I, Boschi-Pinto C, Biloglav Z et al (2008) Epidemiology and etiology of childhood pneumonia. Bull World Health Organ 86:408–416
29. Sato M, Hosoya M, Kato K et al (2005) Viral shedding in children with influenza virus infections treated with neuraminidase inhibitors. Pediatr Infect Dis J 24:931–932
30. Sloots TP, Mackay IM, Bialasiewicz S et al (2006a) Human metapneumovirus, Australia, 2001–2004. Emerg Infect Dis 12:1263–1266
31. Sloots TP, McErlean P, Speicher DJ et al (2006b) Evidence of human coronavirus HKU1 and human bocavirus in Australian children. J Clin Virol 35:99–102
32. Talbot TR, Poehling KA, Hartert TV et al (2005) Seasonality of invasive pneumococcal disease: temporal relation to documented influenza and respiratory syncytial viral circulation. Am J Med 118:285–291
33. Talbot HK, Crowe JE J., Edwards KM et al (2009) Coronavirus infection and hospitalizations for acute respiratory illness in young children. J Med Virol 81:853–856
34. van den Hoogen BG, de Jong JC, Groen J et al (2001) A newly discovered human pneumovirus isolated from young children with respiratory tract disease. Nat Med 7:719–724
35. van der Hoek L (2007) Human coronaviruses: what do they cause? Antivir Ther 12:651–658
36. van der Hoek L, Pyrc K, Berkhout B (2006a) Human coronavirus NL63, a new respiratory virus. FEMS Microbiol Rev 30:760–773
37. van der Hoek L, Sure K, Ihorst G et al (2006b) Human coronavirus NL63 infection is associated with croup. Adv Exp Med Biol 581:485–491
38. Viboud C, Alonso WJ, Simonsen L (2006) Influenza in tropical regions. PLoS Med 3:e89
39. Walter ND, Grant GB, Bandy U et al (2008) Community outbreak of *Mycoplasma pneumoniae* infection: school-based cluster of neurologic disease associated with household transmission of respiratory illness. J Infect Dis 198:1365–1374
40. Weber MW, Mulholland EK, Greenwood BM (1998) Respiratory syncytial virus infection in tropical and developing countries. Trop Med Int Health 3:268–280
41. Weinberg GA, Hall CB, Iwane MK et al (2009) Parainfluenza virus infection of young children: estimates of the population-based burden of hospitalization. J Pediatr 154:694–699
42. Whiley DM, Sloots TP (2006) Sequence variation can affect the performance of minor groove binder TaqMan probes in viral diagnostic assays. J Clin Virol 35:81–83
43. Williams BG, Gouws E, Boschi-Pinto C et al (2002)Estimates of world-wide distribution of child deaths from acute respiratory infections. Lancet Infect Dis 2:25–32
44. Woo PC, Lau SK, Chu CM et al (2005) Characterization and complete genome sequence of a novel coronavirus, coronavirus HKU1, from patients with pneumonia. J Virol 79:884–895
45. World Health Organization (2005) Make any mother and child count. Geneva.
46. Wu B, Liu H, Huang J et al (2009) Pulmonary cryptococcosis in non-AIDS patients. Clin Invest Med 32:E70–E77
47. Zilberberg MD, Shorr AF (2009) Fungal infections in the ICU. Infect Dis Clin North Am 23:625–642

Chapter 6
Blood-Borne Disease

Dominic Dwyer and William Rawlinson

Abstract The use of nucleic acid tests (NAT) has dramatically enhanced the detection and management of blood-borne viruses. Some of the very first applications of NAT using polymerase chain reaction (PCR) were in the identification of HIV in blood. The recognition of other blood-borne viruses, in particular hepatitis B and hepatitis C, further prompted the application of molecular tests to clinical medicine. It was the ease of transmission of viruses through blood and blood products that was one of the main stimuli for the development of nucleic acid testing (NAT) in virology. The improvements in technology leading to automation, reduction in contamination, quantitation and increased sensitivity have enhanced this development and are likely to continue and expand, and so further improve patient management – hopefully with reductions in cost. The introduction of these more sensitive NAT assays has further reduced the likelihood of the acquisition of blood-borne viruses during transfusion or transplantation. The development of commercial viral load assays, or quantitative NAT, has revolutionised for example the clinical management of HIV infection and of other systemic diseases (e.g. cytomegalovirus, Epstein-Barr virus, HHV-6, BK polyomavirus) that are a particular clinical issue in immunocompromised individuals (Table 6.1). Also, resistance genotyping tests can now be performed by sequencing the relevant drug targets e.g. HIV – most commonly the reverse transcriptase and protease regions, but now including integrase and envelope gp 41. Infections in immunocompromised patients may occur in the absence of classical symptoms and the use of PCR to diagnose, in a targeted fashion, the most common infections has revolutionised treatment of these infections in the immunocompromised patient. Knowledge of the most likely agents – for example, CMV in the first 3 months following bone marrow transplant, is often used because such reactivation infection is extremely common (and often fatal) in this population. The availability of polyoma virus PCR means a clinical transplant service is able to monitor their patients for polyoma virus and adjust the level of immunosuppression

D. Dwyer (✉)
Centre for Infectious Diseases and Microbiology, Westmead Hospital,
Westmead, NSW 2145, Australia
e-mail: dominic_dwyer@wmi.usyd.edu.au

M. Schuller et al. (eds.), *PCR for Clinical Microbiology*,
DOI 10.1007/978-90-481-9039-3_6, © Springer Science+Business Media B.V. 2010

appropriately. The availability of rapid diagnosis for fungi or Pneumocystis means transplant clinical services can often reduce the use of antimicrobial prophylaxis for these agents. Therefore, as a result of these rapid diagnostic nucleic acid testing services, clinical services can directly improve patient therapy, thereby reducing side effects and improving outcomes as well as saving money.

Keywords Virus · Immunocompromised · Transplant · Diagnosis · Quantitative PCR

6.1 PCR Use in the Diagnosis of Blood-Borne Viral Diseases

Dominic Dwyer

6.1.1 Introduction

The ease of transmission of viruses through blood and blood products was one of the main stimuli to the development of nucleic acid testing (NAT) in virology. The improvements in technology leading to automation, reduction in contamination, quantitation and increased sensitivity have enhanced this development.

Following the recognition in 1984 of human immunodeficiency virus (HIV) as the cause of AIDS, serological assays to detect HIV-specific antibodies were developed as the main screening and diagnostic tests. However, their inability to detect either early HIV seroconversion or to measure the severity of infection drove the demand for NAT. In fact, some of the very first applications (published in 1987) of NAT using polymerase chain reaction (PCR) were in the identification of HIV in blood. The recognition of other new blood-borne viruses, in particular hepatitis C, as well as the knowledge of the already identified hepatitis B, further prompted the application of molecular tests to clinical medicine.

6.1.2 Qualitative NAT

NAT for blood-borne viruses has gone through various applications. Initially it was characterised by qualitative PCR directly on blood or other tissues. In HIV, this was done on whole blood or different blood components, e.g. plasma (reflecting 'free' RNA virions) and peripheral blood mononuclear cells (reflecting mainly integrated provirus DNA). Importantly, qualitative NAT has reduced the 'window period', the time between infection and HIV seroconversion, to less than 2 weeks. It has

D. Dwyer (✉)
ICPMR, Westmead Hospital, The University of Sydney, Sydney, NSW 2006 Australia
e-mail: dominic_dwyer@wmi.usyd.edu.au

also allowed the resolution of incomplete serological results, as exemplified by the evolving Western blot in acute HIV infection. PCR for integrated HIV DNA is invariably positive in people who are HIV seropositive; it remains useful in confirming indeterminate serological results and determining the presence of infection in babies born to HIV seropositive mothers (where maternal HIV antibodies may circulate for up to 18 months).

Qualitative PCR can also be used to detect HIV in different body tissues, an important tool in the early days of understanding HIV pathogenesis.

6.1.3 NAT in the Blood Supply

In addition to confirming infection in an individual, testing is needed to protect the blood supply and identify asymptomatic infections in blood donors. This was initially done by rejecting individuals with a clinical or social history that put them at risk, people with surrogate markers of blood-borne virus infections e.g. abnormal liver function tests, or people with serological evidence of infection. Despite these assessments, transmission of blood-borne viruses through the blood supply still occurred. The introduction of more sensitive NAT has further reduced the likelihood of the acquisition of blood-borne viruses during transfusion or transplantation. Of course, the availability of testing methodologies varies around the world, and there certainly is inequality in the access to NAT.

Other viruses may also be transmitted by blood or organ transplantation, although these may not reflect the most common modes of transmission. Examples include arboviruses (West Nile virus infection has been associated with transfusion in the USA), influenza A H5N1 infection, SARS, and other novel viruses such as hepatitis G, TTV, EBV, CMV, BK/JC, Dengue etc. NAT for these pathogens are available for both individual patient diagnosis and protection of the blood supply, but overall they are perhaps less clinically relevant than HIV, hepatitis B and hepatitis C.

6.1.4 Quantitative NAT

The development of commercial viral load assays, or quantitative NAT, has revolutionised the clinical management of HIV infection (Table 6.1). The plasma HIV load, expressed in HIV RNA copies/mL plasma or log copy number, allows the clinician to do a number of things. A high viral load, when combined with the CD4+ T cell count, allows the prognosis to be discussed with the patient by determining the likelihood of progressing to either AIDS or death within a defined time period. It facilitates discussion of transmission, as it is known that the higher the HIV load, the more transmissible the virus is to others. The HIV plasma load also guides when antiretroviral therapy should be started and perhaps most importantly, assesses the response to antiretroviral therapy. Reducing the HIV load with antiretroviral therapy not only provides clinical benefit, but also provides a public health benefit by reducing the likelihood of ongoing HIV transmission. The HIV plasma load is used

Table 6.1 Commercial kits, methods and technologies for HBV, HCV and HIV detection

Target	Company	Assay name and version	Dynamic range	Instrument	Technology
HCV viral load	Abbott Molecular	RealTi*me* HCV	12–100,000,000 IU(copies)/mL	*m*2000 system	Real-time PCR
	Bayer	Siemens Versant HCV RNA 3.0 Assay	615–7,690,000 IU/mL	Versant 440 Molecular System	Branched DNA
	Bayer	Siemens Versant HCV RNA 1.0 kPCR Assay	Unavailable	Versant kPCR Molecular System	Real-time PCR
	Roche	COBAS Ampliprep/COBAS Taqman HCV Test	43–69,000,000 IU/mL	COBAS Ampliprep/COBAS Taqman	Real-time PCR
	Roche	COBAS Taqman 2.0 (with HighPure)	25–391,000,000 IU/mL	COBAS Taqman	Real-time PCR
	Abbott Molecular	HCV Genotype II	Viral load and genotypes 1–6 and 1a or 1b	*m*2000 system	Real-time PCR
HBV viral load	Abbott Molecular	RealTi*me* HBV	10–1,000,000,000 IU(copies)/mL	*m*2000 system – *m*24sp and *m*2000rt	Real-time PCR
	Abbott Molecular	RealTi*me* HBV	10–1,000,000,000 IU(copies)/mL	*m*2000 system – *m*2000sp and *m*2000rt	Real-time PCR
	Bayer	Siemens Versant HBV DNA 3.0 Assay	2,000–100,000,000 copies/mL	Versant 440 Molecular System	Branched DNA
	Bayer	Siemens Versant HBV DNA 1.0 kPCR Assay	Unavailable	Versant kPCR Molecular System	Real-time PCR
	Qiagen	Artus HBV Assay	0.02–100,000,000 IU/mL	Multiple platforms and kits	Real-time PCR
	Roche	COBAS Ampliprep/COBAS Taqman HBV Test V2.0	54–110,000,000 IU/mL	COBAS Ampliprep/COBAS Taqman	Real-time PCR
	Roche	COBAS Amplicor HBV Monitor	60–38,000 IU/mL	Cobas Amplicor	PCR

Table 6.1 (continued)

Target	Company	Assay name and version	Dynamic range	Instrument	Technology
	Roche	Amplicor HBV Monitor Test, v2.0	190–7,604,563 IU/mL	Cobas Amplicor	Microwell plate format PCR
HIV viral load	Abbott Molecular	RealTi*me* HIV-1	40–10,000,000 IU(copies)/mL	*m*2000 system	
	Bayer	Siemens Versant HIV-1 RNA 3.0 Assay	50–500,000 copies/mL	Versant 440 Molecular System	Branched DNA
	Bayer	Siemens Versant HIV-1 RNA 1.0 kPCR Assay	35–11,000,000 copies/mL	Versant kPCR Molecular System	Branched DNA
	Biomerieux	Nuclisens EASYQ HIV-1 V2.0	10–10,000,000 copies/mL	Nuclisens EasyQ	NASBA Real-time PCR
	Roche	Ampliprep/Cobas Taqman HIV-1	40–10,000,000 copies/mL	COBAS Ampliprep/COBAS Taqman	Real-time PCR
		TAQman HIV-1 (with HighPure)	47–10,000,000 copies/mL	COBAS Taqman	Real-time PCR
HIV Genotyping	Abbott	ViroSeq HIV-1 Integrase Genotyping Kit	An 864 base pair region of the entire IN gene is amplified and sequenced	ABI PRISM 3100 OR 3130xl	Sequencing based-mutations in the integrase region of the pol gene in RNA
	Bayer	Siemens TruGene HIV-1 Genotyping Assay	A 1038 base sequence of HIV-1 DNA is determined	PCR sequencing based run on the OpenGene DNA System	Sequencing based-mutations in the protease and part of the RT region of HIV

to monitor the response to new antiviral agents, in both treatment experienced and treatment of naive individuals. A rise in HIV plasma load is a surrogate for HIV drug resistance, as well as other therapy-related problems such as compliance and drug absorption, and is an indication for changing therapy.

Following evidence of the clinical value of HIV plasma load assays, quantitative NAT has also proved clinically useful in addressing treatment and antiviral drug resistance issues in hepatitis C and B infection. The management of other systemic diseases (e.g. cytomegalovirus, Epstein-Barr virus, HHV-6, BK polyomavirus) that are a particular clinical issue in immunocompromised individuals is also guided by quantitative NAT on blood, and will be enhanced as multiplex quantitative assays become available.

6.1.5 Genotyping

Genotyping for the detection of mutations associated with HIV drug resistance and clinical failure of therapy is now 'standard of care' in HIV management. A resistance genotyping test is most commonly performed by sequencing the relevant drug targets – most commonly the reverse transcriptase and protease regions, but now including integrase and envelope gp 41. Genotyping is recommended prior to the commencement of the first antiviral regimen to ensure that the individual is not already infected with drug resistant virus. Resistance testing is also indicated when people are failing antiretroviral therapy. HIV phenotype assays are also available, but their cost precluded their routine use in clinical practice and methods have been excluded from this section. The same principles apply to other blood-borne viruses, and assays are available for the detection of hepatitis B and CMV drug resistance mutations. As hepatitis C-specific therapies evolve, it is likely that genotyping will assist clinicians in the best use of antiviral drugs in both acute and chronic hepatitis C.

6.1.6 Molecular Epidemiology

Molecular epidemiology studies are undertaken with HIV, hepatitis B and hepatitis C, and to a lesser extent with the other blood-borne viruses. Here, sequencing of the viral genome allows comparison of a patient's viral sequences with those in large public sequence databases, allowing one to determine in which viral subgroup the patient's virus falls. Molecular subtyping of viral strains assists with understanding how viruses move in a community, and indicate potential subtype-specific responses to antiviral therapy and clinical outcomes. NAT has also been used in medico-legal cases, where HIV or other blood-borne viruses have been transmitted to other individuals

Another molecular advance is the use of NAT to determine the presence of human gene alleles associated with adverse drug reactions. This has been most clearly demonstrated in HIV-infected individuals with abacavir hypersensitivity in the presence of HLA-B5701. This test is now performed as part of 'standard of care' in HIV-infected individuals where abacavir therapy is being considered.

The use of NAT has dramatically enhanced the detection and management of blood-borne viruses. This is likely to continue and expand, and so further improve patient management – hopefully with reductions in cost.

In Australia, assays for the detection of HIV and HCV require approval by the Therapeutic Goods Administration [see Therapeutic Goods Regulations 1990. Schedule 3. Therapeutic goods required to be included in the part of the Register for registered goods (regulation 10)]. The legislative basis for the regulation of in-vitro diagnostic devices (IVDs) (see In vitro diagnostic devices – definitions and links at http://www.tga.gov.au/devices/definitions.htm) involves inclusion on the Australian Register of Therapeutic Goods (ARTG). The ARTG is a database of therapeutic goods approved for supply in Australia. IVDs for the diagnosis of patients infected with the HIV or with the HCV are required to be registered in the ARTG. If the application is acceptable, the Conformity Assessment Branch (CAB) will advise the sponsor of the evaluation fee, any additional data required and the number of test kits required for evaluation. All data for evaluation must be sent to the CAB who will coordinate the evaluation. The TGA undertakes the evaluation of kit integrity while the quality and efficacy aspects are evaluated by the National Serological Reference Laboratory (NRL) located in Melbourne.

Every IVD approved for listing or registration in the ARTG is subject to standard and specific conditions under Section 28 of the Act. http://www.tga.gov.au/docs/pdf/dr4/dr4app04.pdf

IVDs approved for use as screening or as supplemental tests for the diagnosis of infection with Human Immunodeficiency Virus [HIV] (viral load assays exepted) and also for supplemental tests for the diagnosis of infection with Hepatitis C Virus [HCV] may be supplied to authorised laboratories only.

Sponsors and manufacturers also have ongoing responsibilities once a product has been approved for supply in Australia. All IVDs supplied in Australia (including exempt IVDs) are subject to post market requirements and IVD users can report any problems relating to the safety, quality or performance of an IVD supplied for commercial use.

Extract from *http://www.tga.gov.au/docs/pdf/dr4/dr4v1s2.pdf*:

> Conditions of Registration HIV Test Kits
> Test kits will be entered in the ARTG and specified as being suitable for routine screening or supplemental purposes. The conditions relating to the registration will specify the appropriate category.
> Sponsors will be advised by the TGA of the marketing approval, which will include details of the ARTG registration number, conditions of approval and the certificate of registration.
>
> HCV Test Kits
> Test kits are categorised as being suitable either for routine screening or for supplemental purposes only, and the condition of entry in the ARTG will specify the category of supply.

There is no restriction to the supply of HCV test kits approved for screening once entered in the ARTG. However, it is Commonwealth policy that all HCV test kits approved for use as supplemental assays and those using newer technology (such as polymerase chain reaction, branched DNA amplification or procedures currently in developmental stages) be supplied only to laboratories approved by State/Territory health authorities.

Use of Unapproved HIV/HCV Test Kits
Provisions exist for the supply of unapproved HIV/HCV test kits either for research use or under the Clinical Trials / Special Access Schemes in the Therapeutic Goods Act 1989. Refer to Chapter 1.24 Access to Unapproved Therapeutic Devices. . ..
Sponsors are permitted to supply their kits for research purposes prior to the evaluation procedures having been completed. The kits may only be supplied to bona fide research institutes as well as to the designated testing laboratories. Such products cannot be used for diagnostic purposes, i.e. specimens must not be identifiable. . ..

6.2 PCR Use in the Diagnosis of Infection in the Immunocompromised Host

William Rawlinson

6.2.1 Background

The prevention of infection is dependent upon existing (innate) and responsive (humoral and cell mediated) immune functions. All these host functions are critical to preventing infection in the host. However, in most clinical settings we see immune compromise as immunodeficiencies resulting from abnormalities of humoral and cell-mediated components. Ranulocytopaenia results in predominantly bacterial infections and neutropaenia is associated with infection by bacterial species predominantly, although the types of bacterial species differ slightly. Changes in humoral immunity resulting from absence of antibody predominantly cause bacterial infection, whilst impaired cellular immunity results predominantly in infection with intracellular organisms – bacteria such as *Nocardia*, mycobacteria including, atypical mycobacteria, *Pneumocystis jirovecii* and fungal infections. All viral infections have an intracellular component, as viral reproduction is critically dependent upon intracellular infection, and therefore impaired cellular immunity is particularly associated with infection with viruses – either due to reactivation of latent herpesviruses or the more severe infections from exogenous viruses such as respiratory or gastrointestinal agents.

The patient groups demonstrating immunocompromise are many. However, we predominantly focus on immunocompromised patients who present often to hospitals, with severe immune compromise such as results from organ transplantation,

W. Rawlinson (✉)
SEALS Virology Division, Prince of Wales Hospital, Randwick, NSW 2031, Australia
e-mail: w.rawlinson@unsw.edu.au

haematological transplantation (bone marrow transplants are a highly immunocompromised population), compromised immune systems as a result of therapy – such as cancer chemotherapy, and immune compromise resulting from infection with human immunodeficiency virus (HIV) with the development of acquired immune deficiency syndrome (AIDS). As increased numbers of transplants are done each year, and additional transplants (such as bone marrow matched unrelated donor transplantation) are performed, then the number of immune compromised patients has increased. Furthermore, more aggressive immune suppression associated with some of these transplants and longer survival of transplant patients with better medical and surgical therapy, has meant that the number of severely immune compromised patients has increased significantly. The annual cost to the health care system of transplantation transplant complications are also significant – transplants cost $65,000 to $75,000 per transplant with an annual average cost of care of $11,000 (www.transplant.org.au). This cost is significantly increased by post-transplant viral infection – for example, liver transplant recipients who develop cytomegalovirus (CMV) disease have significantly (24–49%) higher costs than patients with no, or asymptomatic CMV infection.

Diagnosis of these infections in immunocompromised patients is critical to proper therapy. Empirical therapy is often begun urgently, due to the tendency of such infections in the immune compromised host to progress, but such empiric therapy often has side effects, requires the patient to remain in hospital, and is costly. Therefore more targeted therapy is needed. Diagnosis, particularly using nucleic acid tests, is critical to allow narrowing of therapy to the minimum required to give the best clinical outcomes. Importantly, much infection may go undiagnosed due to the need for urgent empirical therapy. The febrile neutropenic patient with an undiagnosed infection and treatment with multiple antibiotics, often including antifungal and antiviral agents, is a recurring clinical scenario. Up to one in five infections will remain undiagnosed, and it is apparent that treatment for these different infections, bacterial, fungal or viral is completely different. Therefore diagnosis of infection in the immune compromised host, utilising nucleic acid tests such as PCR is critical.

6.2.2 Issues to Consider in the Diagnosis of Infection in the Immunocompromised Host

The source of the organism in transplant infection may be exogenous or endogenous. The exogenous organisms are those commonly seen in other hosts – for example influenza during winter season, respiratory syncytial virus (RSV) in lung transplant recipients, bacterial agents in neutropenic patients that are normal colonisers of the skin – such as *Staphylococcus aureus* and nosocomial agents introduced via central venous catheter related infections (such as coagulase negative staphylococci, *Staphylococcus aureus*, *Candida albicans*, *Candida parapsilosis*). These agents are often antibiotic resistant as they may be acquired in hospital, and the immunocompromised patient often is treated with multiple antibiotics.

Significant infection exposures of transplant recipients may be derived from the donor organ, preexisting latent infections (herpesviruses, HIV, HCV) in the recipient or donor, the hospital environment (e.g. nosocomial MRSA), and community exposures (e.g. influenza). These have contributed to reducing long term patient survival through death from infection and cancer (ANZDATA and ANZOD transplant data in www.transplant.org.au) and improved diagnosis is vital in reducing these infections.

Diagnosis may be straightforward, as evidence of infection (for example an infected centrovenous catheter) is clinically apparent, a swab can be taken, and the organism will often grow, albeit that antibiotics may inhibit culture in some cases. Endogenous infections, arising from reactivation of latent viruses such as the herpesviruses (HSV1, HSV2, EBV, CMV, HHV6, HHV7, HHV8), the polyoma viruses (BK/JC virus) may be more difficult to detect. This is because serological assays are often of little use in this setting, as IgG and IgM assays, or even avidity assays, are not accurate because immune suppression reduces production of IgM. Therapy for these infections is often toxic (for example ganciclovir for CMV) and decisions regarding prophylaxis versus pre-emptive therapy often need to be made. It is here that nucleic acid tests, particularly quantitative PCR is extremely useful in deciding whether a (viral) pathogen is reactivating, whether the level is increasing (suggesting active infection), and whether the level continues to increase on therapy, suggesting antiviral resistance has developed.

The clinical presentation of the infected immunocompromised patient is often atypical. Infections may occur in the absence of classical syndromes – fever is often used to indicate active bacterial infection, although this may be absent in the immunocompromised patient. Different pathologic agents (bacterial, viral and fungal infecting agents), may present in a very similar fashion. For example, pneumonia with diffuse, bilateral lung infiltrates in the immunocompromised bone marrow transplant recipient may be due to a virus (RSV, CMV, influenza, adenovirus, parainfluenza virus, VZV, HSV), *Mycoplasma pneumoniae*, *Pneumocystis jirovecii*, bacterial and fungal agents although the latter would normally result in localised rather than diffuse infiltrates. Sampling of the respiratory tract in this setting, by bronchoalveolar lavage with or without bronchial biopsy is often useful, when accompanied by specific PCRs for these agents.

Neurological infections, with encephalitis and/or meningitis can also result in non-specific symptoms (confusion, headache, seizures) and symptoms may initially be non-specific. Infections of the CNS with reactivated HSV (HSV encephalitis), *Toxoplasma gondii* (cerebral toxoplasmosis) or intracellular organisms such as *Listeria monocytogenes* or *Cryptococcus neoformans* may all result in this syndrome. Fungal infection may also cause non-specific symptoms, although again focal infection with cryptococcus, aspergillus or *Nocardia* or mycobacteria species may occur. The obvious difficulty with neurological infection in the immunocompromised patient is that access to clinical samples (cerebrospinal fluid or very rarely brain biopsy) is limited when compared with other organ system infections.

Gastrointestinal infections are occasionally troublesome in immunocompromised patients, although long-standing, ongoing infections are usually due to a small number of agents – *Strongyloides stercoralis*, *Clostridium difficile*, and cytomegalovirus (CMV) can infect the lower gut resulting in ongoing diarrhoea. Oesophagitis due to CMV, HSV, or candida may occur. A more serious problem can be for a hospital unit treating immunocompromised patients, where infection with one of the epidemic viruses such as norovirus or astrovirus results in an acute, severe gastroenteritis which may be associated with widespread infection (including of the staff) that can cause severe fluid and electrolyte depletion. Infections of other parts of the gut such as oropharyngeal gingivostomatitis, severe mucositis and oesophagitis may occur although often specific agents (such as HSV in gingivostomatitis, candida in oesophagitis) are limited to a small number of agents. Again, in all these cases, directed PCR to identify the agent will result in successful treatment strategies. Treatment of infections in immunocompromised patients is critically dependent upon accurate diagnosis. The use of PCR to diagnose, in a targeted fashion, the most common infections has revolutionised treatment of these infections in the immunocompromised patient. Knowledge of the most likely agents – for example, CMV in the first 3 months following bone marrow transplant, is often used because such reactivation infection is extremely common (and often fatal) in this population. This means that other agents then become more likely in the individual patient on prophylaxis, and nucleic acid testing can be more directed towards these other agents. The laboratory services are of enormous benefit in directing the approach to treatment taken by a clinical service. The availability of routine CMV viral loads for example, done two to three times weekly means a clinical service is able to undertake pre-emptive treatment with ganciclovir, rather than relying on prophylaxis. The availability of polyoma virus PCR means a clinical transplant service is able to monitor their patients for polyoma virus and adjust the level of immunosuppression appropriately. The availability of rapid diagnosis for fungi or Pneumocystis means transplant clinical services can often reduce the use of antimicrobial prophylaxis for these agents. As a result of these rapid diagnostic nucleic acid testing services, clinical services can directly improve patient therapy, thereby reducing side effects and improving outcomes.

Chapter 7
Enteric Infections

James Buttery

Abstract Acute gastroenteritis presenting with diarrhoea and or vomiting is extremely common in childhood, and, is extremely difficult to differentiate from community acquired infection. Additionally the aetiology of community acquired acute gastroenteritis varies depending upon setting (developed or developing), age, season and testing modality. In all settings, rotavirus remains the most common cause of gastroenteritis in childhood. Norovirus infections remain important in both developed and developing settings. It is also a major cause of nosocomial and community outbreaks in all ages. Additionally, bacterial and parasitic infections are important as a cause of gastrointestinal disease for residents and tourists in developing countries. Traditionally, the diagnosis of enteric pathogens has been based upon microscopy, culture and antigen detection. However, these methods are relatively insensitive, especially for small viral pathogens including norovirus and sapovirus, and for some protozoa. The advent of molecular diagnostics has marked a significant improvement in the diagnosis of enteric infections, identifying a greater number of episodes in many of the populations studied.

Keywords Viruses · Bacteria · Protozoa · Diagnosis · Gastroenteritis · PCR

7.1 Introduction

Acute gastroenteritis presenting with diarrhoea and or vomiting is extremely common in childhood, and responsible for an estimated 2.5 million deaths annually due to dehydration [6]. The overwhelming burden of this mortality is borne in children under 5 years of age in developing countries, but even in developed nations, acute gastroenteritis is a major cause of morbidity and disruption to families and the

---

J. Buttery (✉)
Enteric Virus Research Group, Murdoch Childrens Research Institute, Royal Children's Hospital, Parkville, VIC 3052, Australia
e-mail: jim.buttery@mcri.edu.au

M. Schuller et al. (eds.), *PCR for Clinical Microbiology*,
DOI 10.1007/978-90-481-9039-3_7,

health care system. In Australia, an estimated 17.2 million cases (95% CI 14.5, 19.9 million) of gastroenteritis occur each year, resulting in 41,000 admissions to hospital [4]. Of these, 20,000 are estimated to occur in the highest risk age group, children less than 5 years of age, with those between 6 and 24 months at especial risk [3]. These figures are consistent with international data, suggesting that only a small proportion of cases present even to primary health care, with most treated at home [2].

In childhood, it is extremely difficult to differentiate between community acquired gastroenteritis pathogens based upon clinical presentation.

While seizures with gastroenteritis strongly indicate Shigella as a likely cause, most Shigella gastroenteritis patients do not fit. At the bedside level, there is also little difference in the way children are treated. Oral rehydration therapy remains the cornerstone of gastroenteritis management. Even in acute bacterial gastroenteritis, antibiotics are rarely used. Bowel antispasmodic agents may be used in many adult cases, but are contra-indicated in children. Outside the context of gastroenteritis, gastro-intestinal pathogens cause ubiquitous disease such as *Helicobacter pylori* and gastritis, as well as severe life-threatening conditions including Haemolytic Uraemic Syndrome (entero-haemorrhagic *E. coli*) and gastric cancer (*H. pylori*).

The decision to test for gastroenteritis pathogens has been based upon several factors, including:

- to allow assessment of organism epidemiology
- to detect potential food-borne outbreaks that may require intervention
- for unusual presentations (such as significant bloody diarrhoea)
- to enable cohorting in hospitals to minimise nosocomial transmission
- for nosocomially acquired cases to detect hospital outbreaks

The aetiology of community acquired acute gastroenteritis varies depending upon setting (developed or developing), age, season and testing modality. Traditionally, rotavirus has been the most common cause of gastroenteritis in childhood in all settings. The more severe the childhood gastroenteritis presentation, the more likely rotavirus is to be the cause [1]. With the introduction of oral rotavirus vaccine into the National Immunisation Program in Australia in 2007, it is expected to decrease in importance there, although definitive effectiveness data is still pending. Worldwide, as vaccine access remains limited, it is still the most important pathogen, killing an estimated 600,000 children annually [6]. Norovirus, the most common cause of sporadic gastroenteritis in adulthood, is the second most common pathogen in children in developed settings, and remains important in the few developing settings it has been studied [7]. It is also a major cause of nosocomial and community outbreaks in all ages. In developing settings bacterial and parasitic infections assume greater importance in gastrointestinal disease for residents and tourists. Common bacterial pathogens include species of *Salmonella, Campylobacter, Shigella* enterotoxigenic *E.coli*, *Vibrio* and *Yersinia*. *Pleisomonas* and *Aeromonas* as well a range of toxin producing organisms may also result in acute diarrhoea. *Clostridium difficile* toxin production is an important cause of nosocomial infection. Symptoms lasting more than 2 weeks should prompt a request to look for parasitic infection particularly *Cryptosporidium* and *Giardia*. A travel history may require more extensive culture and microscopy for less common pathogens. Microscopy is still useful to assess the degree of inflammation present and to detect haemorrhagic diarrhoea.

Seasonally, winter and spring are peak periods for viral pathogens (especially rotavirus), whilst summer heat can favour bacterial pathogens as well as adenovirus.

The diagnostic ascertainment of enteric pathogens has long been based upon microscopy, culture and antigen detection. These methodologies have had the disadvantage of poor sensitivity – especially for small viral pathogens including norovirus and sapovirus as well as less robust protozoa including giardia and cryptosporidia. The 'detection threshold' of these tests probably increased the likelihood that when they were positive, that the organism detected was causing the acute symptoms. The advent of molecular diagnostics has seen the proportion of episodes with a detectable pathogens climb towards 80%, depending upon the populations studied [5]. New pathogens including bocavirus and torovirus have been appreciated, as well as a reordering of relative proportions of known pathogens. However, they have also seen an increase in co-infections as well as asymptomatic detection [8]. While this will serve to enrich our understanding of how humans interact with these organisms, we may need to exercise caution for some positive results until appropriate population studies with or without quantitative interpretation can inform the true significance of these results.

References

1. Bernstein DI (2009) Rotavirus overview. Pediatr Infect Dis J 28(3 Suppl):S50–S53
2. de Wit MA, Kortbeek LM, Koopmans MP et al (2001) A comparison of gastroenteritis in a general practice-based study and a community-based study. Epidemiol Infect 127:389–397
3. Galati JC, Harsley S, Richmond P, Carlin JB (2006) The burden of rotavirus-related illness among young children on the Australian health care system. Aust N Z J Public Health 30:416–421
4. Hall G, Kirk MD, Becker N et al (2005) Estimating foodborne gastroenteritis, Australia. Emerg Infect Dis 11:1257–1264
5. Huh JW, Kim WH, Moon SG, Lee JB, Lim YH (2009) Viral etiology and incidence associated with acute gastroenteritis in a 5-year survey in Gyeonggi province, South Korea. J Clin Virol 44:152–156
6. Parashar UD, Hummelman EG, Bresee JS et al (2003) Global illness and deaths caused by rotavirus disease in children. Emerg Infect Dis 9:565–572
7. Sdiri-Loulizi K, Ambert-Balay K, Gharbi-Khelifi H et al (2009) Molecular epidemiology of norovirus gastroenteritis investigated using samples collected from children in Tunisia during a four-year period: detection of the norovirus variant GGII.4 Hunter as early as January 2003. J Clin Microbiol 47:421–429
8. Subekti D, Lesmana M, Tjaniadi P et al (2002) Incidence of Norwalk-like viruses, rotavirus and adenovirus infection in patients with acute gastroenteritis in Jakarta, Indonesia. FEMS Immunol Med Mic 33:27–33

Chapter 8
Sexually Transmitted Infections

Suzanne M. Garland, David W. Smith, Sepehr N. Tabrizi, and David M. Whiley

Abstract Widespread screening for *Chlamydia trachomatis, Neisseria gonorrhoea* and other STIs, adds a significant burden to the health care budget, but is important for disease prevention and epidemiology. NAATs are now widely used, and have become the preferred tools for detecting and screening for STIs. This is largely due to their enhanced sensitivity and specificity over more traditional methods, and the ability to process non-invasive test samples such as urine samples or self-collected vaginal swabs. However, in order to provide accurate diagnostic results NAATs must be rigorously evaluated for specificity, reproducibility, analytic and clinical validity, and cost before they can be recommended for screening of general populations.

Keywords Sexually transmitted infection · Nucleic acid amplification test · Virus · Bacteria · Protocol · Diagnosis · Clinical

8.1 Introduction to Sexually Transmitted Infections

Sexually transmitted infections (STIs) are infections of antiquity, references to diseases being made as long ago as in the Bible. STIs cut across all socio-economic groups in society, but particularly the poor and the underprivileged. Consequently, STIs are seen to increase in incidence in times of social disruption, particularly that of war.

Those pathogens included as STIs include viruses, as well as bacteria, protozoa and parasites. The incidence of various STIs varies from country to country, as well as within different social groups. It is noteworthy that STIs are particularly common amongst those under 25 years of age. They are diverse in clinical presentation,

S.M. Garland (✉)
Microbiology and Infectious Diseases Department, Royal Women's Hospital, Parkville, VIC 3052, Australia; Department of Obstetrics and Gynecology, University of Melbourne, Melbourne, VIC, Australia; Murdoch Children's Research Institute, Parkville, VIC 3052, Australia
e-mail: Suzanne.Garland@thewomens.org.au

M. Schuller et al. (eds.), *PCR for Clinical Microbiology*,
DOI 10.1007/978-90-481-9039-3_8, © Springer Science+Business Media B.V. 2010

many are asymptomatic or non-specific, yet have the propensity to cause significant morbidity and mortality, if they go unrecognized and untreated.

In particular in the pregnant population, STIs can result in adverse outcomes for the pregnancy (eg spontaneous abortion, premature birth, low birth weight for gestation) as well as cause congenital abnormalities or disease. Consequently, where there are cheap, sensitive screening and confirmatory assays available, plus appropriate and effective intervention in the pregnant woman to prevent or reduce adverse outcomes, antenatal screening for certain STIs is recommended.

Definitive diagnoses for various STIs in some instances has been difficult because pathogens have been difficult to culture by traditional means (eg HCV, HPV), hampered by transport requirements and necessity for specialized media due to the fastidious nature of the organisms (eg *N.gonorrhoeae, Haemophilus ducreyi)* or required cell culture (eg HSV, CMV). However with the advent of molecular biology tools, particularly applied to self collected samples, definitive diagnoses have become more readily available. Techniques such as polymerase chain reaction (PCR) still require careful collection, transport and handling at the clinical level as well as in the laboratory to ensure that there is no DNA contamination. These techniques being much more sensitive and rapid (results within hours of receipt in the laboratory) offer great advantages over traditional diagnoses (particularly for those with poor access to medical services or those not compliant with screening and invasive medical examinations). However they are more expensive, require scientific expertise, and currently do not give antimicrobial sensitivity results.

8.2 Clinical Aspects and Epidemiology

A definitive diagnosis of any STI is important, as prompt and accurate diagnosis can reduce both short and long term morbidity in the patient and have an important public health impact by reducing further transmission.

New, more sensitive and rapid diagnostic tools, particularly nucleic acid amplification tests (NAATs) are having a significant impact on diagnoses, particularly for those pathogens that were difficult to grow, or susceptible to transport conditions. [6]. However as with any diagnostic assay, NAAT shortcomings as well as their strengths need to be recognised, particularly the need for confirmatory assays [21]. The microbiology laboratory is an important source of information for appropriate test selection, particularly as the variety of testing options increase.

It is noteworthy that any individual exposed to or diagnosed with a STI should be tested for other STIs relevant to the local epidemiology, as such infections tend to be transmitted simultaneously [2]. In addition their partner(s) (and in the case of a pregnant woman, their infant) should also be investigated and treated appropriately.

8.2.1 Clinical Presentation

The clinical presentation of STIs in women ranges from urethritis, cervicitis, vaginitis, pelvic pain, and inflammatory disease, to genital ulcers and lesions, although the majority actually occur asymptomatically. In women, long-term complications include infertility, ectopic pregnancy and chronic pelvic pain. When symptomatic, males are more likely to note lesions on the genital skin, or complain of urethral discharge, or dysuria. If STIs become complicated in males, they tend to present with epididymitis or orchitis.

Tables 8.1 and 8.2 outline the various bacterial, parasitic and viral pathogens associated with STIs denoting the areas of infection involved, the symptomatology, common clinical signs, as well as incubation period. Infection with herpes simplex virus and *Chlamydia trachomatis* have been specifically described in detail as more common examples of viral and bacterial infections, respectively.

8.2.2 Genital HSV

The most common cause of genital ulcers in Australia is herpes simplex virus (HSV), but both syphilis and chancroid can present in a similar fashion: thus diagnostic tests are required to differentiate these infections. A history of recent travel may be helpful for more exotic STIs such as chancroid or syphilis.

The prevalence of genital herpes in Australia is around one in seven sexually active adults [12]. Typically genital herpes is caused by HSV-2, although HSV-1 is becoming more common. This reflects changing epidemiology with young people not picking up HSV-1 orally as children, but then being infected genitally. The clinical presentation of primary infection is characterised by vesicular or ulcerative lesions on the external genitalia, which may be associated with systemic symptoms of fever, headache, malaise and myalgia. More commonly the infection goes unrecognised or is totally asymptomatic: here type specific serological assays can be of value. The cervix and urethra may be involved, with vaginal and urethral discharge and dysuria being present. Tender inguinal lymphadenopathy may also occur. Genital recurrences are more common with HSV-2, with 90% having further episodes in the first 12 months, compared with 55% for HSV-1 [14]. Lesions may be atypical, presenting as painful fissures or redness. Intermittent subclinical shedding occurs and results in transmission to sexual partners, as well as neonates during delivery. Neonatal HSV, although rare can have devastating consequences, resulting in death, as well as permanent neurological sequelae [5].

8.2.3 *Chlamydia trachomatis*

Chlamydia trachomatis is the most common bacterial STI and is associated with cervicitis, urethritis, endometritis, salpingitis and perihepatitis in females (Fitz Hugh Curtis syndrome) and urethritis and epididymitis in males. In contrast, lymphogranuloma venereum (LGV) is an invasive infection associated with specific serological variants (LGV L1-3) of *C.trachomatis*. It is usually imported, so seen in travellers and men who have sex with men (MSM), presenting as a genital ulcerative lesion, with suppurative inguinal lymphadenopathy or proctitis.

Table 8.1 Bacterial and protozoan pathogens

Pathogens	Incubation period	Area of infection, transmission pattern	Symptoms and clinical manifestations
Chlamydia trachomatis (D-K)	7–14 days (3–90 days)	Columnar epithelium of cervix, urethra, fallopian tubes, conjunctiva of eye, lungs, anal canal. Genital secretions (transmission)	Most often asymptomatic. Genital tract discharge, cervicitis, urethritis, pelvic inflammatory disease (PID), conjunctivitis urinary frequency, Abnormal bleeding, sticky eye, pneumonitis (infants with infection). Complications include: ectopic pregnancy, infertility, chronic pelvic pain. In the pregnant woman, adverse outcomes include spontaneous abortion, postpartum (or postabortal) endometritis and salpingitis, premature rupture of the membranes and (rarely) intrauterine fetal infection.
Chlamydia trachomatis (LGV)	7–14 days (3–30 days)	Genital epithelium and mucus membranes, anal canal, lymph node, Genital secretions (transmission)	Genital ulceration, lymphatic buboes, proctitis genital ulceration
Haemophilus ducreyi (chancroid)	1 day–2 weeks, usually 3–5 days	Genital epithelium and mucus membranes Genital secretions (transmission)	Genital tract discharge, pain, ulceration
Klebsiella granulomatis (donovanosis)	weeks to months 7–112 days	Genital epithelium and mucus membranes Genital secretions (transmission)	Genital tract discharge, pain, ulceration
Neisseria gonorrhoeae	2–7 days	Cervix, , urethra, Fallopian tubes, eye, rarely: joints Genital secretions (transmission rarely blood borne)	Many women are asymptomatic Cervicitis, urethritis PID, conjunctivitis, rarely: septicaemia, septic arthritis

Table 8.1 (continued)

Pathogens	Incubation period	Area of infection, transmission pattern	Symptoms and clinical manifestations
Mycoplasma genitalium	average 2–3 weeks	Cervix, urethra and possibly fallopian tubes Genital secretions (transmission)	May be asymptomatic, genital tract discharge, pain, Cervicitis, urethritis, Possibly PID.
Treponema pallidum	10–90 days (average 21 days)	Genital epithelium and mucus membranes Genital secretions and blood-borne (transmission)	May be asymptomatic, genital ulceration, rash, genital ulcerations untreated complications of CVS, CNS
Trichomonas vaginalis	4–28 days, usually 7 days	Genital mucus membranes of vagina and epithelium of the bladder Genital secretions and rarely fomites (transmission)	May be asymptomatic or genital discharge, vaginitis urinary frequency

Table 8.2 Viral pathogens

Virus	Incubation period	Area of infection, transmission pattern	Clinical manifestations and symptoms
Adenoviruses	2–8 days	Respiratory tract, gastrointestinal tract, eyes, bladder: specifically urethra, throat	Urethritis: urinary frequency, dysuria or discharge (also pharingitis, gastroenteritis, keratoconjunctivits)
Cytomegalovirus (CMV)	3–12 weeks	Respiratory tract, liver, genital tract. Transmitted through saliva, urine, genital secretions, blood (viraemia)	Largely asymptomatic, fever, nonspecific skin rash, cervical lymphadenopathy, (occasionally hepatitis)
Hepatitis B virus (HBV)	45–180 days with an average of 60–90 days	Liver: blood borne infection causing hepatitis. Genital secretions as well as blood can transmit infection sexually. Once infected may become chronic carrier	Asymptomatic, hepatitis: fever, jaundice, nausea, vomiting
Hepatitis C virus (HCV)	Ranges from 2 weeks–6 months, most commonly 6–9 weeks. HCV antibody tests become (+) 2–3 months after exposure	Liver: blood borne causing hepatitis Once infected may become chronic carrier	Asymptomatic, hepatitis: fever, jaundice, nausea, vomiting
Herpes simplex virus (HSV) types 1 and 2	2–8 days typically with a range of 1–26 days	Genital skin and mucous membranes and sensory ganglia: oropharynx, uncommonly viraemia, meningitis, and encephalitis	Genital or other mucocutaneous vesicles or ulcers, genital discharge, fever general malaise, myalgia, headache

Table 8.2 (continued)

Virus	Incubation period	Area of infection, transmission pattern	Clinical manifestations and symptoms
Human immunodeficiency virus (HIV1, HIV2)	Period from infection to the primary sero-conversion illness 3–8 weeks, with development of anti-HIV antibodies being 3–weeks–3 months Time from HIV infection to the diagnosis of AIDS ranges from 9 months–20 years or longer, with a median of 12 years	Blood borne. Transmitted in genital secretions as well as blood	Fever, rash, loss of weight, secondary opportunistic infections
Human papillomavirus (HPV)	For genital warts, 3 weeks–8 months, the average being 3 months	Infects genital skin and mucus membranes. Low-risk genotypes (6, 11) cause the majority of genital warts or condylomata accuminata. High risk genotypes (especially 16 and 18) cause cervical, vulvar, anal dysplasia, as well as their respective cancers	Most often asymptomatic. Genital warts, abnormal genital bleeding in association with neoplasia.

Detection of *Chlamydia* using traditional methods lack sensitivity and these have been largely replaced by NAATs. Even then, one needs to be very careful about checking for molecular changes in the organisms, as recently described in isolates lacking the cryptic plasmid circulating in Scandinavia [15] (see Section 8.5). Serological tests are not recommended as they cannot reliably distinguish between acute and past infection, even when testing acute and convalescent sera. Moreover they need to be interpreted against the background carriage rate of positive serology. The main value of serology lies in infants with chlamydia pneumonitis or a woman with pelvic inflammatory disease with no further positive detections.

8.3 Commonly Recognised STI Agents

David W. Smith

STIs are caused by a range of diverse organisms that exploit the environment, nature and necessity of sex in human society. They span bacteria, viruses, fungi and parasites linked only by their mode of transmission.

Bacteria continue to feature as the cause of some of the most common and significant STIs throughout the world. Interestingly all are relatively fastidious and each poses its own diagnostic difficulties. For these reasons, PCR based diagnostics have been embraced as the method of choice.

Neisseria gonorrhoeae is a member of a genus that contains one other well known pathogen, *N. meningitidis*, and a large number of other species that are commonly found as commensals of humans. Within the genus, interchange of genetic material is common, leading to some of the difficulties experienced with PCR-based diagnostics. However, culture-based diagnosis is limited by the need for specialised media, and the difficulties of maintaining organism viability during storage and transport.

Chlamydia trachomatis is an obligate intracellular organism that is classified into 15 or more serological varieties (or serovars) that fall into three main biological groupings: Types A, B, Ba and C cause trachoma, types D-K cause the common STIs, while types L1, L2 and L3 cause the chronic infection lymphogranuloma venereum. The organism exists in two forms: the reticulate bodies that are the fragile form in which it grows within cells, and the elementary bodies which are the hardier forms that are able to be transmitted. As expected, culture requires growth

David W. Smith (✉)
Division of Microbiology, PathWest Laboratory Medicine, QEII Medical Centre, Nedlands, WA 6909, Australia; Faculty of Life and Physical Sciences and Faculty of Medicine, University of Western Australia, Nedlands, WA 6009, Australia
e-mail: David.Smith@health.wa.gov.au

in cells and has only been reliably achieved with the genital types and even then, is difficult and highly specialised.

Antigen detection tests are directed at proteins either of the reticulate body, or at the major outer membrane protein (MOMP) of the elementary body, though they lack both sensitivity and specificity. Diagnosis now relies heavily on the nucleic acid detection tests, which are most commonly directed at the sequences coding for the MOMP or at those within a multi-copy cryptic plasmid. The MOMP sequences can also be used to determine the individual serovar.

Treponema pallidum is a spiral-shaped bacterium related to other treponemal and non-treponemal spirochaetes that cause human disease (such as yaws, pinta and leptospirosis) and others that are commensal organisms of the mouth. Like many of the other sexually-transmitted bacteria they are highly adapted to their special environments within the human body, and do not grow well elsewhere. Equally difficult are *Haemophilus ducreyi*, which causes chancroid, and *Klebsiella granulomatis* (formerly *Calymmatobacterium granulomatis*), which causes granuloma inguinale. For both, the conventional diagnostic methods are difficult, insensitive and technically challenging.

The family of mycoplasmas are all highly fastidious, slow growing bacteria that are found commonly in the genital tract of healthy individuals. *Mycoplasma genitalium* appears to be a cause of cervicitis, urethritis and possibly endometritis. The pathogenic role of the ureaplasmas is less certain, while *M. hominis* and several other species are probably only commensal organisms. Culture-based detection is difficult and is not commonly performed, so it is expected that with the increasing information gained from molecular diagnosis and typing we will gradually learn more about the significance of this group of bacteria.

A number of viruses are transmitted predominantly or significantly by sexual contact. The most common of these are human papilloma viruses (HPV) and HSV. HPV has been known to be the cause of both genital warts and genital tract cancers in males and females for several decades, but the more recent application of molecular detection methods has provided much more detailed information about the range of genotypes and their relative importance. That has assisted in the determination of risk to individual women as well as providing important tools to guide vaccination programs and to measure their success.

HSVs are the best known of the herpesvirus family of viruses. Following primary infection with any member of this family, the virus remains in the body for life in a latent state, from which it periodically reactivates and is secreted either asymptomatically or in association with disease. HSV-2 is the virus that most commonly causes genital disease, varying from florid blisters and ulcers of primary infection, to milder recurrent blistering and/or ulceration (Fig. 8.1). Recurrent disease follows reactivation of latent virus from its reservoir within the nerve roots, usually of the sacral spinal cord, with virus travelling back to the skin or mucosal surface supplied by the involved nerve. Lesions can be quite subtle (e.g. areas of reddening or splitting) or virus excretion can occur without lesions. HSV-1 is more commonly a cause

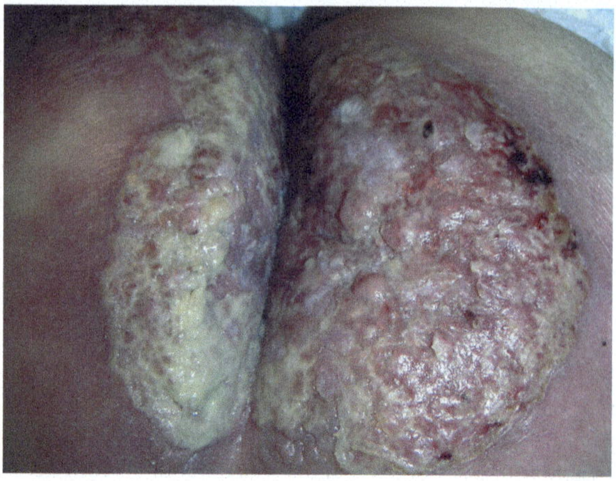

Fig. 8.1 Illustration of extensive chronic perianal ulceration in an immunosuppressed patient with aciclovir/penciclovir resistant HSV. The virus could only be detected by PCR (not culture) and was characterised by thymidine kinase gene sequencing. It significantly improved with topical cidofovir

of mouth ulcers (as primary disease) or cold sores (as recurrent disease). The newer molecular diagnostic tests have also increased our recognition of HSV-2 as a cause of benign meningitis, sometimes recurrent, that may occur with or without evidence of active genital lesions.

One other member of the herpesvirus family that occasionally causes disease in the genital region is varicella-zoster virus. While it is not a sexually transmitted virus, both the primary disease (chickenpox or varicella) and the recurrences (shingles or zoster) can cause genital lesions that may be confused with herpes simplex (Fig. 8.2). Cytomegalovirus and Epstein-Barr virus may be found in the genital tract and can be transmitted in this way from mother to child or between sexual partners, but do not cause genital disease.

Molluscum contagiosum is a pox virus that causes self-limiting infections of skin and mucosal surfaces. Infection can occur anywhere on the body where it is able to enter via minor trauma, which is common during sexual activity.

There are a number of other sexually transmitted viruses that either do not or only rarely produce genital disease, of which HIV is the best known and most important. Also sexual transmission of hepatitis B is common in many parts of the world, especially where there are high carrier rates and, though less common, sexual transmission of hepatitis A, hepatitis C and hepatitis D all occur. There are many other examples of viruses that may be sexually transmitted, including exotic viruses such as Lassa fever.

Although the bacteria and viruses account for the major STIs of humans, the yeast *Candida albicans*, can also be acquired or maintained by sexual transmission, but more commonly it results from spread from reservoirs in the gastrointestinal tract.

Trichomonas vaginalis is a protozoan parasite that is largely maintained in humans by sexual transmission, and is probably one of the most common STIs of

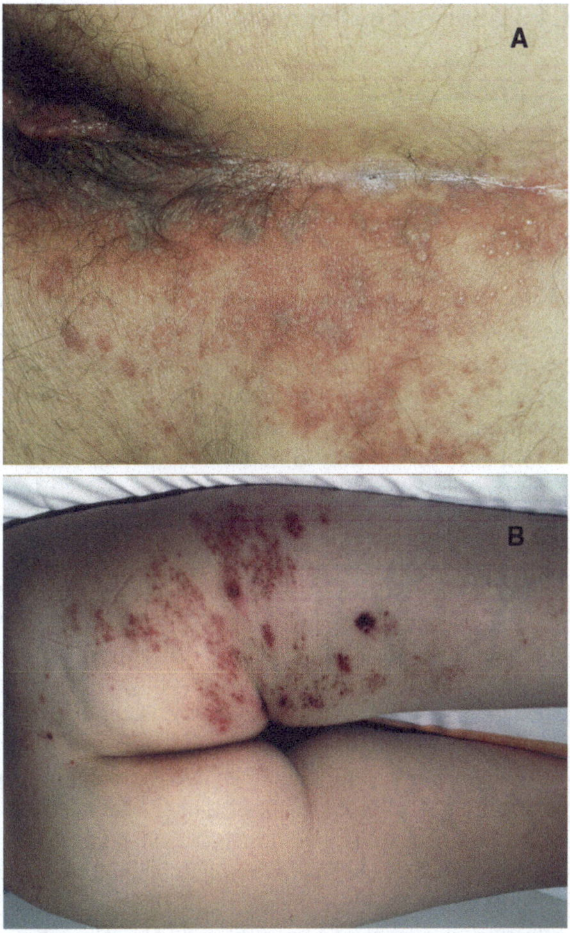

Fig. 8.2 (**a**) Illustration of perianal VZV. The blisters are typical of HSV and VZV, but there is a clear unilateral distribution that stops at the midline. (**b**) Genital VZV in a 21-year-old woman who had been told she had genital herpes. The lesions have a dermatomal distribution and do not cross the midline

women. In symptomatic women it is usually present in large amounts on the vagina and it can be found by direct microscopy and or culture, although these methods lack sensitivity. Newer molecular methods of detection provide a more sensitive and timelier means of diagnosis.

The ectoparasites *Sarcoptes scabiei* (scabies) and *Phthirus pubis* (pubic lice) are spread by the intimate contact associated with sex, though both can occur elsewhere on the body.

A number of infectious agents that are primarily gut pathogens can also be spread by orogenital or oroanal contact. One of these, hepatitis A, has already been mentioned, but others include various *Shigella* species, *Campylobacter* species, *Cryptosporidium parvum, Giardia lamblia*, *Entamoeba histolytica,* and a range of less common organisms.

8.4 Laboratory Diagnosis

Sepehr N. Tabrizi

8.4.1 Specimens and Transport

Conventional diagnostic assays for the detection of STIs generally require fastidious transport conditions to preserve viability and usually a clinician to obtain specimens from the site of infection. Nucleic acid amplification technologies allow the use of self-collected specimens from material collected further away from the original site of infection that may consequently contain fewer organisms than in conventional specimens. In addition transport conditions are less critical for test performance. In contrast to traditional internal examinations and genital swab collections at the physician's office, such strategies for testing have led to improved partner tracing and universal screening, including difficult to reach populations [10]. Several self-sampling methods such as first-void urine, cervicovaginal lavage, low vaginal swabs, mini menstrual pads, and tampons have been described [1, 3, 7, 16–18, 23]. Urine specimens were among the first self-sampling methods to be utilised and adapted for molecular diagnosis of STIs and have become an acceptable and widely applicable method of sampling in both men and women [3]. Tampons and vaginal swabs provide a convenient sampling method for detection of STIs in women. Due to their larger surface area, tampons collect a better specimen in cases where detection of multiple targets are desired, or where there are low amounts of pathogen. Self-collected swabs also have been utilised in collecting anal specimens from men who have sex with men (MSM) for successful detection of STIs [8, 19] and can provide a convenient sampling method for screening of LGV strains among this population. Modified menstrual pads have also been utilised in non-invasive collection of genital and urethral cells [1]. Though not used to a great extent for the detection of STIs due to the availability and convenience of swab and urine sampling, vaginal lavage is another method for obtaining vaginal secretions.

In addition to urine, assessment of STI infection in men has also been evaluated using semen. This type of specimen has been used for screening of sperm donors and in prevalence studies where obtaining specimens is not possible by other means, ie assessing STI prevalence in clients of sex workers [19]. Liquid based cytology medium PreservCyt™ (Hologic, Massachusetts, USA) and SurePath™ (BD, New Jersey, USA) has been a source of providing a specimen which can be used for not only a Pap smear but also to test for multiple STIs. Utilisation of such

Sepehr N. Tabrizi (✉)
Microbiology and Infectious Diseases Department, Royal Women's Hospital, Parkville, VIC 3052, Australia; Department of Obstetrics and Gynecology, University of Melbourne, Parkville, VIC 3052, Australia
e-mail: Sepehr.Tabrizi@thewomens.org.au

specimens for NAAT. However would require aliquoting prior to preparation of slides for Pap smear to prevent contamination [20]. Biopsy specimens, either fresh or archival paraffin embedded, can also be utilized for detection of some STIs, in particular human papillomavirus. Blood specimens are also important not only serological assays, but also for NAATs, in particular for detection of HIV, CMV, for HBV and HCV.

8.4.2 Traditional Methods

Conventional methods have provided, and in some instances continue to provide, important clinical information and are still utilised across laboratories. The culture-based method requires reliable media and rapid transportation. Culture can provide important antimicrobial susceptibility information (in particular for *N. gonorrhoeae*) and should be performed when feasible. Immunological methods such as coaggulation and enzyme immunoassay can also be performed; however they are not sensitive and are not used to a great extent. Microscopic examination is particularly useful when clinical symptoms need to be confirmed i.e. urethral discharge from symptomatic male patient showing a Gram negative diplococci is considered diagnostic for *N. gonorrhoeae*. Microscopy is rapid and inexpensive; however, it has a low sensitivity in most clinical cases.

8.4.3 Criteria for Test Selection

Selection of a test by a clinician for detection of STIs depends on its availability, result turnaround time, as well as sensitivity, specificity, negative and positive predictive values. Laboratories may also base the test selection on ease of performance, availability of quality assurance program as well as price of the test. Assays that have been validated against the gold standard and demonstrated equivalent or better performance for detection of STIs are generally selected. Most commercial assays have such evaluation conducted as well as provide quality tested reagents. Therefore use of well described commercial assays is beneficial in obtaining quality results. When the test target is not available commercially, in-house assays are utilised. It is important that assays employed are fully validated and quality controlled. Participation in all quality assurance programs, where available, or alternatively exchange of panels with another laboratory is important to assess ongoing assay performance.

8.4.4 Commercial Assays

A number of commercial assays are available for laboratory diagnosis of STIs. These assays generally include all components necessary to obtain a result including an internal control for assessment of adequacy of specimen collection or inhibitors present in the specimen. Most utilised, FDA approved commercial assays are available for the detection of *C. trachomatis* and *N. gonorrhoeae* including COBAS Amplicor (Roche Applied Science, Australia), APTIMA Combo 2 (GenProbe, Australia), ProbeTec ET (BD, Australia) and RealTime CT/NG (Abbott, Australia). GenProbe also offers a test for *T. vaginalis*. A number of commercial assays are available for the detection of human papillomavirus. Currently two such assays are FDA approved; Digene Hybrid Capture 2 (QIAGEN, Australia) and Cervista (Hologic). A number of assays are also available for genotyping including Linear Array (Roche Applied Science, Australia), InnoLipa (Innogenetics, Australia) and HPV Genotyping RH (QIAGEN, Australia). Commercial assays for the detection of HBV comprise Amplicor Monitor HBV (Roche Applied Science, Australia), Genostics (Abbott), Hybrid Capture 2 (QIAGEN, Australia), and Versant branched DNA (bDNA) signal amplification assay (Bayer Diagnostics). Commercial assays for the detection of HIV include Amplicor HIV Monitor (Roche Applied Science, Australia), Quantiplex (Bayer Diagnostics) and NucliSense HIV 1 QT (bioMérieux). For the detection of CMV a number of real-time PCR assays are available, however the more widely used assays include Amplicor Monitor CMV Test (Roche Applied Science, Australia) and the CMV Hybrid Capture 2 (QIAGEN, Australia).

8.5 PCR Diagnosis of Sexually Transmitted Infections

David M. Whiley

8.5.1 Technical Considerations

There is no doubting the suitability of NAATs for detecting STIs; they very sensitive, do not require viable organism for detection (only the presence are of intact DNA) and can be applied to a whole range of different specimen types (as described above). However, no single test is ever perfect and NAATs do have some limitations. The main problem with NAATs, compared to other methods is that they involve many complex processes to be successfully completed before, a valid result can be obtained. As such, there are many ways in which laboratories perform the technology and there are also a whole range of quality control issues that need to be considered.

David M. Whiley (✉)
Queensland Paediatric Infectious Diseases Laboratory, Sir Albert Sakzewski Virus Research Centre, Queensland Children's Medical Research Institute, Children's Health Service District, Brisbane, QLD 4029, Australia
e-mail: d.whiley@uq.edu.au

8.5.2 Which Technology, Platform, Reaction Mix?

Leaving aside the commercial methods, in-house NAAT STI protocols are almost exclusively based on PCR technology. PCR comes in all shapes and sizes as is highlighted by the protocol chapters. The described protocols comprise real-time PCR using 5′ nuclease probe and hybridisation probe methods, as well as a conventional PCR using ELISA detection. Numerous reaction mixes, detection platforms and cycling conditions are also described. We do not expect that all protocols would be performed as exactly described within a single laboratory. Rather, the diverse range of reactions mixes, cycling conditions and detection technologies are meant to provide a snap-shot of methods used throughout Australia. The key is to choose the system that will best suit the laboratory's needs, and where possible, revalidate assays using different reaction parameters.

In choosing a system, we recommend adhering to the adage of *keeping it simple*. For example, our laboratory first gained access to a Roche LightCycler and so assay development initially concentrated on hybridisation probe-based methods. However, for various reasons, our laboratory ultimately acquired a number of detection platforms, including the QIAGEN Rotorgene, ABI-7500 and Roche 480 instruments. This accumulation of different platforms resulted in using both 5′ nuclease and hybridisation probe formats across different instruments, amplification reagents and cycling parameters. In turn, a whole range of processes, including reaction mix preparation through to staff training, became unnecessarily complex. To circumvent this problem we investigated methods and reagents that could be more universally applied, and soon focussed on 5′ nuclease technology and generic amplification reagents now offered commercially by several companies.

8.5.3 Quality Control Issues

Standard quality control practices must be implemented when testing for STIs by NAAT, including the use of a suitable positive control, negative control, extraction control and inhibition control. These will ensure efficient extraction and amplification of the target template while also checking for contamination. As a general rule, specimens routinely used for NAAT detection of STIs, including urine and genital swabs specimens, are typically straightforward to process and extract. Thus, QC issues should not exceed those observed for other specimen types when using good quality extraction and amplification reagents in well controlled, distinct PCR areas.

Routine quality control procedures unfortunately do have their limitations as is highlighted by recent problems encountered with NAAT detection of both *N. gonorrhoeae* and *C. trachomatis*. These include false-positive and false-negative results caused by diversity in sequence targeted by the assays;

(1) False-positive results caused by cross-reaction with commensal *Neisseria* species have been a major issue for *N. gonorrhoeae* NAATs, with a number of *N. gonorrhoeae* assays, including commercial and in-house methods,

achieving notoriety for this problem [4, 21]. The issue arises from genetic exchange occurring between different *Neisseria* species, with some commensal strains acquiring gonococcal NAAT sequence targets. The problem culminated in the Australian Public Health Laboratory Network (PHLN) developing guidelines to address the problem, including the use of supplementary testing for all specimens providing positive results in *N. gonorrhoeae* NAATs [13].

(2) False-negative results caused by sequence variation in NAAT targets are now being reported for a range of organisms. The most notable example of this is the "Swedish variant" of *C. trachomatis*, which has a previously undescribed 377 bp deletion in the *C. trachomatis* cryptic plasmid. The deleted sequence comprised the sequence targets of both commercial *C. trachomatis* NAATs manufactured by Roche and Abbott and so produced false-negative results in these assays [11]. Similar problems have been experienced in Australia with *N. gonorrhoeae* PCR assays targeting the *cpp*B gene, with *cpp*B-negative isolates prevalent in the Northern Territory and other parts of the country [9]. The Australian PHLN *N. gonorrhoeae* NAAT guidelines state that "*N. gonorrhoeae* PCR assays targeting the *cpp*B gene should not be used in diagnostic laboratories" [13].

As alluded to above, the problem with these target sequence-related issues is that they will not be flagged by routine quality control measures, as none of the routine controls described above would be impacted by the variation. Hence, there will be no indication of an incorrect result. From a public health perspective, sequence-related false-negative results are of considerable concern given they can facilitate unchecked spread of certain clones, as occurred with the *C. trachomatis* "Swedish variant" [11]. To circumvent these problems there are now calls for NAATs to comprise two-sequence targets for each organism [22]. These problems also reinforce the need to participate in quality assurance programs.

References

1. Alary M, Poulin C, Bouchard C et al (2001) Evaluation of a modified sanitary napkin as a sample self collection device for the detection of genital chlamydial infection in women. J Clin Microbiol 39:2508–2512
2. Antibiotic Expert Group & Spicer, J & Victorian Drug Usage Advisory Committee. Antibiotic Guidelines Sub-Committee (2006) Antibiotic guidelines. Therapeutic guidelines: Antibiotic Therapeutic Guidelines Limited, North Melbourne, Vic. http://nla.gov.au/anbd.bib-an40720842
3. Chernesky MA, Jang D, Lee H et al (1994) Diagnosis of *Chlamydia trachomatis* infections in men and women by testing first void urine by ligase chain reaction. J Clin Microbiol 32: 2682–2685
4. Farrell DJ (1999) Evaluation of AMPLICOR *Neisseria gonorrhoeae* PCR using *cppB* nested PCR and 16S rRNA PCR. J Clin Microbiol 37:386–390
5. Garland SM (1992) Neonatal Herpes Simplex: Royal Women's Hospital 10-year experience with management guidelines for herpes in pregnancy. Aust N Z J Obstet Gynaecol 32: 331–334

6. Garland SM, Tabrizi SN (2004) Diagnosis of sexually transmitted infections (STI) using self collected non-invasive specimens. Sex Health 1:121–126
7. Knox J, Tabrizi SN, Miller P et al (2002) Evaluation of self collected samples in contrast to practitioner collected samples for detection of *Chlamydia trachomatis, Neisseria gonorhoeae*, and *Trichomonas vaginalis* by polymerase chain reaction among women living in remote areas. Sex Transm Dis 29:647–654
8. Lister NA, Smith A, Tabrizi S et al (2003) Screening for *Neisseria gonorrhoeae* and *Chlamydia trachomatis* in men who have sex with men (MSM) at male only saunas. Sex Transm Dis 30:886–889
9. Lum G, Freeman K, Nguyen NL et al (2005) A cluster of culture positive gonococcal infections but with false negative *cpp*B gene based PCR. Sex Transm Infect 81:400–402
10. Morton AN, Wakefield T, Tabrizi SN et al (1999) An outreach programme for sexually transmitted infection screening in street sex workers using self administered samples. Int J STD AIDS 10:741–743
11. Ripa T, Nilsson P (2006) A variant of *Chlamydia trachomatis* with deletion in cryptic plasmid: implications for use of PCR diagnostic tests. Euro Surveill 11: http://www.eurosurveillance.org/ew /2006/061109
12. Sasadeusz JJ, Silvers JE, Kent HE et al (2008) Prevalence of HSV 2 antibody in a Melbourne antenatal population attending a tertiary obstetric hospital. Aust NZJ Obset Gynaecol 48:266–272
13. Smith DW, Tapsall JW, Lum G (2005) Guidelines for the use and interpretation of nucleic acid detection tests for *Neisseria gonorrhoeae* in Australia: a position paper on behalf of the Public Health Laboratory Network. Commun Dis Intell 29:358–365
14. Spruance SL, Overall JC Jr, Kern ER et al (1977) The natural history of recurrent herpes simplex labialis: implications for antiviral therapy. N Engl J Med 297:69–75
15. Stevens MP, Tan SE, Horvath L et al (2008) Absence of a *Chlamydia trachomatis* variant, harbouring a deletion in the cryptic plasmid, in clients of a sexually transmissible infection clinic and antenatal patients in Melbourne. Commun Dis Intell 32:77–81
16. Tabrizi SN, Paterson B, Fairley CK et al (1997) A self-administered technique for the detection of sexually transmitted diseases in remote communi-ties. J Infect Dis 176:289–292
17. Tabrizi SN, Paterson BA, Fairley CK et al (1998) Comparison of tampon and urine as self-administered methods of specimen collection in the detection of *Chlamydia trachomatis, Neisseria gonorrhoeae* and *Trichomonas vaginalis* in women. Int J STD AIDS 9:347–349
18. Tabrizi SN, Fairley CK, Chen S et al (2000) Evaluation of patient administered tampon specimen for *Chlamydia trachomatis* and *Neisseria gonorrhoeae*. Sex Transm Dis 27:133–137
19. Tabrizi SN, Skov S, Chandeying V et al (2000) Prevalence of sexually transmitted infections amongst clients of female commercial sex workers in Thailand. Sex Transm Dis 27:358–362
20. Tabrizi SN, Stevens MP, van Leeuwen MT et al (2009) Effect of ThinPrep preparation on human papillomavirus detection and genotyping in rectal samples by PCR. J Clin Microbiol 47:227–229
21. Whiley DM, Garland SM, Harnett G et al (2008) Exploring 'best practice' for nucleic acid detection of *Neisseria gonorrhoeae*. Sex Health 5:17–23
22. Whiley DM, Lambert SB, Bialasiewicz S et al (2008) False negative results in nucleic acid amplification tests: do we need to routinely use two genetic targets in all assays to overcome problems caused by sequence variation? Crit Rev Microbiol 34:71–76
23. Wiesenfeld HC, Heine RP, Rideout A et al (1996) Vaginal introitus: a novel site for *Chlamydia trachomatis* testing in women. Am J Obstet Gynecol 174:1542–1546

Part III
PCR Protocols for Bacterial Pathogens

Chapter 9
Bordetella pertussis and *Bordetella parapertussis*

Leanne Sammels

9.1 Acceptable Specimens

Nasopharyngeal aspirates, nasal swabs or throat swabs may be processed for *Bordetella pertussis/parapertussis* detection. Sputum, endotracheal tube aspirates, and BAL fluids are also occasionally processed for *B. pertussis/parapertussis* detection.

9.2 Sample Extraction

The Roche MagNA Pure TNA Isolation Kit is used with the MagNA Pure Instrument to purify high quality total nucleic acid (TNA) from 16 or 32 samples. The TNA is extracted from 200 μl of specimen and is eluted into a 50 μl volume. This method of isolation meets the standards required for qualitative PCR analysis on the Roche LightCycler version 1 and 2 instruments. The isolation procedure is based on magnetic bead technology. The specimens are lysed by incubation with a buffer containing high chaotropic salt and Proteinase K. Magnetic glass particles are added and TNA's contained in the specimen are bound to their surface. Unbound substances are removed by several washing steps, and then purified TNA is eluted.

9.3 Primers and Probe Sequences

The oligonucleotides used in this assay are complementary to repetitive elements Insertion Sequence (IS) 481 and IS1001 of *B.pertussis* and *B.parapertussis* respectively (Table 9.1) [1].

L. Sammels (✉)
Department of Microbiology, PathWest Child and Adolescent Health Service,
Princes Margaret Hospital, Perth, WA 6008, Australia
e-mail: Leanne.Sammels@health.wa.gov.au

M. Schuller et al. (eds.), *PCR for Clinical Microbiology*,
DOI 10.1007/978-90-481-9039-3_9,

Table 9.1 Primer/Probe stock solutions are made up to a concentration of 20 μM and stored at –20°C for up to 6 months

Name	Sequence	Length	T_m (°C)
PERT-FOR	GAT TCA ATA GGT TGT ATG CAT GGT T	25	55.3
PERT-REV	TTC AGG CAC ACA AAC TTG ATG GGC G	25	61.8
PERT-FLU	TCG CCA ACC CCC CAG TTC ACT CA-FLU	23	63
PERT-640	Red640-AGC CCG GCC GGA TGA ACA CCC-PH	21	62
PARA-FOR	CAC CGC CTA CGA GTT GGA GAT	21	60.2
PARA-REV	CCT CGA CAA TGC TGG TGT TCA	21	60.1
PARA-FLU	GTT CTA CCA AAG ACC TGC CTG GGC-FLU	24	65
PARA-705	Red705-AGA CAA GCC TGG AAC CAC TGG TAC-PH	24	60.7

9.4 PCR Amplification and Product Detection

The Roche Faststart DNA Master^{Plus} HybProbes kit is used in conjunction with the appropriate primers/probes and heat-labile UNG to prepare the master mix. Immediately prior to making the master mix, combine entire volume of 1 tube of "Reagent 1a" (contains DNA-polymerase enzyme) with entire volume of 3 tubes of "Reagent 1b" (contains, reaction buffer, $MgCl_2$ and dNTP′s {contains dUTP instead of dTTP}). Label this tube "Reagent 1". Make up *Bordetella pertussis/parapertussis* master mix as shown in Table 9.2.

Table 9.2 Composition of master-mix used in the real-time PCR assay

Reagent used	Master Mix (μl)
Reagent 1	400
PCR grade H_2O	830
UNG (1 unit/μl)	50
PERT-FOR Primer 20 μM	20
PERT-REV Primer 20 μM	40
PARA-FOR Primer 20 μM	20
PARA-REV Primer 20 μM	40
PERT-FLU Probe 20 μM	20
PERT-LCR640 Probe 20 μM	20
PARA-FLU Probe 20 μM	20
PARA-LCR705 Probe 20 μM	40
Total	1500

9.5 PCR Amplification and Product Detection

Cycling conditions for the assay are shown in Table 9.3. Perform data analysis as per LC Instrument operator manual. Make sure colour compensation is enabled and use Second Derivative Maximum method. Fluorescence analysis is performed in channel F2/Back-F1 (for LC instrument version 1) or 640/Back-530 (for LC instrument version 2) for *B. pertussis* and in channel F3/Back-F1 (LC1) or 705/Back-530 (LC2) for *B. parapertussis.*

A sample is determined as DETECED when the LC software calculates a crossing point in the quantification analysis screen and when a characteristic melt curve with a Tm of ~63°C for *B. pertussis* and ~69°C for *B. parapertussis* is produced in the melt curve analysis screen. The positive control should produce a result which fits the above criteria for the assay to be valid and the negative control should produce no amplification or melt curve for the assay to be valid.

9.5.1 Quality Control and Validation Data

A *B. pertussis and B. parapertussis* positive control (isolated from clinical specimens in our laboratory) is included in every assay. A negative control (saline) is extracted in each MagNAPure TNA isolation assay and the eluate used as the No Template Control (NTC) in the subsequent PCR assays as a means of testing for cross contamination either during the TNA extraction or PCR setup.

An internal control DNA is added into the Lysis/Binding buffer prior to specimen extraction. The purified TNA isolated from the specimen can then be assayed for the presence of the internal control DNA as a means of checking the quality of the extracted TNA.

Uracil DNA Glycosylase (UNG) is used in conjunction with dUTP in the master mix to eliminate PCR "carry-over" contaminations from previous DNA synthesis reactions.

Participation in the RCPA Quality Assurance Program for Nucleic Acid Detection of *B. pertussis* is important to ensure consistent and accurate results.

Prior to the introduction of this assay into our routine diagnostics, our laboratory performed a series of "in house" validation experiments. Based on extraction and amplification of a bacterial suspension, the sensitivity of the assay is 0.1–1 CFU of *B. pertussis* or *B. parapertussis* per reaction (derived from 200 μl of original sample).

Of 576 respiratory specimens (NPA's and nasal swabs) from patients with clinical symptoms of a respiratory infection screened for *B. prtussis/parapertussis* using the assay, 32 were positive for *B. pertussis* and 18 were positive for *B. parapertussis*. The *B. pertussis* was confirmed in 27 of the 32 specimens by culture and/or a previously validated conventional PCR assay. The remaining five specimens were not confirmed by alternative assays. Possible reasons include false positive results or the LC assay has greater sensitivity than the alternative assays. *B. parapertussis* LC assay positivity was confirmed in 10 of the 18 samples by culture isolation. No

Table 9.3 Cycling conditions

Cycle program data	UNG and denaturation		Amplification			Melting			Cooling
Analyst mode	None		Quantification			Melting curves			None
Cycles	1		55			1			1
Segment	1	2	1	2	3	1	2	3	1
Target temperature (°C)	30	95	95	55	72	95	40	80	40
Incubation time (min:s)	5:00	10:00	0:10	0:15	0:15	0	1:00	0	0:30
Temperature transition rate (°C/s)	20	20	20	20	20	20	20	0.1	20
Secondary target temperature (°C)	0	0	0	0	0	0	0	0	0
Step size (°C)	0	0	0	0	0	0	0	0	0
Step delay (Cycles)	0	0	0	0	0	0	0	0	0
Acquisition mode	None	None	None	Single	None	None	None	Continuous	None

other non culture based assay for the detection of *B. parapertussis* was available to our laboratory.

Specificity of the assay when tested on a broad range of respiratory pathogens has been previously reported [1, 3]. We evaluated the specificity of the assay by testing clinical specimens positive for a range of bacterial and viral respiratory pathogens and observed no cross-reactivity.

The cross reactivity of the IS481 *B. pertussis* assay with the *B. holmesii* IS481 sequence is well described [1, 3, 4]. Our laboratory tested the assay on two *B. holmesii* ATCC strains and a clinical blood culture isolate confirmed on sequencing. Both strains were "positive" in our *B. pertussis* IS481 LC assay and negative in the *B. parapertussis* IS1001 assay.

9.6 Assay Limitations

The described real-time duplex assay on the LC instrument offers a rapid and sensitive tool for the diagnosis of *B.pertussis* and *B. parapertussis* in respiratory specimens. However, a possible reduction in sensitivity by running the assays as a duplex rather than as two simplex assays need to be considered. The potential for false *B. pertussis* detection due to *B. holmseii* infection must also be kept in mind. A pertussis-like illness associated with *B. holmseii* infection has been previously reported [5]. Some strains of *B. bronchiseptica* also have the IS481 [2] although testing on a limited number of strains using this assay showed no cross-reactivity [1, 3].

References

1. Kosters K, Reischl U et al (2002) Real-time LightCycler PCR for detection and discrimination of *Bordetella pertussis* and *Bordetella parapertussis*. J Clin Microbiol 40:1719–1722
2. Register KB, Sanden GN (2006) Prevalence and sequence variants of IS481 in *Bordetella bronchiseptica*: implications for IS481-based detection of Bordetella pertussis. J Clin Microbiol 44:4577–4583
3. Reischl U, Lehn N et al (2001) Real-time PCR assay targeting IS481 of Bordetella pertussis and molecular basis for detecting *Bordetella holmesii*. J Clin Microbiol 39:1963–1966
4. Templeton KE, Scheltinga SA, van der Zee A et al (2003) Evaluation of real-time PCR for detection of and discrimination between *Bordetella pertussis*, *Bordetella parapertussis*, and *Bordetella holmesii* for clinical diagnosis. J Clin Microbiol 41:4121–4126
5. Yih WK, Silva EA et al (1999) *Bordetella holmesii*-like organisms isolated from Massachusetts patients with pertussis-like symptoms. Emerg Infect Dis 5:441–443

Chapter 10
Brucellosis – The Role of PCR in Diagnosis and Management

Shane Byrne, John Bates, and Jennifer M.B. Robson

10.1 Clinical Implications

Brucellosis is a classic bacterial zoonosis. Common *Brucella* species that are pathogenic in humans and their usual animal reservoirs include *B. melitensis* in sheep and goats, *B. abortus* in cattle, and *B. suis* in swine. The recent description of Brucella from marine mammals has extended its ecologic range.

Worldwide, *B. melitensis* is the most significant species with high endemnicity in countries of the Mediterranean, Middle East, East Asia, and Central and South America. In July 1989 Australia was declared bovine brucellosis *(B.abortus)* free. Less than 50 locally acquired cases of brucellosis are reported annually. Most cases today occur in those involved in the recreational pursuit or processing of feral pigs which are associated with a significant export market of sanglier particularly to European countries. The causative agent is *B. suis.* All isolates characterised to date have been *B. suis* bv1. A small number of imported cases in migrants, travellers and visitors from endemic areas and laboratory acquired infections also occur from time to time. In this situation *B. melitensis* or *B. suis* is usually responsible.

Human brucellosis may be acute, relapsing or chronic in presentation and also can be associated at any stage with focal complications. Serology is the mainstay of diagnosis for acute disease and is cross reactive across the species for all except the rare *B.canis*. The identification of *Brucella* isolates can be time consuming and difficult and can pose a hazard to laboratory personnel. The organism is biochemically inert, notoriously difficult to identify and often misidentified because it is not included in the data base of many microbial identification kits.

In our experience the utility of PCR include confirmation of the identification of bacterial isolates, detection of the organism in tissues involved in focal suppurative disease and occasionally as a tool in the diagnosis of acute and relapsing

J.M.B. Robson (✉)
Sullivan Nicolaides Pathology, Cnr WhitmoreSt & Seven Oaks St, Taringa, QLD 4068, Australia
e-mail: jrobson@snp.com.au

M. Schuller et al. (eds.), *PCR for Clinical Microbiology*,
DOI 10.1007/978-90-481-9039-3_10, © Springer Science+Business Media B.V. 2010

disease which is commonly bacteremic. The literature describes a role for detection of *Brucella* by PCR in the blood in identifying those patients that evolve to chronic disease but this has not been our experience to date.

10.2 Assay Descriptions

The *Brucella* spp. LightCycler real time PCR has been developed to enable rapid speciation of the three most commonly encountered *Brucella* species, namely *B.suis*, *B.meltitensis* and *B.abortus*. The method does not detect the other three known species of *Brucella*, and if these are suspected, conventional *Brucella* PCR with gel based detection should be employed [2]. The PCR has been optimised for use on boiled extracts of either colonies or broth culture; direct detection from samples has not been fully evaluated, however positives have been obtained from human tissue and fluids. If a result is negative and suspicion remains that the isolate is a *Brucella*, the gel based PCR method should also be employed. Combination of the real time and the gel based methods covers most biovars of the *Brucella species*. A genus specific TAQman PCR is under development but not yet validated.

10.2.1 Brucella spp. LightCycler Real-Time PCR

This assay utilises a common shared forward primer located within the IS711 element of the *Brucella* genome in combination with unique reverse primers for each of the three species variously located in species or biovar specific chromosomal loci. Reactions are carried out in three separate capillaries (one for each of the three species) and each capillary mix includes a specific pair of LCRed640 labeled HybProbes for each of the three species to serve as the detection mechanism for PCR product formation, and to improve specificity. The method is based upon the work of Redkar et al. [1]. Modifications to the published method include an alteration in the shared Brucella-F primer located in IS711, whereby the base A, needs to be inserted after position 11, and the *B. abortus* FRET probe 1, at position 19 has been changed from C to G.

10.2.2 Gel Based PCR Method

The principle of this test is the amplification of fragments of two homologous genes omp2a and omp2b. These genes code for a major outer membrane protein or porin of *Brucella* spp. Although the omp2 locus is well conserved it displays enough genetic polymorphism to enable differentiation of six *Brucella* species. The omp2 locus in all species (except *B.ovis* which lacks omp2b) is comprised of the two closely

Table 10.1 Identification of *Brucella* spp. based on patterns of amplification

	PCR fragments generated				*KpnI* RFLP of 900 bp fragment		
Sample	200	600	720	900	900 only	700+200	900+700+200
B. melitensis	+	+		+		+	
B. suis	+	+		+	+		
B. neotomae	+	+		+			+
B. abortus	+	+	+				
B. canis		+		+			
B. ovis				+			

related genes omp2a and omp2b. The two genes are oriented in opposite transcriptional directions and primers have been selected that bind to sites present on both genes. The method allows differentiation of *B. abortus*, *B. canis* and *B.ovis* directly. Coupling the test to an additional RFLP using the restriction enzyme *KpnI* and a specific site on a PCR fragment amplified from the omp2b gene, enables indirect differentiation of the other three species, *B.suis*, *B.melitensis* and *B.neotomae*.

One shared primer, 2ab acts as a starting point on both omp2a and omp2b; three other primers are used to generate up to three PCR products. 2ab200 combining with 2ab generates a 200 bp fragment. 2b600 combining with 2ab generates a 600 bp (or 720 bp for *B. abortus*) fragment. Primer 2a900 combining with 2ab generates a 900 bp fragment. The test as published in the literature [2] was designed as a multiplex test, however given that in most instances the 900 bp PCR product will require RFLP testing it is more advantageous to perform the test as three singleplex reactions. The method has only been validated against the following species/biovar combinations; *B. melitensis* bv1,2; *B. abortus* bv1,5; *B. suis* bv1; *B. neotomae*; *B. canis* and *B. ovis*. Identification of *Brucella* spp based on patterns of amplification are shown in Table 10.1.

10.3 Safety

Live cultures of the organism *Brucella spp.* should be handled with extreme caution. This organism is one of the most common causes of laboratory acquired infection. Infection with the organism produces a severe febrile disease. Sample manipulation prior to DNA extraction should be performed in a Class 2 Biohazard cabinet, preferably within a PC3 containment laboratory.

10.4 Acceptable Specimens

Bacterial isolates, Blood (EDTA), Serum, Buffy coat, fresh tissue (liver, spleen, thyroid, bone, synvoial tissue and fluid).

10.5 Extraction Procedure

(i) *Bacterial preparation:* Any accepted method for DNA purification would be appropriate. A simple and effective method for bacterial colonies is to select sufficient colonies to fill a 1 μl inoculating loop and transfer bacteria to 400 μl of TE (e.g. Promega 10 mM Tris–HCl 1 mM EDTA, pH 7.9–8.1). The bacterial suspension is boiled for a minimum of 10 min to kill the bacteria and to facilitate the lysis of the bacteria as a template for the reaction

(ii) *Blood and Tissue preparation:* Total genomic DNA can be extracted from blood samples using the MagNA Pure LC DNA Isolation Kit 1 together with the MagNA Pure LC instrument (Roche Diagnostics GmbH, Manheim, Germany) as described by the manufacturer. Tissue specimens can also be extracted manually using the High Pure PCR Template Preparation Kit (Roche Diagnostics GmbH, Manheim, Germany).

10.6 Primer and Probe Sequences

10.6.1 Brucella *spp. LightCycler Real Time PCR*

Primer/Probe name	Sequence 5′–3′
Brucella-F-Corrected	CATGCGCTATGATCTGGTTAC
B.suis-R	ACCGGAACATGCAAATGAC
B.meli-R	AGTGTTTCGGCTCAGAATAATC
B.abortus-R	GGCTTTTCTATCACGGTATTC
*B.suis*1	5′-CCCAAGCGATAATGCATTCACC(Fluo)-3′
B.suis2	5′-(Red640)CCGCATAAGTAGGGTCTAAGCCG(Phosphate)-3′
*B.melitensis*1	5′-GGTAAGCTATTCCAATCTCGCTATTG(Fluo)-3′
*B.melitensis*2	5′-(Red640)TAATGGCGTCTATTGGATATTACTGCT(Phosphate)-3′
*B.abortus*1	5′-GCCCTAGAACGCCTTTCGGAAGG(Fluo)-3′
*B.abortus*2	5′-(Red640)CAGATTAAGCCGAAACGGCCCC(Phosphate)-3′

10.6.2 Primer/Probe Sequences: Brucella *spp. Gel Based PCR*

Primer name	Sequence 5′–3′
BRU-2ab	ACTGACGGATCCGCGCTCAGGCGGCCGACGCAA
BRU-2a900	ACTGACTTCGAATTGCCTTTTCGGGGGCAATGA
BRU-2b600	ACTGAAGCTTAGCCGTCGATGTGGTAGT
BRU-2ab200	ACTGACTTCGAAACCAGCCATTGCGGTCGGTAC

10.7 PCR Amplification Conditions and Product Detection

10.7.1 Brucella spp. LightCycler Real Time PCR

Three mastermixes are prepared substituting the appropriate Reverse primer and HybProbe pair for the appropriate species and 2 μl of DNA extract from each sample for testing is then added to the appropriate capillary of the *B. suis*, *B. melitensis* and *B. abortus* mix capillaries.

Component	Final [ ]	Volume/1 test (μl)
LightCycler DNA Master Hyb Mix – 'Red 1' Tube	1X	2.0
Kit $MgCl_2$ 'Blue 2' Tube	2.5 mM	3.2
Brucella-F-C Primer	10 pmol/20 μl	1.0
'Applicable Reverse Primer'	10 pmol/20 μl	1.0
Probe 1	0.2 μM	2.0
Probe 2	0.2 μM	2.0
dH_2O	–	6.8
DNA extract		2.0
Total		20

Brucella spp. LightCycler Speciation Program: Pre-Incubation: 95°C for 10 min, Amplification: 45 cycles of 95°C for 10 s, 58°C for 12 s, 72°C for 15 s followed by Melt Curve at 45°C for 5 s raised to 95°C for 5 s and cooling at 40°C for 2 min.

Positive results are determined by three means: Firstly, an amplification curve which is of the same general shape as the positive control for that mastermix should be evident for samples which are positive. Positive samples will generate a value in the 'Crossing Point' column. This value may be more or less than the positive control, if the value is significantly more than the positive control value and is near to the last five cycles of the reaction, it may be a false positive or simply a sample with very low concentrations of *Brucella* DNA. Confirmation is sought from the third method of analysis – melting curves. A valid melting curve for positive samples should generate a Tm value within the following values:

B. suis	66–67°C
B. melitensis	65–66°C
B. abortus	89–90°C

10.7.2 Brucella spp. Gel Based PCR

For the Singleplex Format:

Component	Final [ ]	Volume/1 test (μl)
GeneAmp 10X Buffer II	1X	5.0
GeneAmp $MgCl_2$	3 mM	6.0
Amersham dNTP's	200 μM	0.5
F-Primer	25 pmol/50 μl	2.5
R-Primer	25 pmol/50 μl	2.5
AmpliTaq Gold	2.5 U	0.5
dH_2O	–	28
DNA extract		5
Total		50

Three mastermixes are made for the singleplex:

F-Primer		:	R-Primer
Mix 1	2ab	:	2ab200
Mix 2	2ab	:	2b600
Mix 3	2ab	:	2a900

5 μl of appropriate sample template or control template is added to the respective tube. Samples are then transferred to a PE9700 thermalcycler and are run using the following programme: 94°C × 10 min, 35 cycles of 95°C × 30 s, 66°C × 30 s, 72°C × 90 s followed by 72°C × 7 min and 4°C× ∞.

After thermalcycling is completed, samples are loaded onto 1.5% Agarose gels using Promega Loading Buffer in the ratio of 10 μl PCR product to 2 μl of Loading Buffer. A Promega 100 bp DNA Ladder is used at either end of the samples at the ratio of 5 μl of DNA ladder to 2 μl of Loading Buffer. The gels are electrophoresed at 80 V 400 mA for 60 min (minimum). When the electrophoresis is complete the gel is imaged using a UV transilluminator and imaging device.

KpnI RFLP on 900 bp Fragment:

Digestion Mix:

Reagent	Stock concentration	Volume/1 Test (μl)
Roche Surecut Buffer L	10X	2.5
BSA	1 mg/mL	2.5
Roche KpnI RE	10 U/μl	1
dH_2O	–	4
900 bp PCR Product	–	15
Total		25

15 μl of each PCR product to be digested is transferred into the appropriately labeled tube and mixed. The PE9700 thermalcyclers are utilised as incubators by setting a single program of 37°C for ∞ time. Digestions are carried out for at least 2 h and preferably overnight. When the digestion is complete, samples are loaded

into a number 1.5% Agarose gels using well mixed Promega Loading Buffer in the ratio of 10 μl Digest Mix to 2 μl of Loading Buffer. A Promega 100 bp DNA Ladder is added at either end of the samples at the ratio of 5 μl of DNA ladder to 2 μl of Loading Buffer. The gels are electrophoresed at 80 V 400 mA for 60 min (minimum). The gel is imaged using a UV transilluminator and imaging device.

Identification of *Brucella* spp is based on patterns of amplification as previously described in Table 10.1.

10.8 Quality Control

No template controls as well as DNA extracts prepared from isolates of *B. suis, B. abortus and B. melitensis* are used. In addition *B. abortus* complement fixing antigens (manufactured by Serion Immunodiagnostica GmbH, Wurzburg, Germany), conventionally available as sources of antigen in complement fixing serological assays, can be used in the absence of culture derived *B.abortus* DNA.

10.9 Validation Data

Both assays were validated by testing historical culture confirmed *B. suis* and *B. melitensis* patient isolates as well as cattle derived *B. abortus* isolates. Specificity assessment was determined through testing a panel of non-Brucella fastidous gram negative bacteria and assessment of the primers and probes using electronic sequence databases. Assay sensitivity for the Hyb-probe and gel based method is approximately 25 genome equivalents and 250 genome equivalents respectively in the test volume. Neither of the two marine mammal related species *B. pinnipedialis* and *B. ceti,* nor the most recent species addition, *B. microti* isolated from wild voles, have been evaluated.

References

1. Redkar R, Rose S, Bricker B et al (1997) Real-time detection of *Brucella abortus, Brucella melitensis* and *Brucella suis*. Mol Cell Probes 15:43–52
2. Sifuentes-Rincón AM, Revol A, Barrera-Saldaña HA (1997) Detection and differentiation of the six *Brucella* species by polymerase chain reaction. Mol Med 3:734–739

Chapter 11
Burkholderia cepacia Complex and *Burkholderia gladioli*

Timothy M. Kidd

11.1 Summary of Methods

The PCR assays described herein can be used to reliably determine if an organism belongs to the *Burkholderia cepacia* complex or *Burkholderia gladioli*. Further characterisation of the mostly commonly encountered *B. cepacia* complex species can be achieved using a range of species-specific PCR assays.

11.2 Background

The *B. cepacia* complex comprises a group of at least ten closely related bacterial species. These organisms are recognised as significant respiratory pathogens in persons with cystic fibrosis (CF), but are also capable of causing a range of infections amongst patients with impaired immunity. Chronic CF-related *B. cepacia* complex infection has been associated with reduced survival, increased loss of lung function and life threatening systemic infection. Significant evidence of person-to-person spread has led to the establishment of strict cohort segregation policies [2]. Infection with these species is also considered as a significant contraindication to lung transplantation in many centres. Pulmonary infection caused by *B. gladioli* has also been previously reported in persons with CF, chronic granulomatous disease, and acquired immune deficiency syndrome [2]. Phenotypic identification methods have limited utility in distinguishing the *B. cepacia* complex from *B. gladioli* and other nonfermenting Gram-negative organisms. Commercial identification systems are also widely considered as unreliable for the definitive identification of these organisms. Molecular identification and species characterisation is now widely regarded as the preferred approach [2].

T.M. Kidd (✉)
Queensland Paediatric Infectious Diseases Laboratory, Sir Albert Sakzewski Virus Research Centre, Queensland Children's Medical Research Institute, Children's Health Service District, Brisbane, QLD 4029, Australia; Microbiology Department, Pathology Queensland – Central Laboratory, Herston Hospitals Campus, Brisbane, QLD 4029, Australia
e-mail: Tim_Kidd@health.qld.gov.au

M. Schuller et al. (eds.), *PCR for Clinical Microbiology*,
DOI 10.1007/978-90-481-9039-3_11,

11.3 Acceptable Specimens

This method is designed for the identification of suspected *B. cepacia* complex and *B. gladioli* cultures and has not yet been validated for the detection of these organisms in clinical samples.

11.4 Sample Extraction

From a pure culture prepare 1 ml of a 0.5 McFarlane turbidity equivalent in sterile saline, centrifuge at 10,000 rpm for 3 min, and remove the supernatant. Emulsify the pellet in 20 μl lysozyme (10 mg/ml) plus 100 μl of 0.1 M Tris buffer, and incubate at 37°C for 1–2 h. Genomic DNA from each lysate can then be extracted using the Roche High Pure PCR template preparation kit (Roche Applied Science, Australia) as per the manufacturer's instructions.

11.5 Primer Sequences

Oligonucleotide primer sequences for each assay are listed in Table 11.1. The *B. cepacia* complex and species primers used are complementary to *rec*A gene sequences as described previously [3, 5, 6]. The PCR assay used to identify *B. gladioli* utilises primers targeted at sequences of the 16 s rRNA gene [1].

11.6 PCR Amplification and Product Detection

Each 25 μl *B. cepacia* complex and species specific PCR assay (i.e. Assays 1–9 as listed in Table 11.1) can be performed using a PCR mix containing 2.5 μl of 10X QIAGEN PCR Buffer (QIAGEN Australia), 5.0 μl of QIAGEN Q solution, 0.5 μl of 250 μM dNTP mix (Roche Applied Science, Australia), 20 pmol of each primer (Table 11.1), 1 U QIAGEN Hot Star Taq DNA Polymerase, 2 μl of template DNA, and nuclease free water.

The following thermal cycling parameters should be applied: an initial incubation of 15 min at 95°C; 30 cycles of 94°C for 30 s, the appropriate annealing temperature (Table 11.1) for 45 s, and 72°C for 60 s; followed by a final elongation step of 72°C for 10 min.

B. gladioli specific PCR can be performed using a PCR mix containing 2.5 μl of 10X QIAGEN PCR Buffer (QIAGEN Australia), 0.5 μl of 250 μM dNTP mix (Roche Applied Science, Australia), 33 pmol of each primer (Table 11.1), 1 U QIAGEN Hot Star Taq DNA Polymerase, 5 μl of template DNA, and nuclease free water. The following thermal cycling parameters should be applied: an initial incubation of 15 min at 95°C; 30 cycles of 95°C for 30 s, 54°C for 30 s, and 72°C for 45 s; followed by a final elongation step of 72°C for 10 min.

Amplified PCR product detection can be undertaken using a 2.0% agarose gel containing 2.5 μl of ethidium bromide. Horizontal electrophoresis of 2 μl sample

Table 11.1 Primer sequences, associated annealing temperatures and expected PCR product sizes

Assay number and name	Primer name	Sequence	Annealing temperature (°C)	PCR product size (bp)	Reference
B. cepacia complex PCR	BCC-FWD	TGACCGCCGAGAAGAGCAA	58	1043	[5]
	BCC-RVS	CTCTTCTTCGTCCATCGCCTC			
B. cepacia (genomovar I) PCR	BC-FWD	CAGGTCGTCTCCACGGGT	62	492	[5]
	BC-RVS	CACGCCGATCTTCATACGA			
B. multivorans (genomovar II) PCR	BM-FWD	CGGCGTCAACGTGCCGGAT	62	714	[5]
	BM-RVS	TCCATCGCCTCGGCTTCGT			
B. cenocepacia (genomovar III) *recA* clade A PCR	BCENOA-FWD	GCTCGACGTTCAATATGCC	62	387	[5]
	BCENOA-RVS	TCGAGACGCACCGACGAG			
B. cenocepacia (genomovar III) *recA* clade B PCR	BCENOB-FWD	GCTGCAAGTCATCGCTGAA	60	781	[5]
	BCENOB-RVS	TACGCCATCGGGCATGCT			
B. stabilis (genomovar IV) PCR	BS-FWD	ACCGGCGAGCAGGCGCTT	64	647	[5]
	BS-RVS	ACGCCATCGGGCATGGCA			
B. vietnamiensis (genomovar V) PCR	BV-FWD	GGGCGACGGCGACGTGAA	62	387	[5]
	BV-RVS	TCGGCCTTCGGCACCAGT			
B. ambifaria (genomovar VII) PCR	BAMB-FWD	GTCGGGTAAAACCACGCTG	60	810	[3]
	BAMB-RVS	TCCGCAGCCGCACCTTCA			
B. anthina (genomovar VIII) PCR	BANTH-FWD	TACGGTCCGGAATCGTCG	61	473	[6]
	BANTH-RVS	CGCACCGACGCATAGAAT			
B. gladioli PCR	BGLAD-FWD	AGRGTTYGATYMTGGCTCAG	54	463	[1]
	BGLAD-RVS	CGAAGGATATTAGCCCTC			

aliquots should be performed at 100 V for 1 h. The presence of a band of the correct size (see Table 3.12) is recorded as a positive result.

11.7 Quality Control and Validation Data

A total of at least four controls should be included in each of these procedures. These should include two contamination controls and two organism controls. The first contamination control should consist of control template DNA and water which replaces the PCR mix. In the second contamination control water is used instead of template DNA. No PCR amplification should be detected in either of contamination controls. The organism controls should contain positive and negative control organism DNA which adequately tests the performance of the assay used. Each of these methods has been validated against an extensive range of reference and clinical isolates derived from national and international sources. With little exception the methods exhibited expected results for each of the isolates analysed [4].

11.8 Assay Limitations

The *B. cepacia* complex PCR assay offers a rapid and reliable strategy for the definitive and collective identification of isolates belonging to the *B. cepacia* complex. Furthermore, analysis of *B. cepacia* complex PCR positive isolates using the listed *B. cepacia* species specific PCR assays (Assays 2–9) can provide definitive species identification. Nevertheless, species specific assays are not currently available for all of the currently described *B. cepacia* complex species (e.g. *Burkholderia dolosa, Burkholderia pyrrocinia*).

References

1. Bauernfeind A, Scheider I, Jungwirth R et al (1999) Discrimination of *Burkholderia multivorans* and *Burkholderia vietnamiensis* from *Burkholderia cepacia* genomovars I, III, and IV by PCR. J Clin Microbiol 37:1335–1339
2. Gilligan PH, Kiska DL, Appleman MD (2006) Cumitech 43: Cystic Fibrosis Microbiology. ASM Press, Washington, DC
3. Henry DA, Mahenthiralingam E, Vandamme P et al (2001) Phenotypic methods for determining genomovar status of the *Burkholderia cepacia* complex. J Clin Microbiol 39:1073–1078
4. Kidd TJ, Bell SC, Coulter C (2003) Genomovar diversity amongst *Burkholderia cepacia* complex isolates from an Australian adult cystic fibrosis unit. Eur J Clin Microbiol Infect Dis 22:434–437
5. Mahenthiralingam E, Bischof J, Byrne SK et al (2000) DNA-based diagnostic approaches for identification of the *Burkholderia cepacia* complex, *Burkholderia vietnamiensis*, *Burkholderia multivorans*, *Burkholderia stabilis*, and *Burkholderia cepacia* genomovars I and III. J Clin Microbiol 38:3165–3173
6. Vandamme P, Henry D, Coenye T et al (2002) *Burkholderia anthina* sp. nov. and *B. pyrrocinia*, two additional *Burkholderia cepacia* complex bacteria, may confound results of new molecular diagnostic tools. FEMS Immunol Med Microbiol 33:143–149

Chapter 12
Chlamydia and Lymphogranuloma Venereum (LGV)

Sepehr N. Tabrizi

12.1 Summary of Methods

This protocol describes two real-time PCR assays, each using a different probe chemistry. The *Chlamydia trachomatis* assay uses a hybridisation probe format, comprising two primers and a pair of hybridisation probes, and targets the *C. trachomatis omp*1 gene. The LGV assay uses a 5′ nuclease-based real-time PCR system, comprising two primers and one probe, and targets LGV-specific sequences on the *pmp* gene [3].

12.2 Acceptable Specimens

Acceptable specimens are urine, urethral swab, endocervical swab, dry swab (including throat, rectal and vaginal), tampon [1] cells collected in liquid based cytology medium PreservCyt™ (Hologic, Bedford, MA, USA) and SurePath™ (BD, Franklin Lakes, NJ, USA).

12.3 Unacceptable Specimens

Samples other than those indicated above, would need to be validated prior to their use.

S.N. Tabrizi (✉)
Microbiology and Infectious Diseases Department, Royal Women's Hospital, Parkville, VIC 3052, Australia; Department of Obstetrics and Gynecology, University of Melbourne, Parkville, VIC 3052, Australia
e-mail: Sepehr.Tabrizi@thewomens.org.au

M. Schuller et al. (eds.), *PCR for Clinical Microbiology*,
DOI 10.1007/978-90-481-9039-3_12, © Springer Science+Business Media B.V. 2010

12.4 Sample Extraction

Perform the following pre-extraction steps prior to extraction:

- Swab – rotate 15 times in 400 μl of phosphate buffered saline (PBS).
- Tampon samples – This sample should be transported in 10 ml PBS. Using a sterile glove, twist tampon and collect the resultant cell suspension. Pellet at 4000×*g* 10 min and resuspend cell pellet in 500 μl of PBS.
- Liquid cytology medium – Aliquot 1 ml of PreservCyt and SurePath media and pellet at 15,000×*g* for 10 min and resuspend cell pellet in 200 μl of PBS.

Extraction of 200 μl of processed sample can be performed by the manual QIAamp DNA Blood Mini Kit (QIAGEN, Australia) or automated MagNA Pure LC (Roche Applied Science, Australia) using the DNA Isolation Kit 1 (Roche Applied Science, Australia) according to the manufacturer's instructions. DNA should be eluted in 100 μl final volume and a 5 μl aliquot used for PCR.

12.5 Primers and Probe Sequences

The specific primers and hybridisation probes for the detection of *Chlamydia* targeted the *omp1* gene [2]. Sequences for these were; Primers CtforLC: 5′- AAA AGA TAG CGA GCA CAA AG-3′ and CT3: 5′- CCA T/CA GAA TTC CGT CGA TCA T-3′ with probes CT-F1S: 5′- CTC CTT GCA AGC TCT GCC-FLUO-3′ and CT-F2S: 5′- LC640-GTG GGG AAT CCT GCT GAA- PHOS-3′.

The real time PCR assay for Lymphogranuloma venereum targeted the *pmp* gene [3] using primers LGVf: 5′- CTG TGC CAA CCT CAT CAT CAA-3′ and LGVR: 5′-AGA CCC TTT CCG AGC ATC ACT-3′ with a TaqMan MGB probe; LGVMGB-probe: 5′-6-FAM-CCT GCT CAA CAG T-MGB-3′.

12.6 PCR Amplification and Product Detection

The PCR for *C. trachomatis* and LGV are performed in a separate LightCycler® 480 Multiwell Plate 96. Set up each PCR by the addition of 20 pmol of each primer, 4 pmol of each probe, 1× LightCycler® 480 Probes Master (Roche Applied Science, Australia) and 5 μl of DNA, in a final reaction volume of 20 μl.

Cycling conditions conducted by LightCycler 480 instrument (Roche Applied Science, Australia) are as follows:

C. trachomatis: Initial 10 min at 95°C followed by 55 cycles of 95°C 1 s, 50°C 10 s, 72°C 10 s, with temperature transition rates of 20°C/s. Real-time detection of PCR products is conducted by acquisition of fluorescence at every cycle. The PCR is followed by melting curve analysis of amplicon with continuous acquisition starting at 40°C and proceeding to 80°C at a linear rate of 0.2°C/s.

LGV: Initial 10 min at 95°C followed by 50 cycles of 95°C 15 s, 60°C 60 s, with temperature transition rates of 20°C/s.

12.7 Quality Control and Validation Data

Extracted DNA from *C. trachomatis* serovar D (strain UW-3/Cx) and serovar L2 (strain 434) obtained from ATCC can be used as positive control. Positive control should be added at 10 copies per reaction. Extracted DNA from *C. trachomatis* negative cell line such as A549 (carcinomic human alveolar basal epithelial cells) equivalent to 200 copies should be used as a negative control.

This assay will detect the presence of *C. trachomatis* or LGV serovar in the clinical specimen at sensitivity of 1–10 copies/reaction. Clinical sensitivity and specificity of these assays are over 95%. Although such measurements are dependent on the gold standard, sensitivity and specificity is achieved when compared with an expanded gold standard composed of consensus results from two to three different nucleic acid amplification assays.

A clinical PPV and NPV of over 95% are achieved using this assay.

12.8 Assay Limitations

Commercial assays are available and are routinely used in diagnostic laboratories. Most commercial assays however do not differentiate between LGV and other serovars of *C. trachomatis*. Vigilance in performance of any assay, including commercial assays, is essential. Recent data from Sweden showed a new CT variant, containing a 377-base pair deletion in the cryptic plasmid [4]. This finding came after the observation of an unexpected decrease in chlamydia infections. Assays targeting the deleted region of the cryptic plasmid resulted in false negative results. Any assay utilised in the laboratory should be monitored for unexpected reduction in prevalence.

References

1. Fairley CK, Chen S, Tabrizi SN et al (1992) Tampons: a novel patient-administered method for the assessment of genital human papillomavirus infection. J Infect Dis 165:1103–1106
2. Holland SM, Gaydos CA, Quinn TC (1990) Detection and differentiation of *Chlamydia trachomatis*, *Chlamydia psittaci*, and *Chlamydia pneumoniae* by DNA amplification. J Infect Dis 162:984–987
3. Morré SA, Spaargaren J, Fennema JS et al (2005) Real-time polymerase chain reaction to diagnose lymphogranuloma venereum. Emerg Infect Dis 11:1311–1312
4. Ripa T, Nilsson P (2006) A variant of *Chlamydia trachomatis* with deletion in cryptic plasmid: implications for use of PCR diagnostic tests. Euro Surveill 11: http://www.eurosurveillance.org/ew /2006/061109

Chapter 13
Chlamydophila

James Branley and Bronwen Roy

13.1 Summary of Methods

Genus specific and species specific sequences should be used. An algorithm was devised to detect firstly the *Chlamydia* genus and then to speciate with *Chlamydophila pneumoniae* and *Chlamydophila psittaci* specific primers.

These have generally been based on the MOMP gene (major outer membrane protein) and the 16 s RNA for genus and genes from the 16 s RNA region for species. A more recently described species specific target for detecting *C. psittaci* is the *inc*A gene. PCR based genotyping for *C. psittaci* is available.

13.2 Clinical Background

Chlamydia (*Chlamydophila*) are obligate intracellular organisms producing oculo-genital disease (*C. trachomata*) respiratory and systemic infections (*C. pneumoniae, C. psittaci*).

C. pneumonia is a primary human pathogen responsible for lower respiratory tract infection. It causes acute and sub-acute respiratory disease, e.g. Pneumonia, sinusitis, bronchitis and exacerbations of obstructive lung disease. It is associated with asthma particularly the adult onset form of the disease. It has been linked inconclusively to some chronic diseases such as acute coronary artery disease (CAD) and neurological conditions such as multiple sclerosis and Alzheimer's disease and has been detected in tissues such as unstable coronary artery plaques and brain tissue involved in chronic neurological conditions.

B. Roy (✉)
Department of Microbiology, Nepean Hospital, Penrith, NSW 2750, Australia
e-mail: royb@wahs.nsw.gov.au

M. Schuller et al. (eds.), *PCR for Clinical Microbiology*,
DOI 10.1007/978-90-481-9039-3_13, © Springer Science+Business Media B.V. 2010

C. pneumoniae is a pathogen of variably reported prevalence. Internationally evidence of infection is common. Community antibody levels are up to 50–80% and 6–20% of community acquired pneumonias may involve infection with this organism. Different yields from diagnostic tests (PCR, culture, immunofluorescence and serology) may partially explain these differences. These tests are not well standardized. Quite conflicting results have been obtained associating *C. pneumoniae* and coronary artery disease, however large trials of treatment with anti Chlamydia antibiotics have not shown a benefit in CAD. Australia lacks evidence for this organism being a widespread pathogen.

C. psittaci is primarily a pathogen of birds with human infection (psittacosis) occurring rarely, usually as a result of direct bird exposure but also inadvertent bird exposure in Australian forested settings. This sporadic infection has been linked to activities such as lawn mowing. Clinical presentation is with a range of systems and organ involvement and often relatively mild respiratory symptoms. The clinical picture is dominated by fever myalgia, headache and sometimes abdominal symptoms. Patient to patient spread is rare. Mortality can be significant in untreated patients particularly the immunosuppressed and pregnant. Extensive lung involvement and multi-system failure may necessitate intensive care support. Treatment is with tetracyclines although the quinolones also are active. Macrolide antibiotics are an alternative but work somewhat more slowly. *C. psittaci* is regarded as a potential biological warfare agent.

PCR is difficult to validate on human specimens due to the rarity of illness outside of an outbreak. Avian specimens provide good control material.

Most positive human clinical specimens have DNA just at the limit of detection. Most targets are single copy genes hence copy number reflects organism number (exception *Inc*A based PCR). Sensitivity of the test is low given low copy number but specificity is high. There can be an inverse relationship between DNA presence and serological response.

Repeat infection is possible as immunity is partial. There are significant veterinary implication for *C. psittaci* both as a zoonosis and an agent of commercial bird loss. As a sporadic and epidemic cause of lower respiratory tract infection it is an important human pathogen.

13.3 Acceptable Specimens

Respiratory material including throat swabs, nose swabs, nasopharyngeal aspirates, bronchoalveolar lavage, whole blood (EDTA), urine and CSF, occasionally environmental/bird specimens.

13.4 Unacceptable Specimens

Wound swabs, drainage fluid, faeces.

13.5 Sample Extraction

Specimens are prepared as follows:

- Swabs are soaked in 200 μl of 0.9% NaCl and vortexed periodically for several hours at room temperature before DNA extraction.
- Urine is centrifuged and the pellet resuspended in 200 μl of 0.9% NaCl before extraction.
- Buffy coat from EDTA blood specimens is treated with Biorad Instagene® (Bio-Rad Hercules, Hercules, CA, USA) in order to lyse cells and inactivate DNAse and RNAse enzymes, and then resuspended in 200 μl of 0.9% NaCl.
- DNA is extracted from the specimens using the QiaAmp DNA® Mini-kit (QIAGEN, Australia), according to the manufacturer's instructions. Prior to all PCRs, bird samples are diluted 100-fold as the target organism is often too concentrated in neat specimens.

13.6 Primers and Probe Sequences

Genus specific primers used in the reaction were targeted to the 23rRNA and produced an amplification product of 168 bp [1]. The sequence for these was: CHL23SUP: 5′-GGG GTT GTA GGG TYG AGR AIA WRR CAT C-3′ and CHL23SDN: 5′-GAG AGT GGT CTC CCC AGA TTC ARA CTA-3′.

The species-specific primers and probes for *C. psittaci* were directed to the *incA* gene (74 bp) [2] and had the following sequence: *C. psittaci* Fi-inca-Cpsi: 5′-GCC ATC ATG CTT GTT TCG TTT-3′ and *C. psittaci* Ri-inca-Cpsi: 5′-CGG CGT GCC ACT TGA GA-3′. The sequence for the *C. psittaci* dual labeled probe was: 5′-FAM-TCATTGTCATTATGGTGATTCAGGA-TAMRA-3′.

Primers for the detection of *C. pneumoniae* are as follows: CPN 90: 5′ GGT CTC AAC CCC ATC CGT GTC GG 3′ and CPN 91: 5′ TGC GGA AAG CTG TAT TTC TAC AGT T-3′ resulting in an amplification product of 197 bp [3].

13.7 PCR Amplification and Product Detection

PCR is performed on a RotorGene 3000 real time PCR machine (QIAGEN, Australia) using primer sets targeting both genus (23S rRNA gene) or *psittaci* species (Inc A gene and 16-23S rRNA interspacer region).

Samples for PCR are prepared in a class 2 laminar flow hood. The reaction mixture contains 5 μl of QIAGEN® SYBR green master mix, including hot start polymerase (QIAGEN, Australia), dNTP mix and SYBR® green, 0.1 μl of primers (3 pmol/ml) and 5 μl of extracted DNA. The cycling protocol is an initial denaturation for 15 min at 95°C followed by 50 cycles of 10 s at 95°C, 15 s at 55°C, and 20 s at 72°C. Amplicons were detected using SYBR® green and melt curve analysis and confirmed with gel electrophoresis. Amplicons are occasionally sequenced to confirm identity. An avian specimen is used as a positive control, as described. Genotyping is performed targeting variation in the OMP gene [4].

For samples testing positive for *C. psittaci* a further PCR was performed using the *IncA* gene.

13.8 Quality Control and Validation Data

DNA concentrations were determined by spectrophotometry and copy numbers calculated by reference to a standard curve that had been constructed using serial dilutions of control DNA similar to a published method [5] was found to repeatable detect 100 fg/μl of *Chlamydia* DNA.

The positive control used was a *C. psitacci* wild type avian strain provided by Professor Timms, Queensland University of Technology, Brisbane, Queensland.

Six hundred and forty human respiratory, urine or blood samples from patients displaying symptoms suggestive of *C. psittaci* were tested from 2005 to 2006. Of these 20 patients tested positive for *C. psittaci*. One of these samples was cultured and confirmed positive by DFA.

13.9 Assay Limitations

No international controls are available for *C. psittaci* due to restrictions on transport of potential bioterrorism agents. Local avian and human strains are available.

References

1. DeGraves FJ, Gao D, Kaltenboeck B (2003) High-sensitivity quantitative PCR platform. Biotechniques 34:106–110, 112–115
2. Geens T, Dewitte A, Boon N et al (2005) Development of a *Chlamydophila psittaci* species-specific and genotype-specific real-time PCR. Vet Res 36:787–797
3. Madico G, Quinn TC, Boman J et al (2000) Touchdown enzyme time release-PCR for detection and identification of *Chlamydia trachomatis, C. pneumoniae*, and *C. psittaci* using the 16S and 16S-23S spacer rRNA genes. J Clin Microbiol 38:1085–1093
4. Menard A, Clerc M, Subtil A et al (2006) Development of a real-time PCR for the detection of *Chlamydia psittaci*. J Med Microbiol 55:471–473
5. Messmer TO, Skelton SK, Moroney JF et al (1997) Application of a nested, multiplex PCR to psittacosis outbreaks. J Clin Microbiol 35:2043–2046

Chapter 14
Coxiella burnetii

John Stenos, Stephen Graves, and Michelle Lockhart

14.1 Introduction

Q fever (infection with *Coxiella burnetii*) is usually a difficult disease to diagnose as it has no pathognomic symptoms or signs. The main features are fever, myalgia and fatigue, often with associated respiratory symptoms (similar to influenza or legionnaires disease) or hepatitis. The patient history often reveals animal contact (mainly cattle, sheep and goats in Australia), as Q fever is an occupational disease of unvaccinated workers in the meat industry, shearers or farmers. Occasionally a tick-bite is recalled. However, most patients have not had close contact with animals and have been infected by the air-borne route. A chest x-ray may show an interstitial pneumonia. A post Q fever fatigue syndrome is recognised in some patients. Laboratory tests often demonstrate a hepatitic pattern of hepatitis with raised transaminases, a normocytic leucocyte count but with lymphopaenia and thrombocytopaenia and significantly raised CRP (>100).

While serology is the standard method for diagnosing Q fever, antibodies are only detectable 7–10 days into the illness and during the acute illness the patient may be seronegative. During this time nucleic acid amplification is the diagnostic test of choice. More laboratories are now offering this test (public and private). A positive result is consistent with acute Q fever (or chronic Q fever if the patient has been unwell for some months). A serum sample should also be sent to the laboratory for Q fever serology.

Q fever is usually treated with doxycycline (or cotrimoxazole in pregnant patients and children less than 7 years).

J. Stenos (✉)
Australian Rickettsial Reference Laboratory, Barwon Biomedical Research, The Geelong Hospital, Bellarine Street, Geelong, VIC 3220, Australia; Division of Verterinary and Biomedical Science, Murdoch University, Murdoch, WA 6150, Australia
e-mail: jstenos@unimelb.edu.au

M. Schuller et al. (eds.), *PCR for Clinical Microbiology*,
DOI 10.1007/978-90-481-9039-3_14, © Springer Science+Business Media B.V. 2010

14.2 Assay Description

The real time assay was designed as a duplex PCR, targeting the Com 1 and IS1111a sites. Although these targets can be amplified individually using the described conditions the assay was optimised for the two targets. Primers and probes were designed using Primer Express (Applied Biosystems) (Table 14.1). These conserved regions appear to amplify all of the Coxiella isolates tested to date. This assay is suitable for any open real time PCR platform, however this assay was optimised on a Rotorgene 3000 (Corbett Research). Extracted DNA from a cultured strain of *C. burnetii* is used as a positive control. Negative controls, including a non-template and extraction controls are incorporated in this assay. We have found that the sensitivity of this assay is between 1 and 10 copy numbers and the sensitivity is relatively high given the two gene targets.

Table 14.1 Master mix reagents for Q fever qPCR

Reagent	Stock	Final concentration	Amount
UDG master mix	2X	1X	12.5 μl
Com1 forward	10 μM	400 nM	1 μl
Com1 reverse	10 μM	400 nM	1 μl
Com1 probe	5 μM	200 nM	1 μl
IS1111a forward	5 μM	200 nM	1 μl
IS1111a reverse	5 μM	200 nM	1 μl
IS1111a probe	2.5 μM	100 nM	1 μl
MgCl	50 mM	1.5 mM	0.75 μl
H_2O			0.75 μl
DNA			5 μl

14.3 Acceptable Specimens

A variety of clinical samples can be used such as blood, serum, bone marrow, biopsies (including those imbedded in paraffin) and cerebrospinal fluid. However this assay has been optimised and validated using uncoagulated blood collected in an EDTA tube.

14.4 Sample Extraction

Buffy coat preparation: Blood samples were spun and the buffy coat transferred to a fresh 10 ml tube. A volume of 2.5 ml of red blood cell lysis solution (Qiagen) was added to each 0.5 ml of buffy coat. This was incubated at 35°C for 5–10 min before centrifugation at 4,500×*g* for 5 min. The resultant white blood cell pellet was washed with 5–10 ml sterile phosphate buffered saline (PBS) and resuspended in a final volume of 0.2 ml PBS.

DNA was extracted using a blood mini Kit spin column (Qiagen). Samples such as biopsies (including those embedded in paraffin wax) can be prepared for DNA extraction as described by the handbook supplied with the kit. Buffer AL (200 μl) and 20 μl of proteinase K was added directly to the buffy coat and incubated with shaking at 56°C for 10 min followed by a pulse centrifuge. Two hundred microliters of 100% ethanol was added, mixed by vortex and pulse centrifuged. The solution was then transferred to a column and centrifuged at 8,000×*g* for 1 min. The column was placed into a fresh collection tube to which 500 μl of WB1 buffer was added, followed by centrifugation at 8,000×*g* for 1 min. The column was then transferred to a fresh collection tube and 500 μl WB2 buffer added, followed by centrifugation at 14,000×*g* for 3 min. The collected liquid was decanted and the column centrifuged at 14,000×*g* for 1 min. The column was placed in a fresh sample tube and 50 μl of EB buffer added. This was allowed to stand at room temperature for 1 min followed with centrifugation at 8,000×*g* for 1 min. The extracted DNA was kept at 4°C until tested.

14.5 Primer and Probe Sequences

The primers and probes for the Com1 OMP gene are:

Com1-F (5′- AAA ACC TCC GCG TTG TCT TCA-3′)
Com1-R (5′- GCT AAT GAT ACT TTG GCA GCG TAT TG-3′)
Com1 probe (5′- FAM AGA ACT GCC CAT TTT TGG CGG CCA BHQ1-3′)
The primers and probes for the IS1111a gene are:
IS1111a-F (5′- GTT TCA TCC GCG GTG TTA AT-3′)
IS1111a R (5′- TGC AAG AAT ACG GAC TCA CG-3′)
IS1111a probe (5′- QUASAR 670 CCC ACC GCT TCG CTC GCT AA BHQ2-3′).

14.6 PCR Amplification and Analysis

The components of the PCR mixture are listed below. A UDG Master mix (Invitrogen) was utilised in this assay in order to minimise amplicon contamination in subsequent runs.

Amplification is performed on a Rotorgene 3000 (Corbett Research) under the following conditions: 50°C for 3 min followed by 95°C for 5 min and then 65 cycles of 95°C for 20 s and 60°C for 40 s. Fluorescence is recorded at the end of the annealing/extension step. A characteristic sigmoid amplification is observed with each respective channel. Confirmation of a successful run is given with the amplification of a chromosomal positive control and the absence of amplicons in the negative controls. Samples are considered positive if they amplify in both assays.

Acknowledgments The IS1111a qPCR assay was originally development by Michael Banazis as part of his Ph.D. studies at Murdoch University, Western Australia and the Com 1 qPCR assay was developed by the Adelaide Q fever research group.

Chapter 15
Diarrheagenic Escherichia coli Pathotypes (DEP) Including Enterohaemorrhagic (EHEC)/Shiga-toxin *E. coli* (STEC)

James Chin, Sam Abraham, Ren Zhang, and Rafat Al-Jassim

15.1 Summary of Methods

Many methods are available for the detection and isolation of diarrheagenic *Escherichia coli* pathotypes (DEPs). A versatile protocol is one that will enable detection from any source material ranging from fecal samples, swabs, and foods or food products that have been processed in a stomacher. The first step is to generate a sample suspension in enrichment medium such as mEC (modified EC broth; CM853, Oxoid, Basingstoke, UK) (Fig. 15.1). Aliquots of the broth are serially diluted following overnight incubation and plated onto selective medium such as BVCC (Tryptose blood agar supplemented with vancomycin-cefixime-cefsulodin [5] and SMAC (Sorbitol-MacConkey). The purpose of plating is to facilitate the

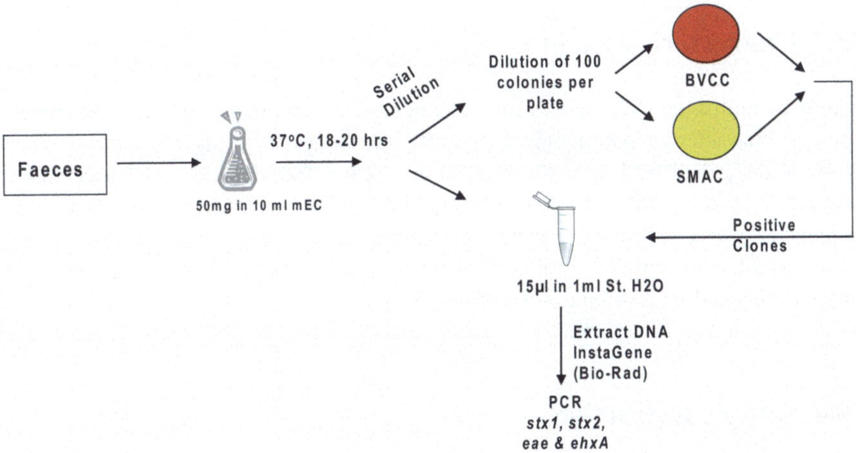

Fig. 15.1 Processing of samples for enrichment of *E. coli* followed by clone identification on selective culture media (based on Hornitzky et al., 2005)

J. Chin (✉)
School of Veterinary Science, University of Queensland, Brisbane, QLD 4072, Australia
e-mail: xion.biotech@gmail.com

M. Schuller et al. (eds.), *PCR for Clinical Microbiology*,
DOI 10.1007/978-90-481-9039-3_15, © Springer Science+Business Media B.V. 2010

isolation of individual clones for further molecular characterization. Alternatively, if the intention was to detect the presence of DEPs, then the sample can be resuspended in sterile PBS and DNA extracted for PCR. However, enrichment followed directly by DNA extraction is also another alternative to plating and would be more likely to yield a positive outcome.

15.2 Organism

To date, six major diarrheagenic *E. coli* (Ec) pathotypes have been described in the literature. These are: enteropathogenic Ec (EPEC); enterotoxigenic Ec (ETEC); enteroinvasive Ec (EIEC); enterohemorrhagic Ec (EHEC) or shiga toxin-producing Ec (STEC) or verocytotoxin-producing Ec (VTEC); enteroaggregative Ec (EAEC) and diffusely adherent Ec (DAEC). Pathotypic assignment of an isolated strain is dependent upon possession of a specific combination(s) of virulence genes encoding adherence proteins and/or toxins [7]. Diarrheagenic Ec (DEC) are of public health significance and an important cause of morbidity and mortality. As fecal contaminants, they cannot only be disseminated via meat, aquatic and vegetable products, but can also pollute waterways in catchment areas adjoining intensive farm operations. Despite these versatile modes of transmission, detection and characterization of DECs is not routinely undertaken in most hospital and veterinary diagnostic microbiological laboratories.

15.3 Characterisation of DEPs

Although multiplex PCR assays for virulence gene combinations have been developed to characterize pathogenic Ec [1, 8], the simplified single multiplex PCR protocol described by Lopez-Saucedo et al. [6] has found widespread acceptance and use. The assay utilizes 14 primer sets (Table 15.1) designed to amplify unique sequences from the following genes: *ial* (invasion associated locus), *lt* (heat labile toxin), *eaeA* (intimin), *bfpA* (bundle forming pilus), *stx2* (shiga toxin2), *st* (heat stable toxin) and *stx1*(shiga toxin1) (Fig. 15.2).

15.4 Sample Extraction

The most expedient way for PCR is to resuspend a bacteria pellet or clone (lifted from a plate with a sterile loop) in 1 ml of sterile Milli-Q water. This is boiled for 1 min and aliquots may be stored frozen for later use.

It is often assumed that good quality DNA is an important determinant for successful PCR amplification. For instance, boiled lysates from enriched cultures generally provide amplifiable DNA templates compared to those prepared by direct

Table 15.1 Primer sequences for the differentiation of four diarrheagenic *E. coli* pathotypes by a single multiplex amplification of virulence/toxin genes (after Lopez-Saucedo et al. [6])

Target gene	Primer sequence	Primer (pmol) in PCR mix
Ial	F: 5′GGT ATG ATG ATG ATG AGT CCA 3′ R: 5′ GGA GGC CAA CAA TTA TTT CC 3′	10.25
Lt	F: 5′GGC GAC AGA TTA TAC CGT GC3′ R: 5′ CGG TCT CTA TAT TCC CTG TT3′	5.0
*eae*A	F: 5′ GAC CCG GCA CAA GCA TAA GC3′ R: 5′CCA CCT GCA GCA ACA AGA GG3′	3.88
*bfp*A	F: 5′AAT GGT GCT TGC GCT TGC TGC3′ R: 5′ GCC GCT TTA TCC AAC CTG GTA3′	2.5
*stx*2	F: 5′GGC ACT GTC TGA AAC TGC TCC3′ R: 5′TCG CCA GTT ATC TGA CAT TCT G3′	2.5
st	F: 5′ATT TTT CTT TCT GTA TTG TCT T3′ R: 5′CAC CCG GTA CAA GCA GGA TT3′	6.47
*stx*1	F: 5′CTG GAT TTA ATG TCG CAT AGT G3′ R: 5′AGA ACG CCC ACT GAG ATC ATC3′	3.88

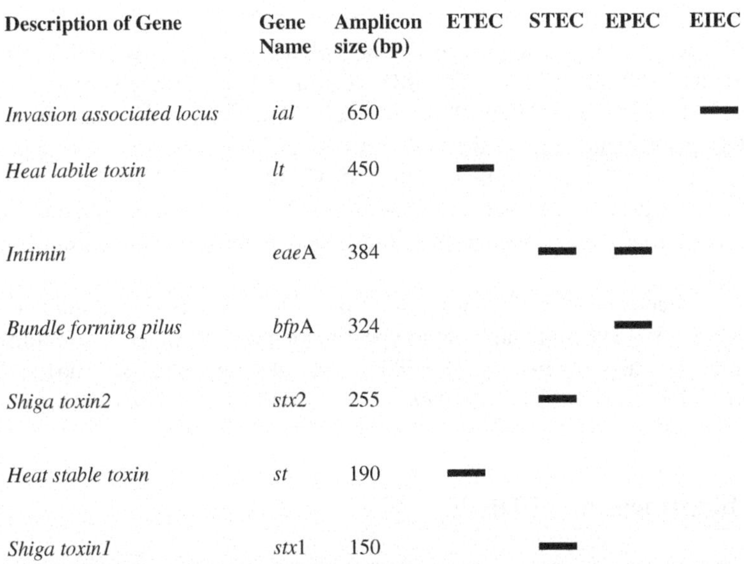

Fig. 15.2 Diagrammatic representation of amplicons resolved by agarose gel electrophoresis following single multiplex amplification of virulence/toxin genes (after Lopez-Saucedo et al. [6])

boiling of sample suspensions. However, the sensitivity of the PCR process is also its Achilles heel as contaminants, particularly from plant products, can interfere with polymerase activity. In these circumstances, an additional kit-based DNA extraction step using InstaGene or Chelex (Biorad) (Figs. 15.1 and 15.3) would ensure successful amplification.

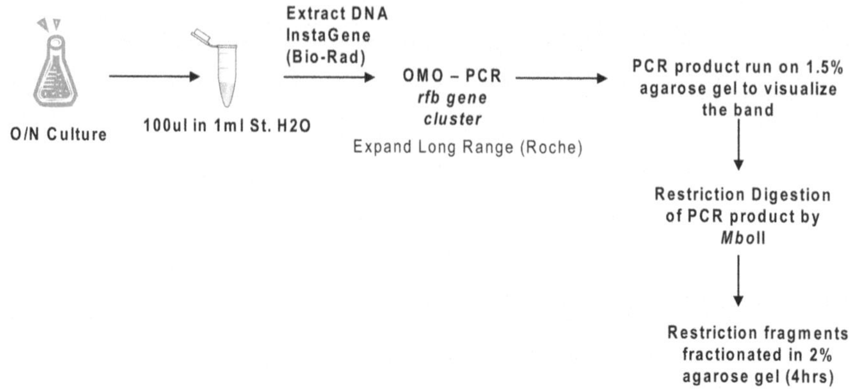

Fig. 15.3 Processing of single *E. coli* colonies for O-molecular (OMO) typing (after Abraham, Ph.D. thesis)

15.5 PCR Amplification and Product Detection

The PCR reaction mix contains in a final volume of 23 μl, Tris-HCl (10 mM, pH 8.3); KCl (50 mM); $MgCl_2$ (2 mM); gelatin (100 μg/ml); glycerol (5% v/v); dATP, dCTP, dGTP and dTTP (200 μM each); Ampli*Taq* polymerase (0.5 U); a mix of 14 primers and 2 μl of the boiled lysate. PCR cycling conditions are: 50°C (2 min, 1 cycle); 95°C (5 min, 1 cycle); 95°C, 50°C and 72°C (45 s each temperature, 40 cycles) and a final extension step (72°C, 10 min). Amplified products can be visualized by agarose electrophoresis and staining with ethidium bromide (Fig. 15.2).

The amplicons generated from the multiplex DEP-PCR is shown diagrammatically in Fig. 15.2. The four major pathotypes are identified by positive amplifications for *lt* (450 bp) and *st* (190 bp) [ETEC]; *eaeA* (384 bp), *stx2* (255 bp) and *stx1* (150 bp) [STEC]; *eaeA* (384 bp) and *bfpA* (324 bp) [EPEC]; and *ial* (650 bp) [EIEC].

15.6 Identification of DEPs

While virulence gene signatures can generally identify the presence of DEC in PCR amplifications of total sample DNA representing an unidentified Ec population, pathotypic assignment is more meaningful with DNA extracted from individual clones grown on selective medium. Unfortunately, there are no hard and fast rules regarding DEP membership because clones with identical virulence gene signatures may be found in isolates belonging to different serogroups or serotypes (Table 15.2). For example, O157H7 is considered to be an STEC but is also a strain type for enterohemorrhagic Ec (EHEC). Within each pathotype, there are strains belonging to different serotypes that may have the same virulence gene signatures but are otherwise considered to be atypical. O165 HNM is regarded

Table 15.2 Published membership list of O-serogroups assigned to different diarrheagenic *E. coli* pathotypes (DEP) (adapted from Riley [7])

DEP	O-serogroup
ETEC	O6, O8, O15, O20, O25, O27, O49, O63, O78, O128, O148, O153, O159, O167, O169
EIEC	O28, O29, O112, O124, O136, O143, O144, O152, O164, O167
EPEC	O55, O86, O111, O114, O119, O125, O126, O127, O128, O142, O157
STEC	O1, O2, O5, O9, O14, O22, O26, O45, O48, O50, O55, O79, O83, O91, O103, O104, O111, O113, O118, O121, O128, O137, O145, O153, O157, O163, O165, O172

as an unusual EHEC/STEC. Traditionally, antibody reagents directed against Ec lipopolysaccharides have been used to differentiate between O-serogroups. More recently, O-serotyping has been replaced by *rfb*-RFLP which involves amplification of the *rfb* gene cluster followed by restriction fragment length polymorphism [3]. We have adapted this procedure successfully in our laboratory to generate a database of O-molecular (OMO) types so that OMO groupings can be readily assigned with clone DNA to identify individual DEP strains.

15.7 O-Molecular (OMO) Typing of DEPs

Figure 15.3 illustrates the processing of samples for OMO-typing. The only requirement in this protocol is the use of Roche Long Range PCR (Roche Applied Sciences, Australia) as some amplicons can be quite large. O-groups are recognized by the restriction fragment patterns generated following *Mbo*II restriction of the *rfb* amplicon (Fig. 15.4). We routinely use BioNumerics (Applied Maths, Belgium) to digitize scanned restriction fragment patterns and these are then cluster assigned to the respective O-molecular groupings. OMO typing provides a rapid means of characterizing DEPs and assists in resolving ambiguities of DECs that belong to different O-serogroups.

High quality DNA preparations are a pre-requisite for OMO-PCR because of the size range of amplicons from 4.2 to 20 kb. DNA extraction can be done according to the method described [2] or with 6% Chelex matrix (Bio-Rad). Briefly, this involves diluting a 100 μl of overnight culture in LB with 1 ml sterile water. The bacteria pellet is suspended in 200 μl of 6% Chelex matrix in TE and incubated at 56°C for 20 min. This is followed by further mixing prior to incubation at 100°C for 8 min, transferred to ice for 5 min and recovery of DNA by centrifugation 11,000 × g for 10 min.

The O antigen gene cluster is amplified by using primers sets JUMPstart 482 (5′-CAC TGC CAT ACC GAC GAC GCC GAT CTG TTG CTT GG-3′) and *gnd* 412 (5′-ATT GGT AGC TGT AAG CCA AGG GCG GTA GCG T-3′). PCR reaction is carried out using Expand Long Range (Roche Applied Sciences, Australia). The PCR reaction mix contains in a final volume of 30 μl, PCR buffer containing 12.5 mM $MgCl_2$ (6 μl), dNTP (500 μM), primer (0.4 μM each), DNA polymerase

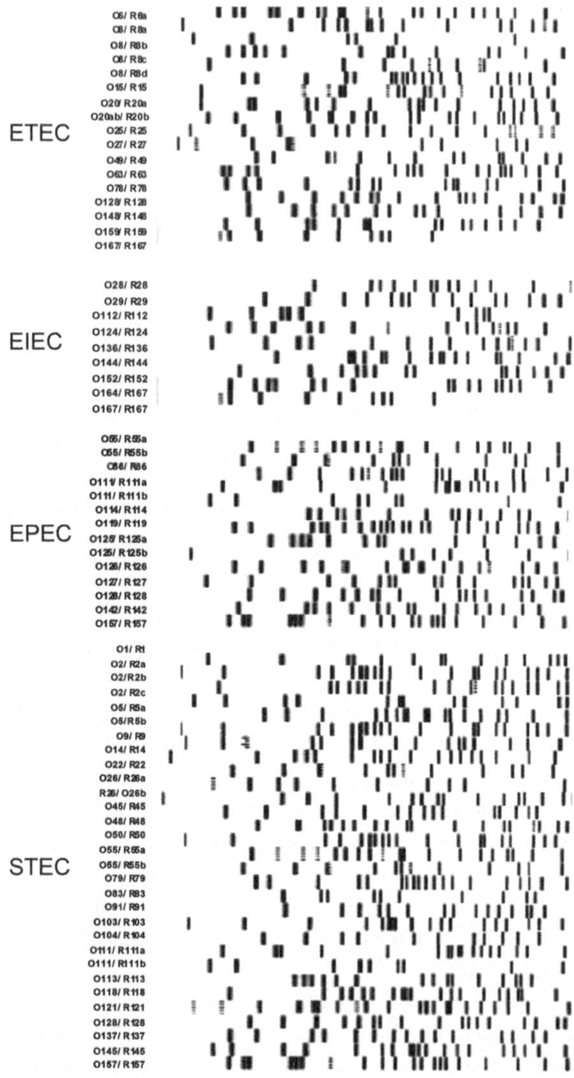

Fig. 15.4 Diagrammatic representation of *E. coli* O-molecular types following *Mbo*II restriction digest of amplicon products generated by long range PCR (Expand Long Template PCR system – Boehringer) (modified from Coimbra et al. [3])

(0.5 μl) and 2 μl of template DNA prepared using Chelex in TE. PCR cycling conditions are: initial denaturation for 94°C (5 min), 10 cycles of 93°C (10 s), 60°C (30 s) and 68°C (15 min). Followed by 20 cycles of 93°C (10 s), 60°C (30 s) and 68°C (15 min) with an increase of 20 s each time for the extension and a final elongation step (68°C for 7 min). The products can be visualized by agarose electrophoresis and staining with ethidium bromide.

Restriction patterns that can be used to define OMO-types are performed by digesting amplicons as follows: 12–18 μl PCR product is digested with 18 U of *Mbo*II (Fermentas) in kit buffer at 37°C for 2 h followed by inactivation of enzyme at 65°C for 20 min. The restriction patterns are visualized following electrophoresis in 2% agarose gel and staining with ethidium bromide, scanned and stored in a database using Bionumerics software.

15.8 Conclusion

Of 174 recognized O-serotypes (181 O-types minus O31, 47, 67, 72, 93, 94, 122 following re-classification according to Bergey's Manual of Systematic Bacteriology), more than a third (61 out of 174 or 35%) are represented by strains belonging to each of the 4 DEP – ETEC (16 O-types), EIEC (10 O-types), EPEC (12 O-types) and STEC (28 O-types). DEC strains belonging to different OMO types within each pathotype can display either a restricted or expanded host range. O157 is a good example, being found not only in humans, but also in a wide range of farm species including cattle, sheep and pigs. While rapid diagnosis of submitted samples can be pathotyped by multiplex PCR, OMO-typing and other molecular techniques such as RAPD and MLST, can also provide invaluable epidemiological information. This is of particular importance because of the zoonotic potential of a number of strains that are pathotypic members of STECs, EPECs and EHECs. The procedures described will be of use in human and veterinary clinical settings as well as in food safety laboratories.

References

1. Chapman TA, Wu XY, Barchia I et al (2006) Comparison of virulence gene profiles of *Escherichia coli* strains isolated from healthy and diarrheic swine. Appl Environ Microbiol 72:4782–4795
2. Coimbra RS, Grimont F, Grimont PAD (1999) Identification of Shigella serotypes by restriction of amplified O-antigen gene cluster. Res Microbiol 150:543–553
3. Coimbra RS, Grimont F, Burguière P et al (2000) Identification of *Escherichia coli* O-serogroups by restriction of the amplified O-antigen gene cluster (rfb-RFLP). Res Microbiol 151:639–654
4. Hornitzky MA, Mercieca K, Bettelheim KA et al (2005) Bovine Feces from Animals with Gastrointestinal Infections Are a Source of Serologically Diverse Atypical Enteropathogenic *Escherichia coli* and Shiga Toxin-Producing *E. coli* Strains That Commonly Possess Intimin. Appl Environ Microbiol 71:3405–3412
5. Lehmacher A, Meier AH, Aleksic S et al (1998) Detection of hemolysin variants of shiga toxin-producing *Escherichia coli* by PCR and culture on vancomycin-cefixime-cefsulodin blood agar. Appl Environ Microbiol 64:2449–2453
6. Lopez-Saucedo C et al (2003) Single multiplex a polymerase chain reaction to detect diverse loci associated with a diarrheagenic *Escherichia coli*. Emerg Infect Dis 9:127–131
7. Riley LW (2004) Distinguishing pathovars from nonpathovars: *Escherichia coli*, in Riley LW Molecular Epidemiology of Infectious Diseases: Principles and Practices. ASM Press, Washington DC, pp 179–206
8. Wu X-Y, Chapman T, Trott DJ et al (2007) Comparative analysis of virulence genes, genetic diversity, and phylogeny of commensal and enterotoxigenic *Escherichia coli* isolates from weaned pigs. Appl Environ Microbiol 73:83–91

Chapter 16
Haemophilus ducreyi and *Klebsiella granulomatis*

Ian M. Mackay

16.1 Summary of Methods

Haemophilus ducreyi is the causative agent of chancroid, and *Klebsiella granulomatis* is the cause of Donovanosis. Both are classified as agents of genital ulcer disease. This method describes a multiplex polymerase chain reaction targeting these organisms in which amplification products are detected by an enzyme linked amplicon hybridization assay.

16.2 Acceptable Specimens

The only samples that are suitable for analysis by this assay include (i) impression smears or press slides, (ii) swabs, and (iii) biopsies.

16.3 Sample Extraction

Biopsy material, scrapings from impression smears and cells from swabs were predigested with proteinase K (2.4 units in 100 μl) at 37°C for 60 min. Digests were extracted using the High Pure PCR Template Preparation Kit (Roche Applied Science, Australia).

I.M. Mackay (✉)
Queensland Paediatric Infectious Diseases Laboratory, Sir Albert Sakzewski Virus Research Centre, Queensland Children's Medical Research Institute, Children's Health Service District, Brisbane, QLD 4029, Australia
e-mail: ian.mackay@uq.edu.au

M. Schuller et al. (eds.), *PCR for Clinical Microbiology*,
DOI 10.1007/978-90-481-9039-3_16, © Springer Science+Business Media B.V. 2010

16.4 Primers and Probe Sequences

The PCR assay for the detection of *H. ducreyi* targeted the 16rRNA gene [4] using primers HD01.1: 5′-CAA GTC GAA CGG TAG CAC GAA G-3′ and HD02.1: 5′-TTC TGT GAC TAA CGT CAA TCA ATT TTG-3′. Amplification products were detected using an immobilized probe HD_P02.1 with the sequence 5′-Biotinyl-CCG AAG GTC CCA CCC TTT AAT CCG A-3′.

Similarly, the primers and probe for the detection of *K. granulomatis* used primers KG01.1: 5′-TCC TCT GCC CAG ACC GAT AAC TTT ATG-3′ and KG02.1: 5′-CCA GGT AGA TAT TGT TGG CGT CA-3′ for amplification, and probe KG_P01.1: 5′-Biotinyl-GCC GTC AGC GCA GCC TAC ACC AGC-3′ for the detection of amplification product [3].

16.5 PCR Amplification and Product Detection

16.5.1 Multiplex PCR Screen

The multiplex PCR combined 5 ml of control or purified specimen with 45 ml of PCR mix (4 mM of $MgCl_2$; 50 mM of KCl; 10 mM of Tris-HCl [pH, 8.3]; 0.2 mM of dATP, dCTP, and dGTP; 0.4 mM of dUTP; and 3 mM of digoxigenin-11-dUTP [Boehringer Mannheim]) containing 2 pmol of each oligonucleotide primer (Sigma-Genosys) and 2 units of PlatinumTaq DNA polymerase (Invitrogen). The primers for *H. ducreyi* were chosen from a previous study [3], whereas the primers for *K. granulomatis were* unique to this study. Amplification commenced with 2-min incubation at 95°C followed by 45 cycles of 20 s at 94°C, 20 s at 62°C, and 30 s at 72°C and final 7-min incubation at 72°C.

Amplicon was detected by agarose gel electrophoresis or by enzyme linked amplicon hybridization assay (ELAHA), a microwell DNA hybridisation method using an enzyme-mediated colour reaction to indicate the presence of specific template amplification [1].

16.5.2 K. granulomatis Confirmation

The *K. granulomatis* PCR primers yield a 328 base pair amplicon from both *K. granulomatis* and other *Klebsiella* species, including *K. pneumoniae, K. oxytoca species, and K. planticola,* and from some *Escherichia coli* strains. Discrimination of these from true *K. granulomatis* templates is achieved by generating a diagnostic 167-base pair fragment following restriction of the amplicon with 5 units of *Hae*III (Promega) in a total volume of 20 ml following the manufacturer's instructions (Figs. 16.1 and 16.2) [3].

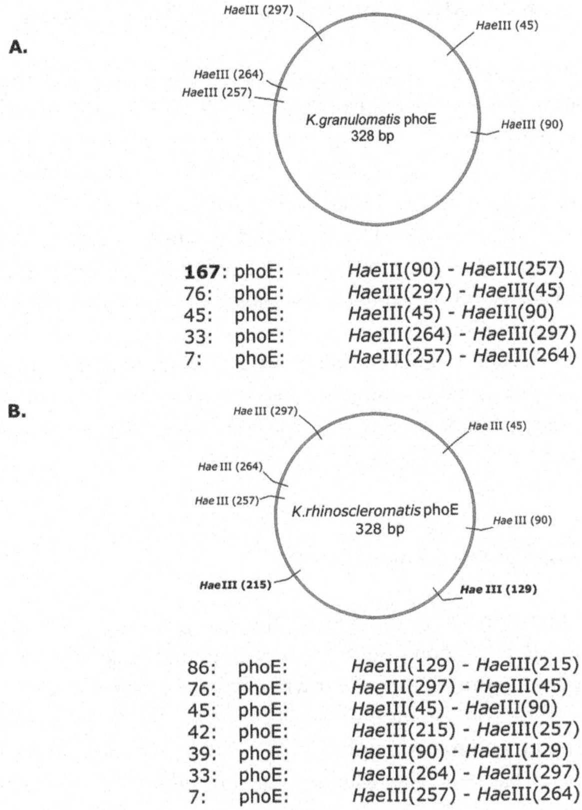

Fig. 16.1 The expected restriction enzyme patterns resulting from *HaeIII* restriction of the 328 bp phoE (phosphate porin) amplicon generated by amplification of *Klebsiella* templates. The 167 bp fragment (**a**) was diagnostic for *K. granulomatis,* and was absent when cross-reacting species were amplified and restricted (**b**). The *K. rhinoscleromatis* sequence is presented as an example of a non-target species of *Klebsiella* (figure adapted from [3])

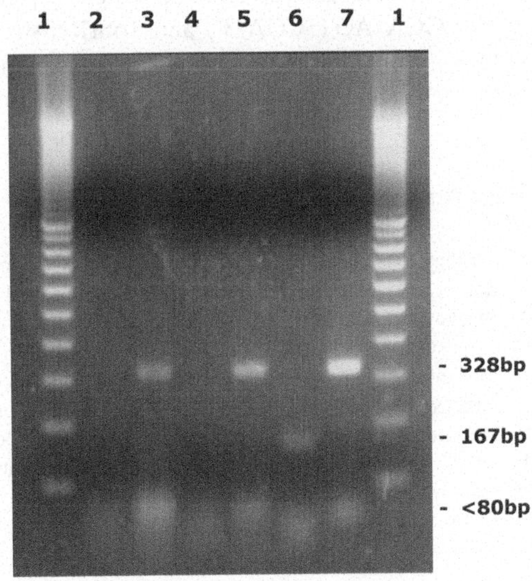

Fig. 16.2 Agarose gel electrophoresis of KG_P01.1-positive amplicon before and after *HaeIII* restriction. Lane 1 – 100 bp DNA molecular weight marker; Lanes 2 and 3 – *E. coli* 626 amplicon following and preceding restriction digestion respectively; Lane 4 and 5 – *K. pneumoniae* amplicon following and preceding restriction digestion respectively; Lanes 6 and 7 – restriction digested and undigested *K. granulomatis* amplicon respectively (figure adapted from [3])

16.6 Quality Control and Validation Data

A previously positive clinical specimen for *H. ducreyi* and *K. granulomatis* may be used for positive control. However this is sometimes difficult to obtain, and synthetic cloned positive controls may be constructed using plasmid vectors as previously described [1]. Distilled water subjected to the full extraction procedure was added to reaction mix and used as a negative control.

Use of the ELAHA increased the limit of amplicon detection by at least 8-fold, compared with agarose gel electrophoresis of an equal amount of amplicon [2]. The ELAHA was capable of detecting 3×10^0 copies of each synthetic control template per 50-μl reaction.

A wide range of related and unrelated nucleic acid templates were tested in the assay and did not show any positive reaction. The only cross-reaction with an unintended template was that for some unrelated enterobacterial targets. A PPV and NPV for this assay were not established.

16.7 Assay Limitations

During the development of this assay, limited clinical material was available and a full validation could not be completed. Use of this procedure therefore needs a qualification to this effect when issuing results.

The confirmation of positive results for *K. granulomatis* is an essential step of the procedure as amplification products of the correct size may be produced with other bacteria. However, only *K. granulomatis* carries the restriction site specific to this organism.

Further confirmation of the presence of *K. granulomatis* may be achieved by amplification of the 16S rRNA gene using primers 16S rRNA 01.1 (5′-CCT AAC ACAT GCA AGT CGA-3′) and 16SrRNA01.2 (5′-CGG TCT GTC AAG TCG GAT-3′). Both strands of amplicon, representing ~500 bases proximal to the 5′ end of the gene should then be sequenced to confirm identity.

References

1. Mackay IM, Metharom P, Sloots TP et al (2001) Quantitative PCRELAHA for the determination of retroviral vector transduction efficiency. Mol Ther 3:801–808
2. Mackay IM, Gardam T, Arden KE et al (2003) Co-detection and discrimination of six human herpesviruses by multiplex PCR-ELAHA. J Clin Virol 28:291–302
3. Mackay IM, Harnett G, Jeoffreys N et al (2006) Detection and discrimination of herpes simplex viruses, *Haemophilus ducreyi, Treponema pallidum*, and *Calymmatobacterium (Klebsiella) granulomatis* from genital ulcers. Clin Infect Dis 42(10):1431–1438
4. Orle KA, Gates CA, Martin DH et al (1996) Simultaneous PCR detection of *Haemophilus ducreyi, Treponema pallidum,* and herpes simplex virus types 1 and 2 from genital ulcers. J Clin Microbiol 34:49–54

Chapter 17
Haemophilus influenzae – Capsulated and Non-typeable

Gerald B. Harnett and Glenys R. Chidlow

17.1 Summary of Methods

The real-time multiplex assay described in this study detects the *bexA* gene which is present in capsulated strains of *Haemophilus influenzae*, the *omP2* gene which is present in all *H. influenzae* strains, capsular genes a-f and equine herpesvirus (EHV) an internal control added to samples during the extraction process. The multiplex assay is structured according to the listing in Table 17.1.

17.2 Background

H. influenzae is a human pathogen which is responsible for cases of meningitis, lower respiratory disease, conjunctivitis, septicaemia and otitis media. *H. influenzae* comprises six serotypes a–f, which are based on the nature of the capsular antigen, and non-capsulated or non-typeable strains. Capsulated strains and non-typeable strains of *H. influenzae* have traditionally been differentiated using a slide agglutination test but concern has been expressed that this method may be less sensitive and specific compared with molecular typing [1, 2]. Capsular typing of *H. influenzae* has been performed with a PCR in which *H. influenzae* types a–f yielded amplification products of differing sizes which could be assessed by gel electrophoresis [1], and a real-time PCR for capsular typing has also been described [4]. *H. influenzae* type b is regarded as the most virulent serotype but its incidence in severe disease has been much reduced by Hib vaccination programmes, leaving other types, particularly type f, causing a higher percentage of serious infections. However, non-typeable strains are present in the throats of most healthy children and there is also *H. influenzae* type b in 3–5% of unvaccinated children although this varies considerably in different populations [3]. The carriage of other capsulated types is not

G.B. Harnett (✉)
Department of Microbiology and Infectious Diseases, PathWest Laboratory Medicine, QEII Medical Centre, Nedlands, WA 6009, Australia
e-mail: Gerry.Harnett@health.wa.gov.au

M. Schuller et al. (eds.), *PCR for Clinical Microbiology*,
DOI 10.1007/978-90-481-9039-3_17, © Springer Science+Business Media B.V. 2010

Table 17.1 The primer and probe sequence sets used in the multiplex PCR method for the identification of *H. influenza* strains

Multiplex	Target gene	Primers and probe sequences	Product size (bp)
1	BexAI-F	5′-GGCGAAATGGTGCTGGTAA-3′	57
	BexAI-R	5′-GAGGCTCAACACCASTCATCAA-3′	
	BexAI-Probe	5′-6FAM- TCCACGCTCATTCG-MGBNFQ-3′	
	BexAII-F	5′-TTGGGATYYTAGGWCGAAATGG-3′	61
	BexAII-R	5′-ACGCCACTGAKCAACCGAAT-3′	
	BexAII-Probe	5′-6FAM-AGCAGGCAAATCAACACT-MGBNFQ-3′	
	OmP2-F	5′-GGTGCATTCGCAGCTTCAG-3′	84
	OmP2-R	5′-TTAYACGACCRCCTAATTCTAC-3′	
	OmP2-Probe	5′-VIC-TTAGTCCCTTCGTTGTTATAA-MGBNFQ-3′	
	EHV-F	5′-GATGACACTAGCGACTTCGA-3′	81
	EHV-R	5′-CAGGGCAGAAACCATAGACA-3′	
	EHV-Probe	5′-Quasar670-TTTCGCGTGCCTCCTCCAG-BHQ2-3′	
2	CapB-F	5′-AGAAGTTTTACTGATGATATGGGTACATCT -3′	79
	CapB-R	5′-GCTCGAAGAATGAGAAGTTTTGTG -3′	
	CapB-Probe	5′-6FAM-TTCGCCATAACTTCATCT-MGBNFQ-3′	
	CapA-F	5′-AAAGCGGAGAGATGTTGTTTTTAAA-3′	73
	CapA-R	5′- AGAAATATCGCACAGGCCAACT-3′	
	CapA-Probe	5′-VIC-CAGCACTAAACTCACTGAAGT –MGBNFQ-3′	
	CapC-F	5′-TTATTATCTGTGTAGATGATGGT-3′	68
	CapC-R	5′-ATATTTACGCTGCCATTT-3′	
	CapC-Probe	5′-Cy5.5-AA+TAA+T+CT+TAG+C+TG+CA+TCT-BHQ3-3′	
3	CapF-F	5′-TGGCAGGAAAAGAAACAGCACTA-3′	69
	CapF-R	5′-GACATTATCTTTCGCATAGAATGCA-3′	
	CapF-Probe	5′-6FAM-CGAAACTAGCAAACAACT-MGBNFQ-3′	
	CapE-F	5′-GGTGAAAACAAACCGCACTTT-3′	73
	CapE-R	5′-ATCTTTAATTACCAGATCCCTTTCAT-3′	
	CapE-Probe	5′-VIC-AATAATCTTAGCTGCATCT-MGBNFQ-3′	
	CapD-F	5′-TGATGACCGATACAACCTGTTTAAA-3′	116
	CapD-R	5′-CCAGAAATTATTTCTCCGTTATGTTG-3′	
	CapD-Probe	5′-Cy5.5-AA+AA+CT+CT+T+CT+TAG+TG+C+TGAAT-BHQ3-3′	

The sets were designed in-house with the exception of those for the *CapE* target which were slightly modified from Maaroufi et al. [4]. The Cy 5.5 probes are LNA probes supplied by Sigma-Aldrich, Australia (+, LNA)

well documented. It is important to monitor the types of *H. influenzae* causing disease since this information will be of potential use in framing future vaccination strategies.

17.3 Acceptable Specimens

This method is designed for the rapid identification of *H. influenzae* cultures and has not yet been validated for the detection of *H. influenzae* in clinical samples.

17.4 Sample Extraction

Bacterial colonies are suspended in 1.0 ml of water and the amount of EHV is added to give a Ct of 30. This is determined by pre-titration of the EHV preparation. The EHV is prepared in large batches and stored as single use aliquots to give consistent readings over a long period. In practice Ct readings exceeding Ct 38 should be considered inhibitory.

17.5 Primers and Probe Sequences

Primer and probe sequences are shown in Table 17.1.

17.6 PCR Amplification and Product Detection

After extraction, 8 μl of supernatant fluid is added to each of multiplex 1, 2 and 3 PCR tubes which contain 0.2 μM of the primers and TaqMan probes listed in Table 17.1, 1x PCR buffer, 0.2 mM of dNTPs, 0.01% BSA and 0.75 units of AmpliTaq Gold DNA polymerase (Applied Biosystems, Australia), in a total reaction volume of 20 μl. The tubes are cycled in a real-time RotorGene 6000 instrument (QIAGEN, Australia) and probe emissions acquired on the appropriate three wavelength channels. For the RotorGene it is important that the tube in position one in the rotor contains all three probes to obtain the correct autogain setting. The cycling programme consists of an enzyme activation and DNA denaturation period of 10 min at 95°C, followed by 40 cycles of 12 s at 94°C, 30 s at 55°C and 20 s at 72°C. Positive controls should be included but can be pooled to limit the number of assay tubes required.

17.7 Quality Control and Validation Data

The EHV TaqMan in Multiplex 1 serves as a control of PCR inhibition and should show a Ct value consistent with the copy number of the EHV target introduced. The *omP2* gene is present in all strains of *H. influenza* including non-typeable strains while the *bexA* gene is present in capsulated strains in one of two clonal forms [5]. Some strains of *H. influenza* type b can lose the *bexA* gene function while retaining the *capA* gene and these are termed b⁻ strains [1]. Multiplexes 2 and 3 identify the capsular types *a–f* if present.

The method has been tested against *H. influenza* reference capsulated types a, b, c, d, e and f. In addition 37 field strains of *H. influenza* type b and 5 of type b⁻, one type c, one type d, two type f, 63 non-typeable strains and 14 *Haemophilus* spp. were tested. The method gave the expected results with all these cultures. It remains to be established whether this multiplex method is suitable for use on clinical samples. Probit analysis carried out on the *H. influenza capB* assay, using closely spaced dilutions of target with 24 replicates per dilution showed that the 95% confidence limit of detection was 417 copies/ml. It is recommended that Ct values for control positive materials be plotted on a Shewhart chart. When about 100 test batches have been processed, the mean Ct value and/or two standard deviations can be determined and influenz to monitor subsequent test batches on a daily basis. When control Ct values fall within these limits they provide a 95% confidence limit for the assay and can be used for Uncertainty of Measurement calculations. There is no need to plot the EHV values and the inhibitory samples are noted by inspection of the individual Ct values.

Acknowledgments We would like to thank Natasha K Ryan who provided valuable technical assistance for this study.

References

1. Falla TJ, Crook DWM, Brophy LN et al (1994) PCR for capsular typing of *Haemophilus* influenzae. J Clin Microbiol 32:2382–2386
2. Gonin P, Lorange M, Delage G (2000) Performance of a multiplex PCR for the determination of *Haemophilus influenza* capsular types in the clinical microbiology laboratory. Diag Microbiol Infect Dis 37:1–4
3. Killian M (2007) Haemophilus. In: Murray P (ed) Manual of Clinical Microbiology, 9th edn. ASM Press, Washington, DC
4. Maaroufi Y, De Bruyne J-M, Heymans C et al (2007) Real-time PCR for determining capsular serotypes of *Haemophilus influenza*. J Clin Microbiol 45:2305–2308
5. Zhou J, Law DKS, Sill ML et al (2007) Nucleotide sequence diversity of the bexA gene in serotypeable *Haemophilus influenza* strains recovered from invasive disease patients in Canada. J Clin Microbiol 45:1996–1999

Chapter 18
Mycobacterium tuberculosis Complex

Maria Globan and Janet Fyfe

18.1 Summary of Methods

Tuberculosis (TB) is an infectious bacterial disease, which is one of the leading causes of morbidity and mortality worldwide. The causative organism belongs to the *Mycobacterium tuberculosis* complex (MTBC). Disease spread between humans generally occurs via aerosols. Due to the slow growth of these bacteria, appropriate molecular methods provide an important tool for rapid diagnosis. We describe a TaqMan real-time PCR assay for the detection of MTBC DNA in clinical samples. The assay does not distinguish between different members of the complex.

18.2 Acceptable Specimens

A range of clinical specimens may be tested using this assay. The most appropriate specimens are from the respiratory tract. Based on the analysis of 572 respiratory specimens, the sensitivity and specificity of the assay compared to culture (Gold Standard) were 87.4% and 97.2% respectively. Specimen types other than respiratory samples may also be appropriate, especially if acid-fast bacilli have been seen on microscopy. Based on the analysis of 604 samples of non-respiratory origin, the sensitivity and specificity of the assay has been estimated to be 69.8% and 95.6%, respectively.

J. Fyfe (✉)
Mycobacterium Reference Laboratory, Victorian Infectious Diseases Reference Laboratory (VIDRL), 10 Wreckyn Street, North Melbourne, VIC 3051, Australia
e-mail: janet.fyfe@mh.org.au

M. Schuller et al. (eds.), *PCR for Clinical Microbiology*,
DOI 10.1007/978-90-481-9039-3_18,

18.3 Sample Extraction

Table 18.1 provides suggested guidelines for pre-extraction procedures for appropriate clinical specimens. DNA is then extracted using the Amplicor Respiratory Specimen Preparation Kit (RSPK, Roche), according to Fig. 18.1.

This crude extract may be further purified and concentrated, where necessary, using the QIAamp DNA Mini kit (Qiagen), following the manufacturer's Protocol for Crude Cell Lysates and Other Samples, with one modification: DNA is eluted from the column using 50 μl of AE buffer (rather than 200 μl).

Table 18.1 Specimen pre-treatment guidelines prior to DNA extraction

Specimen type	Pre-treatment procedure
Mucoid samples i.e. sputum; Bronchial washes and lavages; gastric aspirates	Decontaminate sample with NaOH (4%) and orthophosphoric acid (10%). Resuspend pellet in 1 ml of Ringers solution or PBS
Urine	Centrifuge and resuspend pellet in 1 ml of Ringers or PBS
CSF	Need at least 1 ml. If >1 ml received centrifuge and resuspend in 1 ml Ringers or PBS
Fluids from normally sterile sites i.e. pericardial; pleural; ascitic	Use 1 ml. If >1 ml, centrifuge and resuspend pellet in 1 ml of Ringers or PBS. If mucoid, follow procedure for mucoid samples, but use 2% NaOH
Tissues	Dice up into small pieces; place into a sterile glass bottle containing glass beads. Add 1–2 ml PBS and vortex vigorously for 1–2 min. Remove 1 ml for extraction
Swabs (dry or in transport medium)	Place swab into a sterile glass bottle containing glass beads, break shaft and replace bottle lid. Add 1–2 ml of sterile PBS, vortex vigorously for 1–2 min. Remove 1 ml for extraction
Pus	Dilute sample with sterile Ringers or PBS at a 1:1 ratio. Place contents into a sterile glass bottle with beads, vortex vigorously for 1–2 min to homogenise sample. Remove 1 ml for DNA extraction
Bone marrow; whole blood	Dilute sample with sterile water at a 1:1 ratio to lyse red blood cells. Centrifuge and resuspend pellet in 1 ml of Ringers or PBS
Paraffin – embedded formalin fixed tissue sections (complete extraction procedure)	Place at least 6x 10–20 μm sections into a sterile tube with 1 ml of xylene substitute (limonene). Centrifuge at maximum speed for 5 min. Aspirate and discard supernatant (into a designated discard bottle appropriate for xylene substitute), add 1 ml of absolute ethanol to tissue pellet. Centrifuge at 13,000 rpm/5 min. Aspirate and discard supernatant and resuspend de-paraffinised sample in 20 μl Proteinase K 300 μg/ml (can use reagent from Qiagen kit), and 180 μl Digestion buffer (50 mM Tris-HCl pH 7.5, 10 mM EDTA, 0.5% SDS, 50 mM NaCl). Incubate at 56°C for 24–48 h. Following incubation, proceed to step 5b of the tissue protocol described in QIAmp DNA mini kit handbook. Elute in 50 μl AE buffer

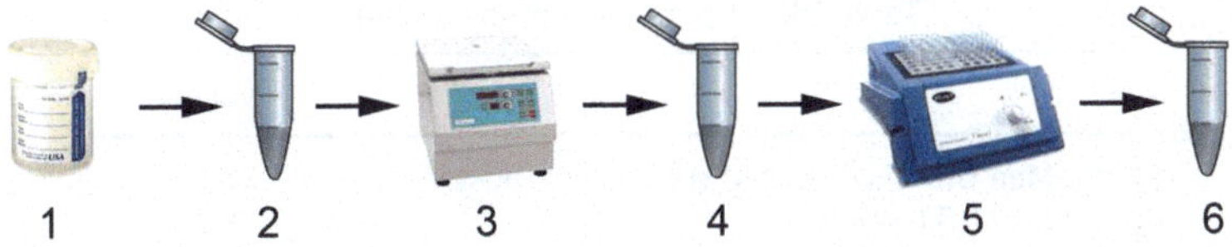

Fig. 18.1 DNA extraction procedure (excluding Paraffin sections see Table 18.1)

A 'reagent only' negative control is included with every run. It consists of 1 ml of phosphate buffered saline (PBS) that is extracted in the same way and at the same time as the other samples.

Positive control samples (including QAP samples provided by the RCPA) should be extracted on a regular basis for quality control purposes.

Note that the extraction of DNA from clinical samples must be performed in a dedicated biosafety cabinet (preferably Class II), using dedicated reagents and equipment that are only used for this purpose and never used for the extraction of DNA from cultured MTBC isolates. A 10% sodium hypochlorite solution is used to wipe out the biosafety cabinet, wipe down pipettes and other equipment following each extraction run. UV decontamination of the cabinet should also follow to prevent any potential contamination of the DNA extracts.

18.4 Primers and Probe Sequences

The primers target the insertion sequence IS*6110*, which is specific to MTBC isolates, and is present in 1–25 copies in 99% of strains. The amplicon detected is 77 bp in size.

IS6110T-F (5′-GGA AGC TCC TAT GAC AAT GCA CTA-3′)
IS6110T-R (5′-GGC TTG CCG GGT TTG AT-3′)

A TaqMan probe fluorogenic 5′nuclease assay is used to detect PCR product.

IS*6110*T-P (5′-6-FAM-AAC GGC CTA TAC AAG AC-MGBNFQ-3′)

The master mix (MM) reaction is described in Table 18.2. The assay is performed on an ABI Prism 7000 Sequence Detection System. Figure 18.2 describes the workflow for setting up a typical run. The assay may be run on any real-time PCR instrument that can simultaneously detect FAM, and VIC fluorescent labels.

Table 18.2 Mastermix for one reaction

Reagent	Volume (μl)
2X TaqMan Universal PCR MM (Applied Biosystems)	12.5
Primer IS*6110*TF (18 μM)	1.25
Primer IS*6110*TR (18 μM)	1.25
Probe IS*6110*TP (5 μM)	1.25
50X Exo IPC DNA – see Table 18.3	0.5
10X Exo IPC mix (VIC)	2.5
Nuclease-free water	0.75
Total	20

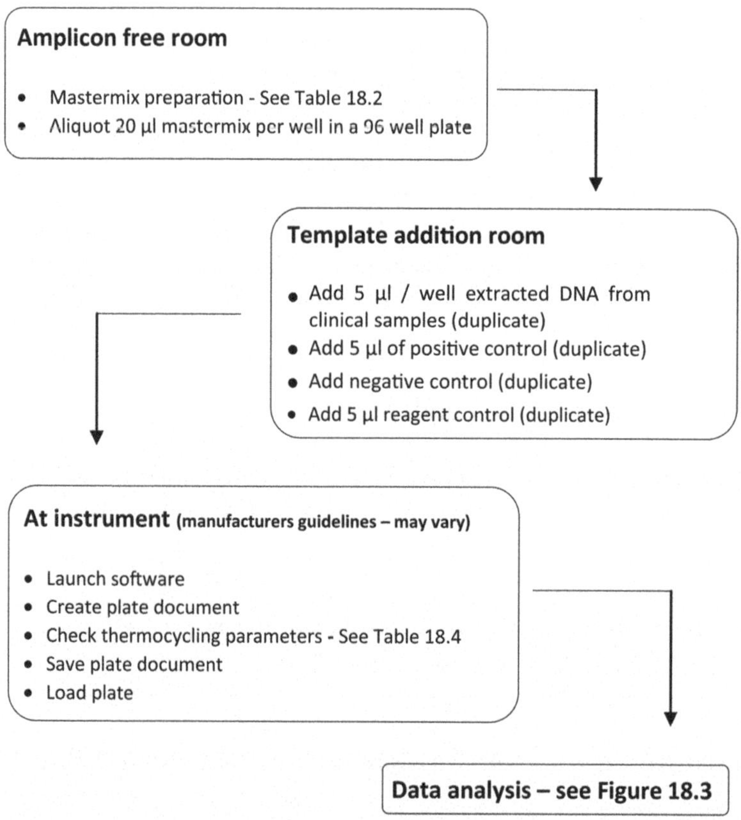

Fig. 18.2 TaqMan assay set up flowchart

18.5 Result Interpretation and Reporting

Refer to Fig. 18.3 for result interpretation. Following 40 cycles, the cycle threshold (Ct) for the positive control should be 30 ± 2 cycles, and negative controls should be below the threshold line.

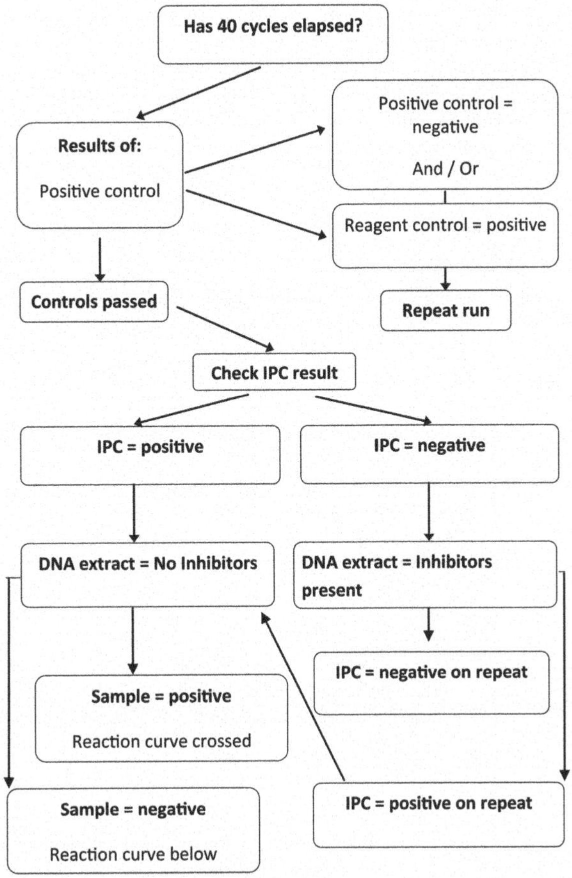

Fig. 18.3 Result interpretation

Report a **positive** specimen as:

***M.tb* complex DNA detected.** With the comment:

This assay shows the presence of *M.tb* complex DNA but does not necessarily indicate the presence of viable organisms. In addition routine culture is recommended.

Report a negative specimen as:

***M.tb* complex DNA NOT detected.** Include comment:

The presence of *M.tb* complex cannot be excluded, routine culture is recommended.

Table 18.3 Controls used

1	Positive control	DNA extracted from a culture of *M. bovis* BCG used at a concentration of approx. 1 pg/μl
2	Negative control	Nuclease free water
3	Reagent only negative control	Prepared during specimen extraction – see Section 3. This tests for the presence of MTBC DNA contamination of any of the reagents used during DNA extraction
4	Internal positive control (IPC)	TaqMan Exogenous Internal Positive Control (IPC) (Applied Biosystems). This tests for the presence of PCR inhibitors in the DNA extract. The IPC reagents consist of two different mixes, one containing template DNA (50X Exo IPC) and the other the primer and probe reagents (10X Exo IPC) specific for the included template

Table 18.4 Amplification conditions

Temperature	Cycles	
95°C	1 cycle for 10 min	
95°C	15s	} 40 cycles
60°C	1 min	

Note: The amplification conditions are instrument and MM dependant. Conditions described are for the TaqMan Universal PCR MM (Applied Biosystems).

Chapter 19
Mycobacteria Other Than *Mycobacterium tuberculosis*

Greg James

19.1 Summary of Methods

The PCR assay targeting the 16S rRNA gene, for the detection and identification of mycobacteria, is modeled from a paper by Boddinghaus et al. [1]. The assay is a nested double amplification PCR utilyzing primers Mb246 + Mb Reverse 247 (R247), to amplify a 590 bp amplicon. Mb246 is a universal primer for bacterial 16SrDNA whereas MbR247 is specific for the genus *Mycobacterium.* An aliquot from the first amplification is reamplified in a second PCR utilizing primers Mb1 and MbR7, designed in-house from published sequence data giving a final product of 469 bp. Mb1 and MbR7 sequence is conserved between all mycobacterial species.

19.2 Introduction

Disease processes caused by *M. tuberculosis* and *M. leprae* are well known, however more than 125 *Mycobacterium* species have been identified and over 60 are capable of causing disease in humans [3]. These mycobacteria species are grouped as mycobacteria other than tuberculosis (MOTT) and are ubiquitous in the environment. MOTT cause opportunist infections and disease in immuno-suppressed and compromised patients and also in immunocompetent hosts. A variety of clinical syndromes can be seen including; chronic pulmonary infections, superficial lymphadenitis, soft tissue and osteoarticular infections and disseminated disease. Treatment of MOTT infection varies with the *Mycobacterium* species present and infection site and may include prolonged multidrug therapy and surgical excision [3]. Traditionally MOTT have been identified using the Runyon system based on growth rates and pigment production. This process requires growth of the organism and as many species are slow growing can result in a lengthy time to identification and commencement of appropriate treatment. Molecular biology methods such as

---

G. James (✉)
Centre for Infectious Diseases and Microbiology Laboratory Services (CIDMLS), Institute for Clinical Pathology and Medical Research (ICPMR), Westmead, NSW 2145, Australia
e-mail: Greg.James@swahs.health.nsw.gov.au

M. Schuller et al. (eds.), *PCR for Clinical Microbiology*,
DOI 10.1007/978-90-481-9039-3_19, © Springer Science+Business Media B.V. 2010

PCR can provide a fast and accurate identification of the *Mycobacterium* species causing the infection and timely implementation of treatment.

All known *Mycobacterium* species can be detected and most identified using PCR targeting the 16SrRNA gene and subsequent DNA sequence analysis. This gene contains genus specific regions and regions of high variability where individual species can be determined.

PCR amplicons are visualized by agarose gel electrophoresis and SYBER SAFE staining. PCR amplicons with an identical length (in nucleotide base pairs) to the positive control are deemed POSITIVE and are characterized to species level by DNA sequence analysis.

19.3 Acceptable Specimens

Culture Isolates: On solid or in liquid media.

Clinical Specimens: Mycobacteria can cause infection and disease in many parts of the body, therefore a variety of specimen types are submitted for diagnosis including; fresh tissue, cerebrospinal fluid, respiratory specimens, and body fluids including uncoagulated blood in EDTA tubes.

19.4 Unacceptable Specimens

Formalin fixed and paraffin embedded tissue samples, clotted blood, serum, swabs, urine and faeces.

19.5 Sample Extraction

Varies with sample type. Culture isolates on solid media can be prepared for PCR by suspending a loop full of bacterial colonies in 1 ml of water and then sonicate for 30 min at 47 kHz with 30 W output in a waterbath sonicator followed by boiling for 20 min and then cooling on ice. Following a pulse centrifugation for 10 s the supernatant can be used in the PCR. Specimens require processing dependant on the site of collection. DNA is purified from liquid media or specimens using commercial automated and manual specimen processing systems such as BioMerieux EasyMag and Roche High Pure kits according to manufacturers instructions.

19.6 Primer Sequences

The primer sequences are;

Mb246	5′ $_{8}$AGA GTT TGA TCC TGG CTC AG$_{28}$ 3′
MbR247	5′ $_{609}$TTT CAC GAA CAA CGC GAC AA$_{590}$ 3′

Mb1 5′ $_{105}$AGT GGC GAA CGG GTG AGT AAC$_{126}$ 3′
MbR7 5′ $_{573}$TTA CGC CCA GTA ATT CCG GAC AA$_{551}$ 3′

Oligonucleotide nucleic acid position based on *Escherichia coli* 16SrRNA gene standard nomenclature.

19.7 First PCR Amplification (Mb246 + MbR247)

The PCR Master Mix (MM) has a final volume of 25.0 μl. The final concentration of reactants in the PCR is: 1xTth buffer (10 mM Tris–HCl pH 8.3, 50 mM KCl, 1.5 mM $MgCl_2$, 0.01% Tween 20, 0.01% Gelatin, 0.01% NP-40), 100 μM dNTP's, 400 nM of Mb246 and MbR247 primers, 0.2 μl (1.8 U) Klen Taq DNA polymerase and 10 μl of sample DNA extract.

Note: See Section 19.14 for sequence of sample addition.

Amplification is performed in a conventional thermocycler (Eppendorf Mastercycler) under the following conditions; 95°C for 5 min followed by 40 cycles of 96°C for 10 s, 68°C for 2 min and 74°C for 3 min. Hold the final extension step at 74°C for 10 min to allow PCR amplicons to be fully extended. Cool the cycler to 25°C, remove the tubes and hold them at 4°C for subsequent nested amplification.

19.8 Nested PCR Amplification (Mb1 + MbR7)

An aliquot from the first amplification is diluted 1:100 in water and 4.0 μl added to the Mb1-MbR7 PCR amplification MM that has a final volume of 50.0 μl and consists of; 1xTth buffer, 50 μM dNTP's, 100 nM Mb1 and MbR7 primers and 0.5 μl (1.0 U) DNAZyme Taq DNA polymerase.

Note: The dilution and addition of each first amplification tube must exactly follow the sequence of additions in the first amplification.

Amplification is performed in a conventional thermocycler (Eppendorf Mastercycler) to the following conditions; 25 cycles of 94°C for 30 s, 68°C for 1 min and 74°C for 15 s. Hold the final extension step at 74°C for 5 min and then cool to 25°C.

19.9 Product Detection

PCR amplified DNA segments are separated by electrophoresis in a 2% agarose gel and visualised using SYBR Safe DNA stain (Invitrogen Life Technologies). Product size is determined by comparison with the Positive Control and a DNA size ladder. The positive PCR DNA is confirmed as belonging to the genus *Mycobacterium* and identified to the species level by DNA sequencing.

19.10 Pre-sequencing Treatment

Prior to DNA sequencing, PCR product is cleaned of unused dNTP's and primer using two hydrolytic enzymes, shrimp alkaline phosphatase and exonuclease 1 according to manufacturer's instructions (ExoSAP-IT, USB Affymetrix).

19.11 DNA Sequencing

Is performed by a commercial service provider and requires approximately 10 ng/μl of DNA per sample and 3 pM/μl of sequencing primer (MbR7). Sequencing reagent variables such as amount of DNA and primer concentrations must be optimized with each provider.

19.12 DNA Sequence Analysis

DNA sequences should be edited using appropriate software (such as Chromas), to produce a clean, easily read sequence of maximum length. Sequences should then be copied into sequence analysis software (such as BioManager) and analyzed using a BLAST search against appropriate databases such as RIDOM (http://www.ridom_rdna.de).

19.13 Reporting

Refer to [2]. A 100% match with the DNA sequence of the type *Mycobacterium* species in the database is required for identification of a *Mycobacterium* species. Further, the DNA sequence identification must correlate with other results such as; High Pressure Liquid Chromatography (HPLC), organism morphology and growth characteristics, conventional biochemical and antimicrobial resistance profiles. The identification must also be discussed with the referring doctor to determine if the identification matches the clinical diagnosis.

19.14 Quality Control

Positive control: A *Mycobacterium* species, rarely if ever encountered in laboratory specimens, of known sequence must be included as a positive control in each PCR assay to ensure the assay is working to specifications and as a control for subsequent DNA sequencing. If the full and correct DNA sequence for the positive control organism is not obtained then the assay is not valid and must be repeated.

Negative controls. As environmental mycobacteria are ubiquitous in the community and in laboratories and the primers used in this assay can detect all known

Mycobacterium species, it is important to include no DNA controls (NDC's) for each sample at each stage of the process (Sample processing and PCR set-up) to determine the possibility of contamination. Add each representative Sample Processing NDC tube first in the same sequence that samples will be subsequently added followed by the first sample. The sample is added diluted 1:100 in water (to dilute possible inhibitory substances) and undiluted. Following addition to the first sample tubes, next add the respective PCR set-up NDC. This tube will show if any contaminating bacteria may have entered tubes during addition of the first sample. The addition process is then repeated for the next sample and so on.

After the last sample and respective PCR set-up NDC has been added, the Positive Control DNA is added to the 'Spike' tubes that contain equal volumes of sample diluted 1:100 and Positive Control diluted to a level that is two dilutions above the detection limit of the assay. Finally the positive control dilution series tubes have Positive Control added from the most dilute to the least dilute to demonstrate the detection limit of the assay and to provide a monitor of assay reproducibility.

Any PCR product formed in NDC's must be DNA sequenced and compared to the sequence of the sample. If the NDC sequence is present in the sample sequence the assay is not valid and must be repeated. Any Spike tube that does not form a PCR product shows that the respective sample was inhibitory. The sample must be re-extracted and re-tested.

19.15 Assay Limitations

The ubiquitous nature of mycobacteria and the potential for environmental mycobacteria to contaminate samples and produce either an incorrect or spurious result or mixed DNA sequence limit the application of this assay for the detection and identification of *Mycobacterium* species directly from clinical specimens to highly experienced staff and reference laboratories.

References

1. Boddinghaus B et al (1990) Detection and identification of Mycobacteria by amplification of rRNA. J Clin Microbiol 28:1751–1759
2. Clinical and Laboratory Standards Institute (2007) Interpretive criteria for microorganism identification by DNA target sequencing; proposed guideline. 27(22):MP18-P
3. Jarzembowski JA, Young MB (2008) Nontuberculous mycobacterial infections. Arch Pathol Lab Med 132:1333–1341

Chapter 20
Mycoplasma genitalium

David M. Whiley

20.1 Summary of Methods

This protocol utilises a 5′ nuclease-based real-time PCR system with two primers and one probe targeting the *Mycoplasma genitalium* MgPa adhesin gene [1].

20.2 Acceptable Specimens

Acceptable specimens are urine and genital swabs.

20.3 Sample Extraction

Swabs were rotated 15 times in 400 μl of phosphate buffered saline (PBS) to dislodge cells and bacteria. 200 μl of PBS or urine were then extracted manually using the High Pure Viral Nucleic Acid kit (Roche Applied Science, Australia), or mechanically using the MagNAPure instrument (Roche Applied Science, Australia), according to the manufacturer's instructions.

20.4 Primers and Probe Sequences

The primer and probe sequences used in this assay targeted the MgPa adhesion gene [1]. Sequences were as follows: Primers; MgPa-355F: 5′-GAG AAA TAC CTT GAT GGT CAG CAA-3′ and MgPa-432R: 5′-GTT AAT ATC ATA TAA AGC TCT ACC GTT GTT ATC-3′ with MGB probe MgPa-380: 5′-FAM-ACT TTG CAA TCA GAA GGT-MGB-3′.

D.M. Whiley (✉)
Queensland Paediatric Infectious Diseases Laboratory, Sir Albert Sakzewski Virus Research Centre, Queensland Children's Medical Research Institute, Children's Health Service District, Brisbane, QLD 4029, Australia
e-mail: d.whiley@uq.edu.au

M. Schuller et al. (eds.), *PCR for Clinical Microbiology*,
DOI 10.1007/978-90-481-9039-3_20,

20.5 PCR Amplification and Product Detection

Perform PCR reactions in 0.2 ml or 0.1 ml reaction tubes. Each reaction mix contains 10 pmol of forward and reverse primer (MgPa-355F and MgPa-432R), 4 pmol of probe (MgPa-380), 1× QIAGEN QuantiTect Probe PCR Master Mix (QIAGEN, Australia), and 5.0 μl of DNA, in a final reaction volume of 25 μl.

Amplification is conducted on a RotorGene 3000 or 6000 instrument (QIAGEN, Australia) with an initial 15 min at 95°C followed by 50 cycles of 95°C for 15 s and 60°C for 60 s (fluorescence acquisition).

20.6 Quality Control and Validation Data

Extracted DNA from a *M. genitalium* isolate or from a *M. genitalium* PCR-positive sample confirmed by sequencing may be used as a positive control. The positive control should be diluted to a concentration providing a cycle-threshold (Ct) value of approximately 30 cycles in the real-time PCR. PCR-grade water is subject to the extraction procedure and is used as a negative control.

This assay will detect presence of *M. genitalium* DNA in the clinical specimen at a sensitivity of 1–10 copies/reaction [1], and has a clinical sensitivity and specificity of greater than 95%. (PPV and NPV data are not available.)

20.7 Assay Limitations

Although *M. genitalium* is a recognized sexually transmitted pathogen, there are still some questions over the value of routine screening given *M. genitalium* DNA may be detected in asymptomatic individuals [2].

References

1. Jensen JS, Bjornelius E, Dohn B et al (2004) Use of TaqMan 5′ nuclease real-time PCR for quantitative detection of *Mycoplasma genitalium* DNA in males with and without urethritis who were attendees at a sexually transmitted disease clinic. J Clin Microbiol 42:683–692
2. Ross JD, Jensen JS (2006) *Mycoplasma genitalium* as a sexually transmitted infection: implications for screening, testing, and treatment. Sex Transm Infect 82:269–271

Chapter 21
Mycoplasma pneumoniae

Rebecca Rockett

21.1 Summary of Methods

The 5′ nuclease-based real-time PCR method described here utilises primers and a TaqMan probe directed to the *Mycoplasma pneumoniae* P1 gene reviewed by Loens et al. ([4], participant 7).

21.2 Background

Mycoplasma pneumoniae is responsible for 10–20% of the cases of community-acquired pneumonia and has been associated with acute exacerbations of asthma [3]. *M. pneumoniae* is also implicated in mild acute respiratory infections, such as sore throat, pharyngitis, rhinitis, and tracheobronchitis [1]. It is generally confined to the respiratory tract with a clinical presentation that may include extra-respiratory symptoms. Infection can occur in all age groups, but is less common in children under the age of five, presenting with more severe clinical symptoms in these subjects [5]. The frequency of infection is generally underestimated in clinical practice because of the lack of specific clinical features, and infection can be confused with viral or bacterial pneumonia.

Diagnosis of *M. pneumoniae* infection is challenging; it is difficult to culture due to the fastidious nature of the pathogen, and there is a high seroprevalence in the population. Also transient asymptomatic carriage may occur. Traditionally, serology has provided the basic strategy for mycoplasma diagnosis in routine clinical

R. Rockett (✉)
Queensland Paediatric Infectious Diseases Laboratory, Sir Albert Sakzewski Virus Research Centre, Queensland Children's Medical Research Institute, Children's Health Service District, Brisbane, QLD 4029, Australia
e-mail: r.rockett@uq.edu.au

M. Schuller et al. (eds.), *PCR for Clinical Microbiology*,
DOI 10.1007/978-90-481-9039-3_21, © Springer Science+Business Media B.V. 2010

practice, and has been improved by the availability of sensitive assays for the detection of specific antibody classes, such as IgM. However, during recent years, PCR has become the method of choice, particularly real-time PCR, and is replacing serology in many laboratories. PCR is most effective in the diagnosis of early acute disease, but PCR together with serology should be considered for the accurate diagnosis of progressing *M. pneumoniae* respiratory infections. Extra-pulmonary disease due to *M. pneumoniae* may be diagnosed by direct pathogen detection alone, but the value of this diagnostic approach is limited by the immunologically mediated pathogenesis of some manifestations [2].

21.3 Acceptable Specimens

Specimens suitable for the detection of *M. pneumoniae* include sputum and bronchoalveolar lavage (BAL) specimens, nasopharyngeal and throat swabs, nasopharyngeal aspirates, tracheal aspirates and pleural fluid specimens.

21.4 Unacceptable Specimens

The method has not been validated for any other specimen type than those listed above.

21.5 Sample Extraction

Swabs are rotated 15 times in 400 μl of phosphate buffered saline (PBS) to dislodge cells and bacteria. BAL specimens may require pre-treatment with proteinase K for 10 min at 37°C.

200 μl of PBS, BAL or NPA specimen are then either extracted manually using a column-based method such as the High Pure Viral Nucleic Acid kit (Roche Applied Science, Australia), or mechanically using the MagNAPure instrument (Roche Applied Science, Australia), according to the manufacturer's instructions.

21.6 Primer and Probe Sequences

The primers used in the reaction were; MP-P1-7-F: 5′-AAA GGA ACA AAC TGA TCC CAC TTC T-3′ and MP-P1-7-R: 5′-CCA GGG CAC ATA ATC CAA CAC-3′. The sequence information for the TaqMan probe was; MP-P1-7-TM: 5′-fam-TCT CCA CCG GGT TCA ACC TTG TGG-bhq-3′ [4].

21.7 PCR Amplification and Product Detection

The QIAGEN QuantiTect kit (QIAGEN, Australia) was used for this assay as per manufacturer's instructions. PCR reaction mix was prepared for 100 reactions using the reagent composition as shown in Table 21.1.

NOTE: These mixes use a total volume of 25 μl: 20 μl of reaction mix + 5 μl of nucleic acid extract are added to each tube.

The assays may be performed on the QIAGEN Rotor-Gene 3000 or 6000 and Applied Biosystems 7500 using the following cycling parameters:

Activation/Denaturation: 1 cycle at 95°C for 15 min, followed by 45 cycles of 95°C for 15 s plus 60°C for 60 s.

Table 21.1 PCR reaction mix for *M. pneumoniae* PCR

Reagent	Volume (μl)
Quantitect Probe master mix	1250.0
MP-Fus5-loens-F (20 pmol/μl)	50.0
MP-Fus5-loens-R (20 pmol/μl)	50.0
MP-Fus5-loens-TM(fam) (20 pmol/μl)	20.0
QIAGEN water	630.0
Total:	2000.0

21.8 Quality Control and Validation Data

10-fold dilutions ranging from 10^{-0} to 10^{-6} were made of a stock suspension of *M. pneumoniae* strain ATCC 29085 (5.8×10^4 CCU/ml) in PBS. Each dilution was tested five-fold using the real-time PCR method described above. The assay detected the 10^{-5} dilution consistently in 95% of all instances. This equated to 1.5 CCU or 15 cells of the organism.

Fourteen BAL specimens and 22 NPA collected from patients diagnosed with acute *M. pneumoniae* infection by presence of specific IgM or increasing IgG titres were available for the validation of this assay. All 36 specimens tested positive for *M. pneumoniae* with this assay. Forty two proven negative BAL and NPA specimens tested negative by the real-time PCR.

A 10^{-4} dilution of the *M. pneumoniae* strain ATCC 29085 described above was used as a positive control in the reaction. Water for injection was extracted using the normal procedure and used as a negative control. Each test run contained 2x positive controls, 5x negative controls and 5x environmental controls consisting of complete mastermix without specimen extract which were left open on the work surface during the preparation of test samples. The environmental controls were used to monitor contamination due to aerosols.

21.9 Assay Limitations

The results of the real-time PCR must be examined in the context of clinical presentation. *M. pneumoniae* may be detected for some time after acute infection following the resolution of clinical symptoms.

References

1. Atmar RL, Greenberg SB (1989) Pneumonia caused by *Mycoplasma pneumoniae* and the TWAR agent. Semin Respir Infect 4:19–31
2. Daxboeck F, Krause R, Wenisch C (2003) Laboratory diagnosis of *Mycoplasma pneumoniae* infection. Clin Microbiol Infect 9:263–273
3. Gil JC, Cedillo RL, Mayagoitia BG (1993) Isolation of *Mycoplasma pneumoniae* from asthmatic patients. Ann Allerg 70:23–25
4. Loens K, Beck T, Ursi D (2006) Two quality control exercises involving nucleic acid amplification methods for detection of *Mycoplasma pneumoniae* and *Chlamydophila pneumoniae* and carried out 2 years apart (in 2002 and 2004). J Clin Microbiol 44:899–908
5. Vervloet LA, Marguet C, Camargos PA (2007) Infection by *Mycoplasma pneumoniae* and its importance as an etiological agent in childhood community-acquired pneumonias. Braz J Infect Dis 11:507–514

Chapter 22
Neisseria gonorrhoeae

David M. Whiley

22.1 Summary of Methods

Specimens tested for *N. gonorrhoeae* by NAAT should be positive in at least two assays (targeting different gonococcal sequences) before a positive result can be reported [2].

This protocol describes two 5′ nuclease-based real-time PCR assays, each comprising two primers and one probe, and targeting the *N. gonorrhoeae porA* pseudogene [1, 4] and *opa* genes respectively [3].

22.2 Acceptable Specimens

Acceptable specimens are urine, urethral swab, endocervical swab, dry swab (including throat, rectal and vaginal).

22.3 Unacceptable Specimens

This assay has not been validated for specimens not included above.

22.4 Sample Extraction

Swabs were rotated 15 times in 400 μl of phosphate buffered saline (PBS) to dislodge cells and bacteria. 200 μl of PBS or urine were then extracted manually using the High Pure Viral Nucleic Acid kit (Roche Applied Science,

---

D.M. Whiley (✉)
Queensland Paediatric Infectious Diseases Laboratory, Sir Albert Sakzewski Virus Research Centre, Queensland Children's Medical Research Institute, Children's Health Service District, Brisbane, QLD 4029, Australia
e-mail: d.whiley@uq.edu.au

M. Schuller et al. (eds.), *PCR for Clinical Microbiology*,
DOI 10.1007/978-90-481-9039-3_22,

Australia), or mechanically using the MagNAPure instrument (Roche Applied Science, Australia), according to the manufacturer's instructions.

22.5 Primers and Probe Sequences

The primers sequences for the detection of *N. gonorrhoeae* targeting the *porA pseudogene* [1, 4] were papTM-F: 5′-CAG CAT TCA ATT TG TTC CGA GTC-3′ and papTM-R: 5′-GAA CTG GTT TCA TCT GAT TAC TTT CCA-3′ and TaqMan probe papTM-P: 5′-FAM- CGC CTA TAC GCC TGC TAC TTT CAC GC -BHQ1-3′.

Primer and probe sequences targeting the *opa* genes [3] were; GCopaF: 5′-TTG AAA CAC CGC CCG GAA-3′ and GCopaR: 5′-TTT CGG CTC CTT ATT CGG TTT AA-3′ with an LNA probe; GCopa-LNA: 5′-FAM-CCG ATA TAA TC+C GTC+ CTT CAA +CAT CAG-BHQ1-3′ (+ indicates position of LNA bases).

22.6 PCR Amplification and Product Detection

Standard reaction mix and cycling conditions are used for both the *N. gonorrhoeae por*A *pseudogene* and *opa* real-time PCR assays. The PCR for *N. gonorrhoeae por*A *pseudogene* and *opa* genes is performed in separate 0.2 ml or 0.1 ml reaction tubes. Each reaction mix contains 10 pmol of forward and reverse primer (papTM-F and papTM-R or GCopaF and GCopaR), 4 pmol of each probe (papTM-P or GCopa-LNA), 1× QIAGEN QuantiTect Probe PCR Master Mix (QIAGEN, Australia), and 5.0 μl of DNA, in a final reaction volume of 25 μl. Amplification is conducted on a RotorGene 3000 or 6000 instrument (QIAGEN, Australia) with an initial 15 min at 95°C followed by 50 cycles of 95°C 15 s, 60°C 60 s (fluorescence acquisition).

22.7 Quality Control and Validation Data

Extracted DNA from a *N. gonorrhoeae* isolate is used as a positive control. Positive control should be diluted to a concentration providing a cycle-threshold (Ct) value of approximately 30 cycles. (Note that the *opa* PCR assay will generally provide Ct values 2–3 cycles lower than the *por*A *pseudogene* assay given the higher copy number of the *opa* target.) PCR-grade water is subject to the extraction procedure and is used as a negative control.

This assay will detect the presence of *N. gonorrhoeae* DNA in clinical specimens at a sensitivity of 1–10 copies/reaction. The clinical sensitivity and specificity of these assays are greater than 95% respectively. Although such measurements are dependent on the gold standard, sensitivity and specificity is achieved when compared with an expanded gold standard composed of consensus results from two to three different nucleic acid amplification assays.

A clinical PPV and NPV of over 95% are generally achieved using these assays.

22.8 Assay Limitations

It is widely recognised that false-positive results have been a considerable problem affecting *N. gonorrhoeae* NAATs. We strongly recommend that users follow the PHLN guidelines for the use and interpretation of nucleic acid detection tests for *Neisseria gonorrhoeae* in Australia [2].

Many commercial assays are available for *N. gonorrhoeae* and are widely used in diagnostic laboratories. In-house PCR assays, including those described above, are generally used as supplementary tests for commercial assays. The reaction mix and cycling conditions described above are those standardised and validated by our laboratory. Alternative conditions may be validated for use in other laboratories [3].

References

1. Goire N, Nissen MD, LeCornec GM et al (2008) A duplex *Neisseria gonorrhoeae* real-time polymerase chain reaction assay targeting the gonococcal *porA* pseudogene and multi-copy *opa* genes. Diagn Microbiol Infect Dis 61:6–12
2. Smith DW, Tapsall JW, Lum G (2005) Guidelines for the use and interpretation of nucleic acid detection tests for *Neisseria gonorrhoeae* in Australia: a position paper on behalf of the Public Health Laboratory Network. Commun Dis Intell 29:358–365
3. Tabrizi SN, Chen S, Tapsall J et al (2005) Evaluation of *opa*-based real-time PCR for detection of *Neisseria gonorrhoeae*. Sex Transm Dis 32:199–202
4. Whiley DM, Sloots TP (2005) Comparison of three in-house multiplex PCR assays for the detection of *Neisseria gonorrhoeae* and *Chlamydia trachomatis* using real-time and conventional detection methodologies. Pathology 37:364–370

Chapter 23
Neisseria meningitidis

Helen V. Smith

23.1 Summary of Methods

Meningococcal septicaemia and meningitis are the most common presentations of invasive meningococcal disease with an average case fatality rate of 10%, although arthritis and pericarditis can also occur. Early antibiotic therapy is strongly recommended, but this decreases the success of conventional diagnostic methodologies, i.e. culture. Culturing should however still be performed as it allows for the monitoring of changes in antimicrobial susceptibility. Molecular detection methods have allowed accurate, rapid diagnosis which enhances patient care and the initiation of public health action among contacts. The increased sensitivity offered by molecular detection and the ability to detect non-viable organisms may be responsible for increasing the diagnosis of laboratory confirmed cases of meningococcal disease by more than 30% [1]. Meningococcal DNA in CSF samples has been detected up to 72 h after commencement of antimicrobial treatment [1].

The target sequence most frequently used for PCR based assays is the *ctrA* gene although some laboratories also use *porA* either singly or as a dual assay with *ctrA*. The *ctrA* gene is involved in capsular biosynthesis and is specific to *N. meningitidis*. The 3′ end is highly conserved across the *N. meningitidis* A, B, C, 29E, W135, X, Y and Z serogroups [2]. This protocol describes a real-time TaqMan assay that targets the *ctrA* gene [1].

23.2 Organism

Neisseria meningitidis is a gram negative diplococcus which is carried in the throat from where it invades the bloodstream. It is transmitted by respiratory droplets. Thirteen serogroups have been identified based on capsular polysaccharide antigens.

H.V. Smith (✉)
Public Health Microbiology Communicable Diseases Forensic and Scientific Services, Queensland Health, 39 Kessels Rd, Coopers Plains, QLD 4108, Australia
e-mail: Helen_Smith@health.qld.gov.au

M. Schuller et al. (eds.), *PCR for Clinical Microbiology*,
DOI 10.1007/978-90-481-9039-3_23,

In Australia greater than 90% of disease is caused by serogroups B and C although the incidence of serogroups C disease has markedly declined since the recent introduction of the meningococcal C vaccine.

23.3 Acceptable Specimens

Specimens accepted for testing are EDTA whole blood, CSF and aspirates from joints. Heparinised or citrated samples can be tested, but EDTA is preferred. Meningococcal detection is possible from punch biopsies from skin lesions but usually insufficient numbers have been tested for method validation and a negative result does not exclude disease.

23.4 Unacceptable Specimens

Plasma and serum samples show a decreased recovery of *N. meningitidis* DNA. A recovery rate from EDTA whole blood of 93% in comparison to 65% from plasma has been reported [3].

23.5 Sample Extraction

The High Pure PCR Template Preparation kit (Roche Applied Science Australia) is used for DNA extraction. 200 μl of blood, CSF or joint fluid are incubated in 200 μl of Binding buffer and 40 μl of proteinase K at 70°C for 10 min. Tissue samples are incubated in 200 μl of Tissue Lysis buffer and proteinase K at 55°C for 1 h. DNA is then extracted according to the manufacturer's instructions and eluted in a final volume of 200 μl.

DNA extractions performed on automated systems such as the MagNA Pure Compact instrument (Roche Applied Science Australia) are also suitable for this assay.

23.6 Primer and Probe Sequences

The primers CtrA-F (5′-GCT GCG GTA GGT GGT TCA A-3′) and CtrA-R (5′-TTG TCG CGG ATT TGC AAC TA-3′) are used to amplify part of the *ctrA* gene which is detected using the hydrolysis probe (5′-6-FAM-CAT TGC CAC GTG TCA GCT GCA CAT-3′) [2].

23.7 PCR Amplification and Product Detection

The PCR reaction mix is set up in a 25 μl volume containing 1X TaqMan® Universal PCR Master Mix, No AmpErase® UNG (Applied Biosytems #4324018), 500 nM of each primer (CtrA-F & CtrA-R), 100 nM of hydrolysis probe and 2.5 μl of DNA extract. Amplification is performed on 7500 Fast Real-Time PCR system (Applied Biosystems) under the following conditions: 95°C for 10 min followed by 40 cycles of 95°C for 15 s and 60°C for 1 min. Fluorescence is measured at the end of the 60°C annealing/extension step. A sample is assumed to be negative if there is no increase in fluorescent signal above the calculated background threshold after 40 cycles [1].

23.8 Quality Control and Validation Data

DNA from CSF and joint fluids is tested for inhibitory substances by running a second PCR tube spiked with 1/10 positive control DNA to 2.5 μl of DNA extract. Inhibition in DNA from whole blood is rare [1] and we do not test for inhibition in these samples. A negative control of molecular biology grade water is included during each DNA extraction and PCR set-up to act as contamination controls. A positive control of 0.3 pg (100 copies) of *N. meningitidis* DNA is included in each PCR run to monitor assay performance.

The assay has a detection limit of 1 copy of target sequence, but at a range of 5–25 copies the assay is considered to be 100% sensitive [1]. For clinical samples, the assay has been estimated to be 96% and 89% sensitive for CSF and whole blood samples, respectively [2, 4]. No cross-reactivity has been reported with other bacteria or viruses [2].

23.9 Assay Limitations

The *ctrA* PCR assay is used as a diagnostic test for meningococcal disease. A negative PCR result does not totally exclude invasive meningococcal disease in a patient with compatible signs and symptoms.

23.10 Serogroup Determination

PCR tests for serogroup determination should be performed from both a confirmatory and epidemiological point of view. The Australian Meningococcal Guidelines recommend that: (i) sample material (CSF or EDTA blood) from which DNA was extracted; and (ii) the actual DNA extract, both be sent to the appropriate National Neisseria Network laboratory [5]. The original sample material is sent to ensure adequate amounts of DNA and uncontaminated DNA preparations are available as required.

References

1. Bryant PA, Hua YL, Zaia A et al (2004) Prospective study of a real-time PCR that is sensitive, specific, and clinically useful for diagnosis of meningococcal disease in children. J Clin Microbiol 42:2919–2925
2. Corless CE, Guiver M, Borrow R et al (2001) Simultaneous detection of *Neisseria meningitidis*, *Haemophilus influenzae* and *Streptococcus pneumoniae* in suspected cases of meningitis and septicaemia using real time PCR. J Clin Microbiol 39:1553–1558
3. Hackett SJ, Carrol ED, Guiver M et al (2002) Improved case confirmation in meningococcal disease with whole blood. TaqMan PCR. Arch Dis Child 86:449–452
4. Porritt RJ, Mercer J, Munro R (2000) Detection and serogroup determination of *Neisseria meningitidis* in CSF by polymerase chain reaction. Pathology 32:42–45
5. Public Health Laboratory Network (2006) Meningococcal infections laboratory case definition http://www.health.gov.au/internet/main/publishing.nsf/ Content/cda-phlncd-mening.htm. Accessed 6 October 2009

Chapter 24
Pseudomonas aeruginosa

Snehal Anuj and David M. Whiley

24.1 Summary of Methods

This protocol describes a duplex real-time PCR assay for the identification of *P. aeruginosa* using two different gene sequences, comprising the *ecfX* and *gyrB* genes. The *ecfX* gene encodes an extra-cytoplasmic function sigma factor, which may be involved in haem uptake or virulence, whereas the *gyrB* gene encodes the DNA gyrase subunit B [5]. Notably, both these genes offer reliable targets for the detection of *P. aeruginosa* by PCR [5, 7]. The benefit of using a duplex assay for a single organism is that it provides simultaneous confirmation of isolate identity, in addition to reducing the potential for false negative misidentification caused by sequence variation in primer or probe targets [7, 12].

24.2 Background

Pseudomonas aeruginosa is an opportunistic pathogen in humans and a major cause of morbidity and mortality in patients with cystic fibrosis (CF) [2, 10]. The organism is generally considered to be easily identifiable using standard microbiology techniques such as phenotypic and biochemical profiles. However, *P.aeruginosa* strains isolated from persons with CF often have unique phenotypes, and the lungs of persons with CF are often co-infected with other non-fermenting Gram-negative bacilli. This can further impede the isolation and identification of *P. aeruginosa* [7, 11] and carries a high risk of misidentification.

At the genomic level, sequence variation and genetic exchange associated with *P. aeruginosa* can also impede molecular identification assays. For example, genetic exchange with other unrelated Pseudomonas species can impact upon the

D.M. Whiley (✉)
Queensland Paediatric Infectious Diseases Laboratory, Sir Albert Sakzewski Virus Research Centre, Queensland Children's Medical Research Institute, Children's Health Service District, Brisbane, QLD 4029, Australia
e-mail: d.whiley@uq.edu.au

M. Schuller et al. (eds.), *PCR for Clinical Microbiology*,
DOI 10.1007/978-90-481-9039-3_24, © Springer Science+Business Media B.V. 2010

specificity of PCR-based methods targeting the 16S rDNA, *oprI* and *oprL* genes [3, 7]. Sequence variation arising from the highly polymorphic nature of the *P. aeruginosa* genome can also impact on these assays, with false-negative results attributed to sequence variation for assays targeting the *toxA* and *algD* genes [5, 7]. Therefore, molecular assays that use single-gene targets are more susceptible to error, including cross-reaction issues with other Gram-negative organisms. The assay described here seeks to address these limitations by targeting two separate genes in a duplex format.

24.3 Acceptable Specimens

This assay was validated for *P. aeruginosa* isolates only.

24.4 Unacceptable Specimens

This assay has been used with other specimen types, including sputa, but was not validated for these samples.

24.5 Sample Extraction

Bacterial DNA was extracted from each isolate for PCR testing using a basic heat treatment method. Briefly, a 1-ml 1.0-McFarland suspension of each isolate in sterile water was heated at 100°C for 10 min. The solution was then vortexed, centrifuged at 3000 rpm for 5 min, and stored at −20°C until further use. The supernatant was used for PCR.

24.6 Primers and Probe Sequences

Primers and probes used in this protocol are shown in Table 24.1.

Table 24.1 Oligonucleotides used for the detection of *P. aeruginosa*

Name	Sequence (5′–3′)	Target	Reference
ecfX-F	CGCATGCCTATCAGGCGTT	ecfX	[1]
ecfX-R	GAACTGCCCAGGTGCTTGC		[1]
ecf-X-TM	yak-ATGGCGAGTTGCTGCGCTTCCT-bhq		[1]
gyrB-F	CCTGACCATCCGTCGCCACAAC	gyrB	[6]
gyrB-R	CGCAGCAGGATGCCGACGCC		[6]
gyr-B-TM	fam-CCGTGGTGGTAGACCTGTTCCCAGACC-bhq		[1]

24.7 PCR Amplification and Product Detection

The PA duplex assay was performed using primers ecfXF, ecfXR, gyrB-F, and gyr-B-R (Table 24.1), and TaqMan probes ecf-X-TM and gyr-B-TM (Table 24.1) for simultaneous amplification and detection of the *P. aeruginosa ecfX* and *gyrB* genes. The expected product size for *ecfX* was 63 bp and for *gyrB* was 220 bp. The *gyrB* primers were previously described [6, 7], whereas the *exfX* primers and both TaqMan probes were designed by Anuj et al. [1] with the aid of Primer Premier Version 5.0 (Premier Biosoft International, Palo Alto, CA) to ensure suitable Tm for TaqMan 2-step cycling and to limit the formation of primer dimers. Candidate primer and probe sequences were also analysed for *P. aeruginosa* sequence conservation and for cross-reactivity with unrelated Pseudomonas species using GenBank's BLAST tool (National Centre for Biotechnology Information, Bathesda, MD).

The reaction mix comprised 12.5 μl of QIAGEN Quantitect Probe Master Mix (QIAGEN, Australia), 0.4 μmol/l of each primer, 0.16 μmol/l of each TaqMan probe, and 2 μl of sample nucleic acid and was made up to a final reaction volume of 25 μl with water. Cycling was performed on an ABI Prism 7500 Detection System (Applied Biosystems, Australia) with an initial hold at 95°C for 15 min followed by 50 cycles at 95°C for 15s and 60°C for 1 min. The gyr-B-TM probe was labelled with carboxyfluorescein (FAM), whereas the ecf-X-TM probe was labelled with a Yakima Yellow fluorophore, enabling the reactions to be distinguished using the ABI7500 FAM and JOE detection channels, respectively.

24.8 Quality Control and Validation Data

The identity of a *P. aeruginosa* isolate was determined by sequencing and was used as a positive control in all assays. The specificity of the assay was tested against 86 unrelated Gram-negative bacilli.

To validate the PCR assay, 91 clinical and environmental isolates were tested by the duplex PCR described above, and compared to results of 3 additional PCR assays targeting the *P. aeruginosa* oprL [9], exotoxinA (ETA) [4], and 16S rDNA genes [8].

Of the 91 isolates tested, 62 ($n = 35$ clinical and 27 environmental) were identified as *P.aeruginosa* (i.e., $\geq$85% identification) by all five PCR assays, and 21 environmental isolates were negative in all five assays. These results were subsequently confirmed by the API 20NE V7.0 kit (bioMérieux), giving 91% concordance between all assays.

However eight isolates, comprising four clinical and four environmental isolates, provided discrepant results between one or more of the PCR assays and the results obtained with the API 20NE V7.0 kit (bioMérieux). Briefly, the 16S DNA PCR gave two false positive results; the *opr*L gave four false-positive results and the ETA PCR gave one false negative result (Table 24.2). All PCR assays identified

Table 24.2 Summary of eight isolates providing discrepant results between one or more tests

	Assay results				
No	API 2ONE	PAduplex	*oprL*	*exoA*	16S DNA
1	*A. xylosoxidans*	Neg	Neg	Neg	Pos
2	*A. xylosoxidans*	Neg	Neg	Neg	Pos
3	*A. xylosoxidans*	Neg	Pos	Neg	Neg
4	*Ralstonia pickettii*	Neg	Pos	Neg	Neg
5	*Shewanella spp.*	Neg	Pos	Neg	Neg
6	*Ps.fluorescens*	Neg	Pos	Neg	Neg
7	*Ps. aeruginosa*	Pos	Pos	Neg	Pos
8	*C. violaceum*	Pos	Pos	Pos	Pos

one clinical isolate as *P. aeruginosa*, yet this was identified as *Chromobacterium violaceum* (profile no. 1150544, 99.5% identification) by the API 20NE kit. Further analysis by sequencing of the bacterial genome confirmed the identification of this isolate as *P. aeruginosa.*

As a result the PA-duplex PCR assay showed a specificity and sensitivity of 100%, using genome sequencing as the standard.

24.9 Assay Limitations

This assay was designed to confirm the identity of *P. aeruginosa* isolates. Although it may also be used to detect the presence of *P. aeruginosa* in other specimens, the result may be influenced by the bacterial load in the sample, and the presence of other, unrelated organisms.

Although the results of the *ecfX* and *gyrB* gene targets were concordant in this study, there is limited sequence information for these genes in the literature and on GenBank, and therefore, we do not believe that these results should be used to advocate testing with a single target only. Rather, our experience suggests that sequence variation is a much broader problem affecting the success of molecular diagnostic assays generally [12], and that the potential for false-negative results should be considered particularly for pathogens, including *P. aeruginosa,* that are known to exhibit extensive genetic diversity. In any event, if only one gene target of the PA-duplex PCR were to provide a positive result, then supplemental testing using a 3rd gene target, such as the *oprL*-PCR, would be warranted.

References

1. Anuj SN, Whiley DM, Kidd TJ et al (2009) Identification of *Pseudomonas aeruginosa* by a duplex real-time polymerase chain reaction assay targeting the *ecfX* and the *gyrB* genes. Diagn Microbiol Infect Dis 63:127–131
2. Hoiby N, Frederiksen B, Pressler T (2005) Eradication of early *Pseudomonas aeruginosa* infection. J Cyst Fibros 4:49–54

3. Jaffe RI, Lane JD, Bates CW (2001) Real-time identification of *Pseudomonas aeruginosa* direct from clinical samples using a rapid extraction method and polymerase chain reaction (PCR). J Clin Lab Anal 15:131–137
4. Khan AA, Cerniglia CE (1994) Detection of *Pseudomonas aeruginosa* from clinical and environmental samples by amplification of the exotoxin A gene using PCR. Appl Environ Microbiol 60:3739–3745
5. Lavenir R, Jocktane D, Laurent F et al (2007) Improved reliability of *Pseudomonas aeruginosa* PCR detection by the use of the species-specific *ecfX* gene target. J Microbiol Methods 70:20–29
6. Motoshima M, Yanagihara K, Fukushima K et al (2007) Rapid and accurate detection of *Pseudomonas aeruginosa* by real-time polymerase chain reaction with melting curve analysis targeting *gyrB* gene. Diagn Microbiol Infect Dis 58:53–58
7. Qin X, Emerson J, Stapp J et al (2003) Use of real-time PCR with multiple targets to identify *Pseudomonas aeruginosa* and other non-fermenting gram-negative bacilli from patients with cystic fibrosis. J Clin Microbiol 41:4312–4317
8. Spilker T, Coenye T, Vandamme P et al (2004) PCR-based assay for differentiation of *Pseudomonas aeruginosa* from other pseudomonas species recovered from cystic fibrosis patients. J Clin Microbiol 42:2074–2079
9. Syrmis MW, O'Carroll MR, Sloots TP et al (2004) Rapid genotyping of *Pseudomonas aeruginosa* isolates harboured by adult and paediatric patients with cystic fibrosis using repetitive-element-based PCR assays. J Med Microbiol 53:1089–1096
10. Tummler B, Kiewitz C (1999) Cystic fibrosis: an inherited susceptibility to bacterial respiratory infections. Mol Med Today 5:351–358
11. Wellinghausen N, Kothe J, Wirths B et al (2005) Superiority of molecular techniques for identification of Gram-negative, oxidasepositive rods, including morphologically nontypical *Pseudomonas aeruginosa*, from patients with cystic fibrosis. J Clin Microbiol 43:4070–4075
12. Whiley DM, Lambert SB, Bialasiewicz S et al (2008) False-negative results in nucleic acid amplification tests: do we need to routinely use two genetic targets in all assays to overcome problems caused by sequence variation? Crit Rev Microbiol 34:71–76

Chapter 25
Rickettsia

John Stenos, Stephen Graves, and Leonard Izzard

25.1 Introduction

Rickettsial diseases are often undiagnosed because the treating doctor doesn't consider it as part of the differential diagnosis, especially in Australia. A travel history is often helpful. For example, scrub typhus (*Orientia tsutsugamushi*) is only present in northern, tropical Australia; returning sick from a southern African game park may be due to African Tick Typhus (*Rickettsia africae*), following tick bites.

Rickettsial diseases are world wide in their distribution, but in Australia their epidemiology is poorly understood.

The key diagnostic clue is the patient having been bitten by a tick, mite or flea within the past 2–3 weeks. The presenting features are fever, rash (usually maculo-papular) and an eschar (the necrotic site on the skin where the ectoparasite attached). The latter may be in anatomical sites not usually examined by the doctor! (e.g. under the scrotum). While serology is the usual diagnostic modality, it is rarely positive during the first week of the illness. A serum taken during this period is still very valuable as it can be paired (tested in parallel) with a later serum sample, to show a seroconversion to a particular rickettsial antigen and confirmation of the disease as rickettsial.

However, during the first week of illness a positive nucleic acid amplification test is a very useful method of diagnosis. Unfortunately, only reference laboratories currently perform this test in Australia.

25.2 Assay Description

Rickettsia of the Spotted Fever Group (SFG) and Typhus Group (TG) are genetically similar. The conserved nature of the citrate synthase (CS) gene of these rickettsiae was utilised as the target of choice of this assay [1]. Conversely, the *Orientia*

J. Stenos (✉)
Australian Rickettsial Reference Laboratory, Barwon Biomedical Research, The Geelong Hospital, Bellarine Street, Geelong, VIC 3220, Australia; Division of Verterinary and Biomedical Science, Murdoch University, Murdoch, WA 6150, Australia
e-mail: jstenos@unimelb.edu.au

M. Schuller et al. (eds.), *PCR for Clinical Microbiology*,
DOI 10.1007/978-90-481-9039-3_25,

tsutsugamushi species appear to be quite heterogeneous in their genetics and finding a suitable gene for designing a qPCR that would amplify the majority of strains proved difficult. The qPCR assay was designed around the 16 s rDNA gene using Primer Express (Applied Biosystems). A positive chromosomal control together with a reagent and extraction controls were utilised in these assays. The sensitivity of the assay was found to be between 5 and 50 copy numbers. The specificity was tested against a number of other Rickettsiales and medically important bacteria and no amplification was observed. These assays were optimised using a Rotorgene 3000 (Corbett Research) but any open PCR platform would support this methodology.

25.3 Acceptable Specimens

A variety of clinical samples can be used with this assay, such as blood, biopsies and cerebrospinal fluid. However the most common specimen is uncoagulated blood in EDTA tubes.

25.4 Unacceptable Specimens

Clotted blood, serum, swabs and urine.

25.5 Sample Extraction

Blood samples were centrifuged at 4500×*g* for 5 min in order to separate the buffy coat from the packed red cells and plasma. The buffy coat was removed with a plastic transfer pipette and added to 5–10 ml of red blood cell lysis solution (Qiagen). This was incubated for 10 min at 37°C followed by centrifugation at 4500×*g* for 10 min. The supernatant was decanted and the pellet rinsed twice with phosphate buffered saline (PBS) before re-suspending in 100 μl of PBS.

Samples were extracted using a mini prep Genomic DNA extraction kit (RBC). The sample was added to 200 μl of GB Buffer, and incubated, mixing at 900 rpm at 70°C for 10 min in a shaking incubator. 200 μl of glacial ethanol was added to the sample and immediately mixed by rapid pipetting or vortexing for 10 s. Following pulse centrifugation, the sample was added to the column and centrifuged at 14,000×*g* for 5 min, with the flow through being discarded. Five hundred microliters of the wash buffer (containing ethanol) was added to the column and centrifuged at 7500×*g* for 30 s. The flow through was discarded and the wash repeated. An additional centrifugation at 14,000×*g* for 3 min was performed in order to dry the column. Preheated elution buffer (50 μl) was added to the center of the column in a new collection tube and allowed to absorb for 2 min prior to centrifugation at 7500×*g* for 30 s. The eluted DNA was stored at −20°C until tested.

25.6 Primer and Probe Sequences

Although there is scope to offer these assays as a duplex the diseases are often clinically and epidemiologically distinct so each assay is performed singularly. Both the CS and 16SrDNA qPCRs utilise the same primer and probe concentrations.

CS-F (5′-TCG CAA ATG TTC ACG GTA CTT T-3′)
CS-R (5′-TCG TGC ATT TCT TTC CAT TGT G-3′)
CS-Probe (5′-FAM TGC AAT AGC AAG AAC CGT AGG CTG GAT G BHQ1-3′)
16SrDNA-F (5′-CTT ATT TGC CAG CGG GTA ATG C-3′)
16SrDNA-R (5′-GGG CCA TGA TGA CTT GAC CTC-3′)
16SrDNA-Probe (5′-FAM CCC ACC TTC CTC CGG CTT AGC ACC BHQ1-3′)

25.7 PCR Amplification and Analysis

The master mixes each contained 200 nM of each oligonucleotide, 2 mM $MgCl_2$, 2 X Platinum qPCR SuperMix-UDG Mastermix (Invitrogen) and 4 μl of DNA in a total volume of 25 μl. Uracil DNA glycosylase readily destroys amplicons incorporating dUTP in subsequent qPCR runs, minimising the chances of carryover contamination.

The reaction tubes are incubated initially at 50°C for 3 min, followed by 95°C for 5 min. Then 60 cycles of 95°C for 20 s, and 60°C for 40 s. Characteristic sigmoid curves are observed with control and samples that are positive. CTs of less than 40 are considered positive.

Reference

1. Stenos J, Graves SR, Unsworth NB (2005) A highly sensitive and specific PCR assay for the detection of spotted fever and typhus group rickettsiae. Am J Trop Med Hyg 73(6):1083–1085

Chapter 26
Streptococcus pneumoniae

Bronwen Roy and Marcel Leroi

26.1 Summary of Methods

Pyogenic bacterial meningitis is a serious and sometimes fatal condition affecting the CNS. Globally, the three major bacterial pathogens causing meningitis are *Haemophilus influenzae* type b (HIB), *Streptococcus pneumoniae* and *Neisseria meningitidis*. Traditional laboratory diagnosis by culture of these pathogens is slow (>36 h) and non-culture based methods such as PCR are increasingly being utilised.

Common PCR targets for *S. pneumoniae* include the virulence factor genes, pneumolysin (*ply*) and autolysin (*lytA*). The real-time PCR assay described below targets the *ply* gene and can be used as a mono- or multiplex assay (in combination with targets specific for *H. influenzae* type b and *N. meningitidis*) on CSF samples [1].

26.2 Acceptable Specimens

CSF and other sterile fluids are the most suitable specimens for testing with this assay.

26.3 Unacceptable Specimens

It is not appropriate to test respiratory specimens with this assay.

B. Roy (✉)
Department of Microbiology, Nepean Hospital, Penrith, NSW 2750, Australia
e-mail: royb@wahs.nsw.gov.au

M. Schuller et al. (eds.), *PCR for Clinical Microbiology*,
DOI 10.1007/978-90-481-9039-3_26, © Springer Science+Business Media B.V. 2010

26.4 Sample Extraction

DNA is extracted from 200 μl of CSF specimen using the QIAamp DNA Mini Kit (Qiagen Pty Ltd, Australia) according to the manufacturer's instructions. If sample volume allows, the CSF is concentrated down and the pellet resuspended in 200 μl of sterile saline. DNA is eluted from the spin column into 100 μl.

26.5 Primer and Probe Sequences

The primers Forward (5′-TGC AGA GCG TCC TTT GGT CTA T-3′) and Reverse (5′-CTC CTT ACT CGT GGT TTC CAA CTT GA-3′) are used to amplify part of the *ply* gene of *S. pneumoniae.* The PCR product is detected by the hydrolysis probe (5′-6′-FAM-TGG CGC CCA TAA GCA ACA CTC GAA-TAMRA-3′). The primer and probe sequences were described previously by Corless et al. [1].

26.6 PCR Amplification and Product Detection

The PCR reaction is set up in a 12 μl volume containing 5 μl of Quantifast Probe PCR mix (containing HotStar Taq *Plus* DNA polymerase, dNTPs and PCR buffer) (Qiagen Pty Ltd, Australia), 150 nM of Forward and Reverse primers, 12 nM of probe, and 5 μl of extracted DNA. DNA is amplified in a Rotorgene 3000 (Corbett Life Science, Australia) thermal cycler under the following conditions: 95°C for 3 min, followed by 50 cycles of 95°C for 15s and 60°C for 30s. Fluorescence is measured at the end of the 60°C annealing/extension step (expressed as a cycle threshold (Ct) value).

26.7 Quality Control and Validation Data

In each PCR run, a negative control water blank is included to monitor for contamination during PCR set-up. A positive control of *S. pneumoniae* DNA (ATCC 49616) is included to monitor the efficiency of PCR amplification. The positive control should have a Ct value between 25–30 cycles. Each clinical sample is tested in parallel with an inhibition control containing 3 μl of sample and 2 μl of positive control DNA. A Ct value >30 cycles or a negative result in this tube indicates partial or complete inhibition, respectively. Inhibited samples are diluted and re-tested to rule out false negative results.

The specificity of the assay was tested on clinical strains of *S. pneumoniae* ($n = 81$) and '*S. viridans*' ($n = 15$), laboratory control strains of *Staphylococcus aureus* (ATCC 29213), *Escherichia coli* (ATCC 25922), *Neisseria gonorrohoeae* (ATCC 49981), and *Listeria grayi* (ATCC 25401) and wild type strains of *Neisseria meningitidis* Type B and C and *Listeria monocytogenes.* The identification of all

S. pneumoniae isolates was confirmed by optochin sensitivity as defined by standard laboratory protocol. The *ply* PCR was specific for *S. pneumoniae* ($n = 81$). All other bacteria, including '*S. viridans*', tested PCR negative.

The analytical sensitivity of the assay is equivalent to 1 CFU/ml of *S. pneumoniae* (ATCC 49619). We are unable to determine the clinical sensitivity of the assay due to a lack of positive CSF samples available for testing. To date one CSF sample ($n = 50$) infected by enterovirus has given a false positive result for *S. pneuomoniae*, however upon repeat testing the sample tested negative.

26.8 Assay Limitations

Although to date the *ply* gene has been detected in all clinically relevant pneumococcal serotypes, it is not specific for pneumococci. It has also been detected in *Streptococcus mitis* and *Streptococcus oralis* isolates [2, 3]. Given the rarity of such '*Streptococcus viridans*' species in meningitis, it assumed that a positive PCR result using our assay represents identification of *S. pneumoniae*, although in culture negative cases it may be prudent to confirm organism identity by use of another molecular target.

References

1. Corless CE, Guiver M, Borrow R et al (2001) Simultaneous detection of *Neisseria meningitidis*, *Haemophilus influenzae*, and *Streptococcus pneumoniae* in suspected cases of meningitis and septicemia using real-time PCR. J Clin Microbiol 39:1553–1558
2. Deutch S, Moller JK, Ostergaard L (2008) Combined assay for two-hour identification of *Streptococcus pneumoniae* and *Neisseria meningitidis* and concomitant detection of 16S ribosomal DNA in cerebrospinal fluid by real-time PCR. Scand J Infect Dis 40:607–614
3. Whatmore AM, Efstratiou A, Pickerill AP et al (2000) Genetic relationships between clinical isolates of *Streptococcus pneumoniae*, *Streptococcus oralis*, and *Streptococcus mitis*: characterization of 'Atypical' pneumococci and organisms allied to *S. mitis* harboring *S. pneumoniae* virulence factor-encoding genes. Infect Immun 68:1374–1382

Chapter 27
Treponema pallidum

Gerald B. Harnett, Julia A. Cattell, Tracy M. Perris, and Glenys R. Chidlow

27.1 Summary of Methods

Treponema pallidum subsp. *pallidum* (*T. pallidum*) is the agent responsible for venereal syphilis which may present with different clinical conditions at various stages of the infection [1]. *T. pallidum* subsp. *pertenue, endemicum* and *carateum* are closely related to *T. pallidum* and cause the diseases of yaws, endemic syphilis or betel and pinta respectively [1, 2]. Syphilitic lesions can appear in the primary, secondary and relapsing stages and in early congenital infection and treponemes are present in these lesion fluids [2]. Serological methods for the diagnosis of syphilis have gradually been improved over a period spanning many years but these are not usually positive in the early stages of infection when rapid diagnosis is important for prompt treatment to prevent transmission of the infection. *T. pallidum* can be detected in lesions fluids by dark-field microscopy and by fluorescent antibody staining of smears but these methods are subjective and may lack sensitivity. Molecular methods can provide sensitive, objective assays which can detect *T. pallidum* in skin lesion materials even when transportation of samples to the laboratory is delayed. Some molecular methods of detection are proposed [3, 4] but it is not clear which provides the best sensitivity. This study describes a previously published nested PCR method [3] with two in-house real-time TaqMan assays applied to a variety of genital samples. Results for positive samples are confirmed by DNA sequencing.

27.2 Acceptable Specimens

Swabs from skin lesions only.

G.B. Harnett (✉)
Department of Microbiology and Infectious Diseases, Pathwest Laboratory Medicine, QEII Medical Centre, Nedlands, WA 6009, Australia
e-mail: Gerry.Harnett@health.wa.gov.au

M. Schuller et al. (eds.), *PCR for Clinical Microbiology*,
DOI 10.1007/978-90-481-9039-3_27, © Springer Science+Business Media B.V. 2010

27.3 Sample Extraction

Lesion swabs are vortexed in PCR grade water and extracted using the QIAGEN QIAmp Viral RNA Mini Kit which is efficient for both RNA and DNA extractions. Equine herpesvirus (EHV) is introduced into the kit lysis buffer to provide a control of DNA extraction and presence of PCR inhibitors.

27.4 Primers and Probe Sequences

See Table 27.1 for details of primer sequences

Table 27.1 The primer sequence sets used in the nested PCR and primer and probe sequences used in the two, in-house, real-time PCR TaqMan assays

Target gene	Primers and probe sequences	Product size (bp)
TyF1-F outer	5′-TGTACCGGACGCTCGTGCCATT -3′	428
TyF1-R outer	5′-TGGCCTTTCCCAACGTCCTCAG -3′	
		364
TyF1-F inner	5′-GCTGCTATCTGCGAGCAATTG-3′	
TyF1-R inner	5′-TCAGTCACTGCATCCCCACTTTCAG-3′	
TprJ-F	5′-TGAAACAGGATCTGGCAGATTTAG -3′	70
TprJ-R	5′-TGGACACTCCGCGAAAAACG -3′	
TprJ-Probe	5′-6FAM-TTATTTCCGTTCGCA-MGBNFQ-3′	
PolA-F	5′-TGCGCGTGTGCGAATG-3′	55
PolA-R	5′-GATGCAATCCATCCGTTTCAC-3′	
PolA-Probe	5′-6FAM-TCAAACGCGACCACA-MGBNFQ-3′	
EHV-F	5′-GATGACACTAGCGACTTCGA-3′	81
EHV-R	5′-CAGGGCAGAAACCATAGACA	
EHV-Probe	5′-6FAM-TTTCGCGTGCCTCCTCCAG-BHQ-3′	

Primers were supplied by Sigma-Aldrich, Australia and TaqMan MGB probes by Applied Biosystems, Australia. The EHV probe was supplied by Biosearch Technologies USA.

27.5 PCR Amplification and Product Detection

Nested PCR: After extraction, 8 μl of eluate is added to the 1st round PCR tubes containing 0.2 μM of the outer primers listed in Table 27.1, 1x PCR buffer, 0.2 mM of dNTPs, 1.5 mM of $MgCl_2$, 0.01% BSA and 0.5 units of AmpliTaqGold DNA polymerase (Applied Biosystems, Australia), in a total reaction volume of 20 μl. After denaturation and enzyme activation for 10 min at 95°C, the tubes are cycled 45 times at 95°C for 30 s, 55°C for 30 s and 72°C for 45 s in a GeneAmp 2700 thermocycler (Applied Biosystems, Australia).

After cycling, 0.5 μl of the first round products is transferred to a second round tube containing PCR mix as above but with 0.2 μM of inner primers (Table 27.1) in a 20 μl reaction volume. These tubes are cycled as above and the PCR products subjected to ethidium bromide gel electrophoresis. Products producing bands of 364 bp are scored as positive and their identity confirmed by DNA sequencing.

Real-time PCR: Extracts (8 μl) are added to tubes of PCR mix containing 0.2 μM of primers and 0.1 μM of TaqMan probe for each of the gene targets (Table 27.1), 1x PCR buffer, 0.2 mM of dNTPs, 4 mM of $MgCl_2$, 0.01% BSA and 0.75 units of AmpliTaq Gold DNA polymerase (Applied Biosystems, Australia), in a total reaction volume of 20 μl. The tubes are cycled in a real-time RotorGene 6000 instrument (Corbett, Australia) and probe emissions acquired by the FAM wavelength channel. The cycling programme consist of an enzyme activation and denaturation period of 10 min at 95°C, followed by 50 cycles of 12 s at 94°C, 30 s at 55°C and 20 s at 72°C.

27.6 Quality Control and Validation Data

A positive control at 10x the detection limit is included with each batch of tests. Water blanks, which had been through the extraction process, are inserted between every fifth sample tube. Ct values for the positive controls are plotted on a Shewhart chart, with lines set at 2 standard deviations from the mean of 100 test batches, to give a 95% confidence limit for the batch of tests. Each sample extract is tested in a real-time assay for EHV and the Ct values indicated whether the DNA extraction has been satisfactory and PCR inhibitors removed.

To validate the assay, samples from 144 patients were tested in all three assays. These comprised mainly genital lesion samples from males but included two male oral swabs and two female perianal samples and a vaginal sample which were all positive for *T. pallidum*. The nested PCR for the *TyF*1 gene detected 65 positive samples each of which was confirmed by DNA sequencing. Samples positive in both of the TaqMan PCRs were also considered as confirmed positives. The *Tpr*J gene PCR detected 68, while the *Pol*A gene PCR detected 62 positive samples. There were no apparent false positive results in any of the assays. Overall, the *Tpr*J TaqMan assay had the best performance having a sensitivity and NPV of 100%. For the *Pol*A TaqMan the sensitivity was 91.2% and NPV was 96.7%. The nested PCR had a sensitivity of 95.6% and NPV of 96.2% and had the advantage that positive results could be confirmed by DNA sequencing if considered necessary.

The limit of detection for the assays was determined by Probit analysis. Replicates of closely spaced dilutions of *T. pallidum* positive sample material were tested in each of the assays and the results processed using the StatsDirect software package. The analytical sensitivities for the nested PCR, the *Tpr*J and *Pol*A TaqMan were 650, 430 and 470 target copies/ml respectively at a 95% confidence level. The reaction efficiencies of the *Tpr*J and *Pol*A TaqMan PCRs were 97% and 107% respectively.

27.7 Assay Limitations

The results of this study indicate that the TaqMan assay for the *Tpr*J gene is the best of the three assays tested, in terms of sensitivity and NPV. It also avoids many of the cross-contamination risks associated with nested PCR assays. The gold standard for the detection of *T. pallidum* is probably rabbit inoculation [1], but this is impractical in a routine clinical microbiology setting. To further validate these PCR methods it would be necessary to compare the PCR results with the serological outcomes of individual patients.

Acknowledgments We would like to thank Natasha K. Ryan who provided valuable technical assistance and A. Levy for the statistical analysis.

References

1. Hay PE, Clarke JR, Strugnell RA et al (1990) Use of the polymerase chain reaction to detect DNA sequences specific to pathogenic treponemes in cerebrospinal fluid. FEMS Microbiol Lett 68:233–238
2. Larsen SA, Steiner BM, Rudolph AH (1995) Laboratory diagnosis and interpretation of tests for syphilis. Clin Microbiol Rev 8:1–21
3. Leslie DE, Assato F, Karapanagiotidis T et al (2007) Development of a real-time PCR assay to detect *Treponema pallidum* in clinical specimens and assessment of the assay's performance by comparison with serological testing. J Clin Microbiol 45:93–96
4. Norris SJ, Pope V, Johnson RE et al (2003) Treponema and other human host-associated spirochetes. In: Murray PR et al (ed) Manual of Clinical Microbiology, 8th edn. ASM Press, USA

Chapter 28
Universal Bacterial Identification by PCR and DNA Sequencing of 16S rRNA Gene

Greg James

28.1 Summary of Method

The assay is a single amplification PCR incorporating three separate reactions. The primers used are broad range primers which recognise conserved sequences within the 16S rRNA gene and amplify the intervening variable regions. The variable portions of the 16S rRNA gene provide unique signatures that can be analyzed to provide an identification of the bacteria species in the sample.

28.2 Introduction

The identification of bacteria culture isolates by analysis of gene sequence is routine in many laboratories. Genes such as; the ribosomal 16S rRNA, 23S rRNA and 16S-23S rRNA internal transcribed sequences, *rpoB* (encoding β subunit of RNA polymerase), *groEL* (encoding heat-shock protein), *gyrB* (encoding β subunit of DNA gyrase) and recombination encoding *recA* are found in virtually all bacteria and have been used for identification using conserved sequences as universal primers in PCR [2]. DNA sequence identification is particularly useful for organisms that are difficult to culture or are slow growing on laboratory media and those that are relatively inert biochemically or produce variable reactions in phenotypic identification schemes.

Detection and identification of bacteria directly from specimens using universal gene targets is compromised due to the presence of other contaminating bacteria in the sample introduced during; collection, specimen processing and PCR set-up. These contaminating bacteria will also produce amplified DNA in the PCR and contribute to a mixed DNA sequence that cannot be interpreted. For this reason PCR assays using universal primers should only be performed on specimens from normally sterile sites and only where bacteria are visible by microscopy. If

G. James (✉)
Centre for Infectious Diseases and Microbiology Laboratory Services (CIDMLS), Institute for Clinical Pathology and Medical Research (ICPMR), Westmead, NSW 2145, Australia
e-mail: Greg.James@swahs.health.nsw.gov.au

M. Schuller et al. (eds.), *PCR for Clinical Microbiology*,
DOI 10.1007/978-90-481-9039-3_28,

organisms are not visible, detection and identification of the causative bacterium may be questionable.

Guidelines for the use of DNA sequencing for the identification of microorganisms have been proposed by Clinical and Laboratory Standards Institute [1]. This publication should be reviewed prior to development of a universal identification system using PCR and DNA sequencing.

The 16SrRNA gene is the most common target used to identify bacterial isolates and the number of sequence depositions into databanks such as GenBank [http://www.ncbi.nlm.nih.gov/GenBank] and RIDOM [http://www.ridom_rdna.de] is increasing at a rapid rate.

28.3 Acceptable Specimens

The assay should only be performed on culture isolates. Detection and identification of bacteria from normally sterile site specimens such as blood, CSF, aspirates from joints and tissue may be attempted when organisms are visible by microscopy or when there is histological evidence of a bacterial infection. When using the assay on specimens the result must clearly state that the assay result must be interpreted along with other clinical and laboratory evidence.

28.4 Unacceptable Specimens

Specimens that are not from normally sterile sites, formalin fixed tissue and paraffin embedded or fresh tissue samples where evidence of infection has not been observed by microscopy.

28.5 Sample Extraction

Varies with sample type. Culture isolates can be prepared for PCR by suspending a colony in 1 ml of water and boiling for 10 min. Following a pulse centrifugation for 10 s the supernatant can be used in the PCR. Specimens require processing dependant on the site of collection. DNA is purified from specimens using commercial automated and manual specimen processing systems such as BioMerieux EasyMag and Roche High Pure kits according to manufacturers instructions.

28.6 Primer Sequences

The primers were derived from Relman [3];

Universal 1 (U1) 5′-ACG CGT CGA CAG AGT TTG ATC CTG GCT-3′
Short Universal 1 (U1R) 5′-GGA CTA CCA GGG TAT CTA AT-3′

Universal 2 (U2) 5′-CGC GGA TCC GCT ACC TTG TTA CGA CTT-3′
Universal 3 (U3) 5′-AGT GCC AGC AGC CGC GGT AA-3′
Universal 4 (U4) 5′-AGG CCC GGG AAC GTA TTC AC-3′
Universal 5 (U5) 5′-TCA AAK GAA TTG ACG GGG GC-3′ (K = G+T)

Primers are used in the following combinations to amplify the 16S rRNA gene;

Culture isolates; U1+U2, U3+U4 and U5+U4.
Specimens; U1R+U1, U3+U4 and U5+U4.

28.7 PCR Amplification

Each master mix (MM) tube contains the components for 10 PCR reactions of 25 μl. Each PCR tube requires 15 μl of MM for a 25 μl PCR. Add 10 μl of sample DNA to each respective PCR tube.

Note: It is advisable, when making up the MMs, to irradiate all tubes and MM components under UV light for 10 min in a biological safety cabinet. As this assay uses universal primers the presence of any extraneous bacterial DNA will introduce contamination to the assay.

Component U1 + U2 PCR MM	1 tube	1 MM
H_2O	5.425 μl	54.25 μl
5x Promega Green GoTaq® Flexi Buffer	5.00 μl	50.0 μl
4.0 mM $MgCl_2$ (25 mM stock)	4.00 μl	40.0 μl
200 μM dNTP's (25 mM stock)	0.20 μl	2.0 μl
350 nM U1 primers (100 μM stock)	0.0875 μl	0.875 μl
350 nM U2 primers (100 μM stock)	0.0875 μl	0.875 μl
Promega Taq DNA polymerase (5 U/μl)	0.20 μl	2.0 μl
Total volume	15.00 μl	150.0 μl

Component U1R + U1 PCR MM	1 tube	1 MM
H_2O	6.475 μl	64.75 μl
5x Promega Green GoTaq® Flexi Buffer	5.00 μl	50.0 μl
3.0 mM $MgCl_2$ (25 mM stock)	3.00 μl	30.0 μl
200 μM dNTP's (25 mM stock)	0.20 μl	2.0 μl
250 nM U1R primer (100 μM stock)	0.0625 μl	0.625 μl
250 nM U1 primer (100 μM stock)	0.0625 μl	0.625 μl
Promega Taq DNA polymerase (5 U/μl)	0.20 μl	2.0 μl
Total volume	15.00 μl	150.0 μl

Component U3+U4 and U5+U4 PCR MM	1 tube	1 MM
H_2O	6.475 μl	64.75 μl
5x Amplification buffer (Promega)	5 μl	50.0 μl
3.0 mM $MgCl_2$ (25 mM stock)	3.00 μl	30.0 μl
200 μM dNTP's (25 mM stock)	0.20 μl	2.0 μl
250 nM (U3), or (U5) primers (100 μM stock)	0.0625 μl	0.625 μl
250 nM (U4) primers (100 μM stock)	0.0625 μl	0.625 μl
Promega Taq DNA polymerase (5 U/μl)	0.20 μl	2.0 μl
Total volume	15.00 μl	150.0 μl

Amplification is performed in a conventional thermocycler (Eppendorf Mastercycler). Initial denaturation at 94°C for 3 min followed by 30 cycles of 96°C for 15 s, 60°C for 1.5 min and 72°C for 2 min. A final extension step at 72°C for 5 min allows all amplicons to be fully extended.

28.8 Product Detection

PCR amplified DNA segments are separated by electrophoresis in a 2% agarose gel and visualised using SYBR Safe DNA stain (Invitrogen Life Technologies). Product size is determined by comparison with the Positive Control and a DNA size ladder.

28.9 Pre-sequencing Treatment

Prior to DNA sequencing, PCR product is cleaned of unused dNTP's and primer using two hydrolytic enzymes, shrimp alkaline phosphatase and exonuclease 1 according to manufacturer's instructions (ExoSAP-IT, USB Affymetrix).

28.10 DNA Sequencing

Is performed by a commercial service provider and requires approximately 10 ng/μl of DNA per sample and 3 pM/μl of sequencing primer (U1, U3 or U5). Sequencing reagent variables such as amount of DNA and primer concentrations must be optimized with each provider.

28.11 DNA Sequence Analysis

DNA sequences should be edited using appropriate software (such as Chromas), to produce a clean, easily read sequence of maximum length. Where possible, sequences from all three PCRs (U1, U3 and U5) should be edited and combined to produce a single continuous read. If not possible, as a minimum, U1 or U3 + U5 must be readable to obtain a result. Sequences should then be copied into sequence analysis software (such as BioManager) and analyzed using a BLAST search against appropriate databases.

28.12 Reporting

Refer to CLSI [1]. The percentage match with DNA sequences deposited in databases required for identification vary with genus and species. Generally a >98% match is required and the DNA sequence identification must correlate with other results such as; gas chromatography fatty acid analysis, organism morphology and growth characteristics, conventional biochemical and antimicrobial resistance profiles. The identification must also be discussed with the referring doctor to determine if the identification matches the clinical diagnosis.

28.13 Quality Control

Positive control: A bacterium species, rarely if ever encountered in laboratory specimens, of known sequence must be included as a positive control in each PCR assay to ensure the assay is working to specifications and as a control for subsequent DNA sequencing. If the full and correct DNA sequence for the positive control organism is not obtained then the assay is not valid and must be repeated.

Negative controls: As universal primers are used in this assay it is important to include no DNA controls (NDCs) for each sample at each stage of the process (Sample processing and PCR set-up) to determine the possibility of contamination. Add each representative Sample Processing NDC tube first in the same sequence that samples will be subsequently added followed by the first sample. The sample is added diluted 1:100 in water (to dilute possible inhibitory substances) and undiluted. Following addition to the first sample tubes, next add the respective PCR set-up NDC. This tube will show if any contaminating bacteria may have entered tubes during addition of the first sample. The addition process is then repeated for the next sample and so on.

After the last sample and respective RCR set-up NDC has been added, the Positive Control DNA is added to the 'Spike' tubes that contain equal volumes of sample diluted 1:100 and Positive Control diluted to a level that is two dilutions above the detection limit of the assay. Finally the positive control dilution series tubes have Positive Control added from the most dilute to the least dilute to demonstrate the detection limit of the assay and to provide a monitor of assay reproducibility.

Any PCR product formed in NDC must be DNA sequenced and compared to the sequence of the sample. If the NDC sequence is present in the sample sequence the assay is not valid and must be repeated. Any Spike tube that does not form a

PCR product shows that the respective sample was inhibitory. The sample must be re-extracted and re-tested.

28.14 Assay Limitations

The ubiquitous nature of bacteria and the potential for exogenous bacteria to contaminate samples and produce either an incorrect or spurious result or mixed DNA sequence limit the application of this assay to highly experienced staff and for practical purposes, the identification of culture isolates.

References

1. Clinical and Laboratory Standards Institute (2007) Interpretive criteria for microorganism identification by DNA target sequencing; proposed guideline. 27(22):MP18-P
2. Drancourt M, Raoult D (2005) Sequence based identification of new bacteria: a proposition for creation of an orphan bacterium repository. J Clin Microbiol 43:4311–4315
3. Relman DA (1993) Universal bacterial 16S rRNA amplification and sequencing. In: Persing DH, Smith TF, Tenover FC, White TJ (eds) Diagnostic Molecular Biology, Principles and Applications. Mayo Foundation Rochester, MN, USA, pp 489–495

Part IV
PCR Protocols for Viral Pathogens

Chapter 29
Adenovirus

Carl Kirkwood

29.1 Summary of Methods

Detection of adenovirus infection can be by monoclonal-based enzyme linked immunosorbent assay (ELISA) or cell culture followed by specific adenovirus detection by immunofluorescent based staining, ELISA or PCR confirmation. Techniques commonly used for adenovirus characterisation include PCR detection followed by the use of restriction enzyme analysis, or type specific probes or primers [1–3]. Multiplex PCR using species specific primers to the fiber gene is a popular method used to conduct adenovirus characterization and typing [3, 4].

29.2 Organism

Adenoviruses are medium-sized (90–100 nm), nonenveloped (naked) icosahedral viruses composed of a nucleocapsid and a double-stranded linear DNA genome. There are 53 described serotypes in humans, which are responsible for 5–10% of upper respiratory infections in children, and many infections in adults as well. The adenovirus genome is linear, non-segmented double stranded (ds) DNA which is between 26 and 45 kbp.

Although epidemiologic characteristics of the adenoviruses vary by type, all are transmitted by direct contact, fecal-oral transmission, and occasionally waterborne transmission. Some types are capable of establishing persistent asymptomatic infections in tonsils, adenoids, and intestines of infected hosts, and shedding can occur for months or years. Other types cause sporadic infection and occasional outbreaks as is seen with adenovirus serotypes 8, 19, and 37 associated with epidemic keratoconjunctivitis. Epidemics of febrile disease with conjunctivitis are associated with waterborne transmission of some adenovirus types, often centering around

---

C. Kirkwood (✉)
Enteric Virus Research Group, Murdoch Childrens Research Institute, Royal Children's Hospital, Parkville, VIC 3052, Australia
e-mail: carl.kirkwood@mcri.edu.au

M. Schuller et al. (eds.), *PCR for Clinical Microbiology*,
DOI 10.1007/978-90-481-9039-3_29,

inadequately chlorinated swimming pools and small lakes. Enteric adenoviruses 40 and 41 cause gastroenteritis, usually in children.

29.3 Sample Preparation and DNA Extraction

Clinical specimens (throat or nasal swabs or faecal samples) are passaged in A549 cells prior to genetic analysis. Adenovirus DNA is extracted from cell culture adapted strains using Qiagen miniElute virus kit (Qiagen, Hilden, Germany) according to the manufacturer's instructions, and eluted into 50 μl of dH_2O.

29.4 Primer Sequences

Primer (sense)	Sequence (5′–3′)	Product size (bp)
A1 (+)	GCTGAAGAAMCWGAAGAAAATGA	1444–1537
A2	CRTTTGGTCTAGGGTAAGCAC	
B1 (+)	TSTACCCYTATGAAGATGAAAGC	670–772
B2	GGATAAGCTGTAGTRCTKGGCAT	
C1 (+)	TATTCAGCATCACCTCCTTTCC	1988–2000
C2	AAGCTATGTGGTGGTGGGGC	
D1 (+)	GATGTCAAATTCCTGGTCCAC	1205–1221
D2	TACCCGTGCTGGTGTAAAAATC	
E1 (+)	TCCCTACGATGCAGACAACG	967
E2	AGTGCCATCTATGCTATCTCC	
F1 (+)	ACTTAATGCTGACACGGGCAC	541–586
F2	TAATGTTTGTGTTACTCCGCTC	

29.5 PCR Amplification and Product Detection

The PCR amplification is performed in a 50 μl volume. This consists of 5 μl of extracted DNA added to a PCR reaction mixture of 45 μl. The reaction mix contains a pool of species specific primers used to identify the various human adenovirus types.

The 50 μl master mix contains 32.25 μl dH_20, 5 μl 10x PCR buffer, 3 μl 25 mM $MgCl_2$, 0.5 μl dNTP (20 mM each), 1 μl Adenovirus primer mix (each at 10 μM), 0.2 μl AmpliTaq (5 Units/μl), 5 μl Template-DNA.

Into each tube 45 μl of the master mix is added followed by 5 μl of extracted adenovirus DNA. After a brief vortex and spin, samples are placed into the thermal cycler. The DNA is denatured by heating at 94°C for 5 min. Then each sample is run for 30 cycles of 94°C for 1 min, 54°C for 45 s, 68°C for 2 min. A final extension of 72°C for 5 min is undertaken.

PCR products are separated in 2% (w/v) agarose gels in 0.5% TBE buffer and stained with ethidium bromide. PCR bands are visualized by UV light, and assigned a specific adenovirus species based on the species specific DNA sizes.

References

1. Gray GC, McCarthy T, Lebeck MG et al (2007). Genotype prevalence and risk factors for severe clinical adenovirus infection, United States 2004–2006. Clin Infect Dis 45:1120–1131
2. Lu X, Erdman DD (2006) Molecular typing of human adenoviruses by PCR and sequencing of a partial region of the hexon gene. Arch Virol 151:1587–1602
3. Selvaraj G, Kirkwood C, Bines J et al (2006) Molecular epidemiology of adenovirus isolates from patients diagnosed with intussusception in Melbourne, Australia. J Clin Microbiol 44:3371–3373
4. Xu W, McDonough MC, Erdman DD (2000) Species-specific identification of human adenoviruses by a multiplex PCR assay. J Clin Microbio. 38:4114–4120

Chapter 30
Cytomegalovirus (CMV)

Jonathan Howard

Two different CMV PCRs are performed on each sample to minimise false negative results.

30.1 CMV gp58 PCR

One method targets the gp58 fragment region of glycoprotein B (gB) gene of CMV. This region is highly conserved and is important in viral infectivity and eliciting immune response [5]. Primers that amplify the gB fragment region of CMV may be more sensitive for the diagnosis of CMV infection with a detection rate of 100% compared to studies targeting major immediate early and late antigen region of CMV [2, 3].

This is a nested PCR requiring two rounds of amplifications using a MyCycler Thermal Cycler (Bio-Rad, Australia) with outer primers (gB1 and gB2) for the first rounds and gB3 and gB4 inner primers for the second round. The first round PCR product size is 150 bp and 100 bp for the second round.

30.1.1 First Round Amplification

Nuclease-free water	6.2 μl
5X GoTaq Hot Start buffer	5.0 μl
25 mM MgCl 2	1.5 μl
dNTPs (1 mM)	5.0 μl
10 μM Primer (gB1)	0.625 μl
10 μM Primer (gB2)	0.625 μl
GoTaq Hot Start polymerase	0.3 μl
Template	5.0 μl

J. Howard (✉)
SEALS Virology Division, Prince of Wales Hospital, Randwick, NSW 2031, Australia
e-mail: jonathan.howard@sesiahs.health.nsw.gov.au

M. Schuller et al. (eds.), *PCR for Clinical Microbiology*,
DOI 10.1007/978-90-481-9039-3_30, © Springer Science+Business Media B.V. 2010

The first round of amplification involved an initial denaturation step at 94°C for 2 min followed by 35 cycles of 94°C for 30 s, 55°C for 30 s and 72°C for 30 s with a final extension at 72°C for 5 min and a 4°C hold cycle.

30.1.2 Second Round Amplification

Nuclease-free water	6.2 μl
5X GoTaq Hot Start buffer	5.0 μl
25 mM $MgCl_2$	1.5 μl
dNTPs (1 mM)	5.0 μl
10 μM Primer (gB3)	1.0 μl
10 μl Primer (gB4)	1.0 μl
GoTaq Hot Start polymerase	0.3 μl
First round PCR product	5.0 μl

The second round of amplification included an initial denaturation at 94°C for 2 min followed by 30 cycles of 94°C for 30 s, 53°C for 30 s and 72°C for 30 s with 5 min final extension at 72°C and a 4°C hold.

30.2 CMV MIE PCR

The second method for detection of CMV targets the major immediate early (MIE). MIE is an early marker of viral replication.

This PCR is nested PCR requiring two rounds of amplifications using a MyCycler Thermal Cycler (Bio-Rad, Australia) with outer primers (CMVOF and CMVOR) for the first rounds and CMVIF and CMVIR inner primers for the second round [4]. The first round PCR product size is 396 bp and 229 bp for the second round.

30.2.1 First Round Amplification

Nuclease-free water	5.7 μl
5X GoTaq Hot Start buffer	5.0 μl
25 mM $MgCl_2$	2.0 μl
dNTPs (1 mM)	5.0 μl
10 μM Primer (CMVOF)	1.0 μl
10 μM Primer (CMVOR)	1.0 μl
GoTaq Hot Start polymerase	0.3 μl
Template	5.0 μl

The first round of amplification involved an initial denaturation step at 94°C for 2 min followed by 30 cycles of 94°C for 45 s, 58°C for 45 s and 72°C for 60 s with a final extension at 72°C for 7 min and a 4°C hold cycle.

30.2.2 Second Round Amplification

Nuclease-free water	6.2 μl
5X GoTaq Hot Start buffer	5.0 μl
25 mM $MgCl_2$	1.5 μl
dNTPs (1 mM)	5.0 μl
10 μM Primer (CMVIF)	1.0 μl
10 μM Primer (CMVIR)	1.0 μl
GoTaq Hot Start polymerase	0.3 μl
First round PCR product	5.0 μl

The second round of amplification included an initial denaturation at 94°C for 2 min followed by 30 cycles of 94°C for 30 s, 58°C for 45 s and 72°C for 45 s with 7 min final extension at 72°C and a 4°C hold.

30.3 Primer Sequences

Primer	Product size (bp)	Target gene	Limit of detection (copies per reaction)
CMVOF 5′ AAG GTT CGA GTG GAC ATG GT 3′	396	MIE	2
CMVOR 5′ CAG CCA TTG GTG GTC TTA GG 3′			
CMVIF 5′ GAG CCT TTC GAG GAG ATG AA 3′	229		
CMVIR 5′ GGC TGA GTT CTT GGT AAA GA 3′			
Reference: [4]			
gB1 5′ GAG GAC AAC GAA ATC CTG TTG GGC A 3′	150	gp58	10
gB2 5′ GTC GAC GGT AGA TAC TGC TGA GG 3′			
gB3 5′ ACC ACC GCA CTG AGG AAT GTC AG 3′	100		
gB4 5′ TCA ATC ATG CGT TTG AAG AGG TA 3′			
Reference: [1, 3]			

All PCR reagents were supplied by Promega (Australia).

These protocols have been used in screening of CMV in the following samples:

Newborn screening cards, fresh placenta, fresh umbilical cord, amniotic fluid, urine, serum, plasma, buffy coats, formalin-fixed, paraffin-embedded tissues.

30.4 Sample Extraction

For the newborn screening cards (NBSC), DNA extraction was employed as previously described [1, 3]. Three disks of 3 mm in diameter were prepared from NBSC using a Wallac 1296-071 DBS Puncher (PerkinElmer, Finland). Carryover DNA contamination was excluded by punching 30 disks from a clean, blank NBSC after punching each sample card. Three blank disks (as negative controls) were processed as test samples to detect any carryover DNA between cards. Blood was eluted from the NBSC by incubating in 45 μl minimal essential media at 55°C for 60 min prior to boiling at 100°C for 7 min. Samples were rapidly cooled, centrifuged at 10,000 × *g* for 3 min and frozen at −80°C for at least 1 h before further testing.

For the other samples, total nucleic acid was extracted using MagNA Pure LC System (Roche, Australia) using MagNA Pure Total Nucleic Acid Extraction kit following the manufacturer's protocol.

30.5 Detection of PCR Products

Products were subjected to electrophoresis on 2% agarose gel containing ethidium bromide (0.5 μg/ml). CMV positive DNA and negative water controls were also included with every reaction.

References

1. Barbi M et al (2000) Cytomegalovirus DNA detection in Guthrie cards: a powerful tool for diagnosing congenital infection. J Clin Virol 17:159–165
2. Barbi M et al (1996) Diagnosis of congenital cytomegalovirus infection by detection of viral DNA in dried blood spots. Clin Diagn Virol 6:27–32
3. Howard J et al (2009) Utility of newborn screening cards for detecting CMV infection in cases of stillbirth. J Clin Virol 44:215–218
4. McIver CJ et al (2005) Development of multiplex PCRs for detection of common viral pathogens and agents of congenital infections. J Clin Microbiol 43:5102–5110
5. Utz U et al (1989) Identification of a neutralizing epitope on glycoprotein gp58 of human cytomegalovirus. J Virol 63:1995–2001

Chapter 31
Dengue Virus

Greg Smith

31.1 Summary of Methods

Dengue Virus (DENV) diagnosis can be performed by isolation of DENV from blood or autopsy samples; demonstration of a four-fold or greater rise in reciprocal IgG or IgM antibody titres to one or more DENV in paired serum samples (acute and convascent); use of the newer rapid diagnostic test kit (also differentiates between primary and secondary dengue infections – these DENV rapid test devices are a solid phase immuno-chromatographic assay for the rapid, qualitative and differential detection of dengue IgG and IgM antibodies to dengue fever virus in human serum, plasma or whole blood); demonstration of DENV antigen in autopsy, serum or CSF samples by immunohistochemistry, immunofluorescence or ELISA methods; demonstration of DENV specific nucleic acid sequences in autopsy, serum, CSF samples by PCR.

31.2 Organism

DENV is a ssRNA positive-strand virus of the family Flaviviridae; genus Flavivirus and there are four serotypes (DEN-1, DEN-2, DEN-3, and DEN-4). The virus has a genome of about 11,000 bases that codes for three structural proteins, C, prM, E; seven nonstructural proteins, NS1, NS2a, NS2b, NS3, NS4a, NS4b, NS5; and short non-coding regions on both the 5′ and 3′ ends.

DENV is the cause of dengue fever and dengue hemorrhagic fever (DHF). These are acute febrile diseases, found in the tropics, and caused by four closely related Dengue virus serotypes within the genus Flavivirus and family Flaviviridae. Epidemics caused by multiple serotypes (hyperendemicity) can occur and each serotype is sufficiently different that there is no cross-protection. It is also known as

---

G. Smith (✉)
Australian Animal Health Laboratory, CSIRO Livestock Industries, Geelong, VIC 3220, Australia
e-mail: Greg.A.Smith@csiro.au

M. Schuller et al. (eds.), *PCR for Clinical Microbiology*,
DOI 10.1007/978-90-481-9039-3_31, © Springer Science+Business Media B.V. 2010

breakbone fever. It occurs widely in the tropics, including northern Australia, some of the Pacific region, the entire Asia/South-east Asian region and South America. Dengue is just as prevalent in the urban districts as in rural areas as it is transmitted to humans by the *Aedes aegypti* or more rarely the *Aedes albopictus* mosquito, which feed during the day and can breed in water sitting anywhere around homes, bush or jungle areas. Dengue may also be transmitted via infected blood products (blood transfusions, plasma, and platelets), but the scale of this problem is currently unknown.

The World Health Organization says around two fifths of the world's population are at risk from dengue and estimates that there may be 50 million cases of dengue infection worldwide every year and the disease is now epidemic in more than 100 countries. Refer to WHO website: http://www.who.int/csr/resources/publications/dengue/Denguepublication/en/

31.3 Acceptable Specimens

Serum is the specimen of choice for dengue virus diagnosis and this allows molecular, antigen (NS1) and serology (IgM) testing to be completed. Samples should be stored and transported at 4°C. In acutely ill patients it is recommended that all three tests be performed. While dengue viral RNA may be detectable up to 10 days post onset of symptoms in some individuals, it is most reliable between days one and five post onset.

31.4 Sample Extraction

This laboratory has employed both automated magnetic bead based kits (*Qiagen EZ1 Virus Mini Kit V2.0, Qiagen*) and silica-gel based kits for use on automated platforms (Qiagen *RNeasy 96 BioRobot 9604 Kit, Qiagen*) or manual centrifugation based approaches (*QIAamp Viral RNA Mini Kit, Qiagen*). The extractions are performed according to the manufacturer's instructions. 5.0 μl of recovered RNA is added to each real-time PCR reaction with a final reaction volume of 25 μl.

31.5 Primer and Probe Sequences

Redundancies have been incorporated into the primer design to account for observed sequence divergence.

Forward primer: 5′AAGGACTAGAGGTTAKAGGAGACCC3′;
Reverse primer 5′CGYTCTGTGCCTGGAWTGATG3′;
Probe 5′$_{\text{FAM}}$TCTGGTCTTTCCCAGCGTCAATATGCTGTT$_{\text{TAMRA}}$3′.

31.6 Amplification Conditions

A dengue group specific assay targeting the conserved 3′-UTR is used routinely [5]. There have been several modifications introduced since the method was published and these are detailed below. Where required samples reactive in this test are tested using additional typing assays to establish the serotype identity of the infecting dengue virus.

The group-specific assay is a real-time PCR assay (TaqMan) employing a FAM-labelled probe. The assay targets the conserved 3′-UTR region of the dengue genome. The *Superscript III Platinum One-Step Quantitative RT-PCR System* (Invitrogen) is used according to the manufacturers instructions with a final concentration of forward and reverse primers of 300 nM and a final probe concentration of 150 nM.

Following a 5 min 50°C reverse transcription step, samples are denatured at 94°C for 2 min followed by 40 cycles of 95°C for 3 s and 60°C for 30 s. Samples with Ct values of <38 and compliant curves are reported as ‘detected’.

31.7 Quality Control and Validation Data

This laboratory uses a synthetic probe control and a synthetic primer control to eliminate the possibility of generating a false positive result due to inadvertent contamination [4]. The current, modified approach employs synthetic DNA probe and primer controls (without a T7 RNA polymerase binding site) in conjunction with an in house developed real-time PCR assay for bovine viral diarrheal virus. The latter serves as both a reverse transcriptase control and an extraction control.

The probe control is comprised of a single synthetic oligonucleotide: 5′-TGC ACC ACC AAC TGC TTA GAT CTG GTC TTT CCC AGC GTC AAT ATG CTG TTA GAA CAT CAT CCC TGC ATC C-3′.

The primer control sequence is: 5′-AAG G AC TAG AGG TTA GAG GAG ACC CAC AGA AGA CTG TGG ATG GCC CCT CAA ATA GCA TCA ATC CAG GCA CAG ATC G-3′.

No cross reactivity was observed when an extensive panel of clinical samples and culture supernatants including; Murray Valley encephalitis virus, Japanese encephalitis virus, West Nile (New York) virus, West Nile (Kunjin virus), Stratford virus, Edge Hill virus, Alfuy virus, Kokobera virus, yellow fever virus, Chikungunya virus, Ross River virus or Barmah Forest virus were examined.

The sensitivity and specificity of the assay in its current form was evaluated using a test panel which included ten viraemic patient serum for each of the four dengue serotypes. The assay detected dengue virus RNA in all 40 of the acutely ill dengue patients but none of ten healthy controls.

The assay has been in use in this laboratory since its introduction in 2001. In that period it has been used extensively during multiple dengue outbreaks in North Queensland and countries within the Western Pacific. A total of 5800 samples have been tested of which there have been 842 reactive samples. These reactive results have been confirmed using either virus isolation, the nested PCR tying assay of Lanciotti et al. [3] or separate real time PCR typing assays for each serotype. The 842 reactive patient samples included 464 dengue 3, 215 dengue 2, 96 dengue 1 and 67 dengue 4 RNA positive samples. The infecting serotype in the majority of these samples were also confirmed by parallel IgM serotyping assays.

31.8 Assay Limitations and Alternative Protocols

As with any nucleic acid based protocol a 'not detected' result should not be considered as evidence that the patient does not have dengue. The viraemic window is short and any request for dengue exclusion testing should be accompanied by antibody and antigen (NS1) testing to definitively exclude dengue infection. Our unpublished protocols for typing individual dengue strains have not been presented here as they have little relevance to the clinical management of patients. However there are a number of published procedures available for that purpose [1, 2].

References

1. Chien LJ, Liao TL et al (2006) Development of real-time reverse transcriptase PCR assays to detect and serotype dengue viruses. J Clin Microbiol 44:1295–1304
2. Das S, Pingle MR et al (2008) Detection and serotyping of dengue virus in serum samples by multiplex reverse transcriptase PCR-ligase detection reaction assay. J Clin Microbiol 46:3276–3284
3. Lanciotti RS, Calisher CH et al (1992) Rapid detection and typing of dengue viruses from clinical samples by using reverse transcriptase-polymerase chain reaction. J Clin Microbiol 30(3):545–551
4. Smith G, Smith I et al (2006) A simple method for preparing synthetic controls for conventional and real-time PCR for the identification of endemic and exotic disease agents. J Virol Methods 135:229–234
5. Warrilow D, Northill JA et al (2002) Single rapid TaqMan fluorogenic probe based PCR assay that detects all four dengue serotypes. J Med Virol 66:4524–4528

Chapter 32
Enterovirus

Jason A. Roberts and Bruce R. Thorley

32.1 Summary of Methods

Human enteroviruses (HEVs) are single stranded, positive sense, RNA viruses in the family *Picornaviridae*. More than 90 serotypes of HEV are classified within four species, A-D, including the poliovirus, coxsackievirus, echovirus and enterovirus prototypes. HEVs cause a wide variety of diseases including mild febrile illness, acute flaccid paralysis and fatal encephalopathy. HEVs are considered a primary cause of aseptic meningitis and a study in the USA reported that HEVs were detected in 4.6% of patients suffering from encephalitis [1].

Given the labour intensive nature of virus culture in mammalian cell lines, detection of HEV in clinical specimens via genetic methods, particularly RT-PCR, has become more prevalent in diagnostic laboratories. The RT-PCR diagnostic detection methods target highly conserved sequence within the 5′ non-translated region (5′NTR) (Fig. 32.1). While this method enables sensitive detection of HEVs from clinical specimens, the serotype cannot be reliably determined from the PCR amplicon, which may be important for full clinical or public health investigation. The serotype identification of enteroviruses is based on primers directed to the VP1 genomic region (Figure 32.1) [2]. While the identification of HEV species A-D is based on the 3′ non-translated region (3′ NTR) [3].

In the National Poliovirus Reference Laboratory a single round real-time RT-PCR (RT-qPCR) assay is used for confirmation of suspected HEV cytopathic effect in cell culture, while a semi-nested assay is used to detect HEV directly from patient samples.

J.A. Roberts (✉)
Poliovirus Reference Laboratory, Victorian Infectious Diseases Reference Laboratory, North Melbourne, VIC 3051, Australia
e-mail: Jason.Roberts@mh.org.au

M. Schuller et al. (eds.), *PCR for Clinical Microbiology*,
DOI 10.1007/978-90-481-9039-3_32,

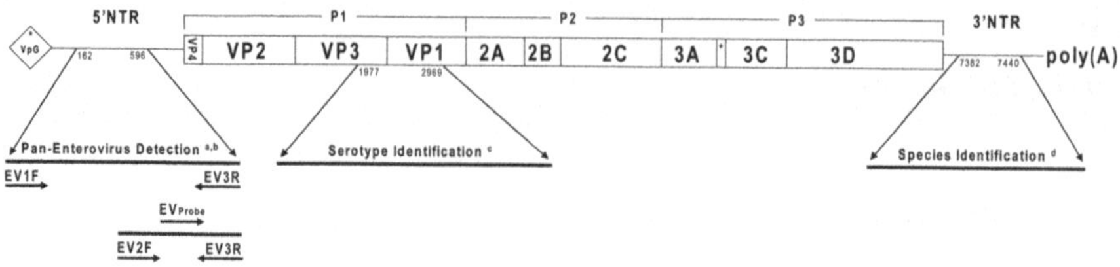

Fig. 32.1 Depiction of the PCR primer annealing sites within the ~7,500 nucleotide enterovirus genome, based on poliovirus type 1 Mahoney, Genbank accession number V01150 (not to scale). A viral encoded protein, VpG, is covalently attached to the 5′ terminus of the genome. [a]Zoll et al. [6], [b]Verstrepen et al. [4], [c]Nix et al. [2], [d]Oberste et al. [3]

32.2 Acceptable Specimens

Faeces, CSF, nasopharyngeal aspirates, cell culture supernatant, throat, vesicle and rectal swabs are all acceptable specimens depending upon the clinical symptoms. Throat, vesicle and rectal swabs should be transported in viral transport media. All specimens should be kept at 4°C until processed. CSF should undergo nucleic acid extraction immediately upon receipt. Faeces and cell culture supernatant may be frozen if processing is delayed for more than 48 hours. The non-enveloped HEV virion is typically shed in faeces for up to six weeks after infection and is therefore the specimen of choice for HEV testing. For optimal testing of faecal material, two specimens collected >24 h apart and within 14 days of the onset of symptoms should be tested due to intermittent virus shedding.

32.3 Unacceptable Specimens

Given the transient nature of HEV viremia, blood samples are not recommended for testing. Tissue samples require specific RNA extraction procedures and also are not recommended.

32.4 Sample Extraction

RNA is extracted from 200 μl of all specimens using the High-Pure Viral Nucleic Acid Extraction Kit (Roche Applied Science Australia). RNA is eluted into 50 μl and stored immediately at −30°C. Cell culture supernatant positive for HEV is freeze-thawed twice and centrifuged at 1500 × *g* for 10 min to pellet cellular debris. RNA extracted cell culture supernatant should be diluted one in five before use. Faecal specimens should be processed in virus maintenance media containing 2% foetal bovine serum and 10% v/v chloroform [5].

32.5 Primer and Probe Sequences

For the semi-nested endpoint PCR, primer pairs EV1F (5′-CAA GCA CTT CTG TTT CCC CGG-3′) (position 162–182) and EV3R (5′-ATT GTC ACC ATA AGC AGC CA-3′) (position 596–577) are used in the first step of the assay and primer pairs EV2F (5′-TCC TCC GGC CCC TGA ATG CG-3′) (position 443–462) and EV3R in the second step. The primer pairs EV2F and EV3R are also used in the RT-qPCR and the product is detected using the EV hydrolysis probe (5′ FAM-AAC CGA CTA CTT TGG GTG TCC GTG TTT C-BHQ1-3′) (Position 535–562) (Fig. 32.1).

32.6 PCR Amplification and Product Detection

32.6.1 Semi-nested Endpoint PCR

The first round pan-EV RT-PCR is performed in a 50 μl volume consisting of 1X SuperScript™ III buffer (containing 0.2 mM each dNTP and 1.6 mM $MgSO_4$) (Invitrogen Life Technologies), 0.4 μM EVIF and EV3R, 2 μl Superscript™ III RT/Platinum® Taq Mix (Invitrogen Life Technologies), 20 U Roche Protector RNAse Inhibitor (Roche Applied Science Australia) and 10 μl of RNA. Amplification is performed on an Applied Biosystems 9700 thermocycler under the following conditions: 45°C for 30 min, 94°C for 2 min, followed by 35 cycles of 94°C for 15 s, 57°C for 45 s, and 68°C for 60 s, and a final extension at 68°C for 5 min.

The second-round pan-EV reaction is performed in a 50 μl volume consisting of 1X Roche FastStart Master mix (Roche Applied Science Australia), 0.2 μM each of primers EV2F and EV3R, and 2 μl of first-round PCR product. Amplification is performed on a GeneAmp® PCR 9700 thermocycler (Applied Biosystems) under the following conditions: 95°C for 2 min, followed by 35 cycles of 95°C for 30 s, 57°C for 20 s, and 72°C for 20 s, and a final extension at 72°C for 7 min. PCR amplicons are separated by electrophoresis in a 2% agarose gel. Products are stained and visualised using a GelDoc 2000 system (BioRad). Final product size is approximately 150 bp, depending on HEV serotype.

32.6.2 TaqMan RT-qPCR

The reverse transcription PCR is performed in a 20 μl volume consisting of 1X SuperScript™ III first strand buffer (containing 50 mM Tris-HCl (pH 8.3), 75 mM KCl and 3 mM $MgCl_2$) (Invitrogen Life Technologies), 0.5 μM of primer EV3R, 0.004 M DTT (Invitrogen Life Technologies), 0.5 mM each of dATP, dCTP, dGTP and dTTP, 160 U Superscript™ III RT enzyme (Invitrogen Life Technologies), 16 U Roche Protector RNAse inhibitor (Roche Applied Science Australia), and 10 μl of template RNA. First strand synthesis is performed on a GeneAmp® PCR 9700 thermocycler (Applied Biosystems) under the following conditions: 25°C for 10 min, followed by 50°C for 30 min and 95°C for 5 min.

The TaqMan qPCR is performed in a 25 μl volume containing 1X TaqMan® Universal PCR Master Mix, No UNG (Applied Biosystems), 0.9 μM each of primers EV2F and EV3R, 0.25 μM EV probe, and 5 μl of template RNA. Amplification is performed on an Applied Biosystems 7500 FAST real time thermocycler under the following amplification conditions: 95°C for 10 min, followed by 45 cycles of 95°C for 15 s and 60°C for 60 s. Fluorescence is measured at the end of the 60°C step. (NOTE: If using TaqMan® FAST Universal PCR Master Mix, no AmpErase® UNG, follow manufacturer's instructions with 45 cycles).

32.7 Quality Control and Validation Data

Sabin poliovirus, obtained from the National Institute for Biological Standards and Control in the United Kingdom, are used as the positive control. Working stocks of 1.0×10^{-2} and 1.0×10^{-4} are used for the qPCR and semi-nested PCR, respectively. Nuclease free water and viral maintenance or transport medium are used as the negative control. Bovine Viral Diarrhoea Virus is included as an internal control to account for PCR inhibition and undergoes the same processing as all samples tested.

The limits of detection for the assays are: (i) 0.84 log $CCID_{50}$ for semi-nested endpoint PCR (Fig. 32.2); (ii) 2.01 log $CCID_{50}$ for TaqMan RT-qPCR using TaqMan Universal Master Mix; and (iii) 2.68 log $CCID_{50}$ for TaqMan RT-qPCR using TaqMan FAST Universal Master Mix (Fig. 32.3).

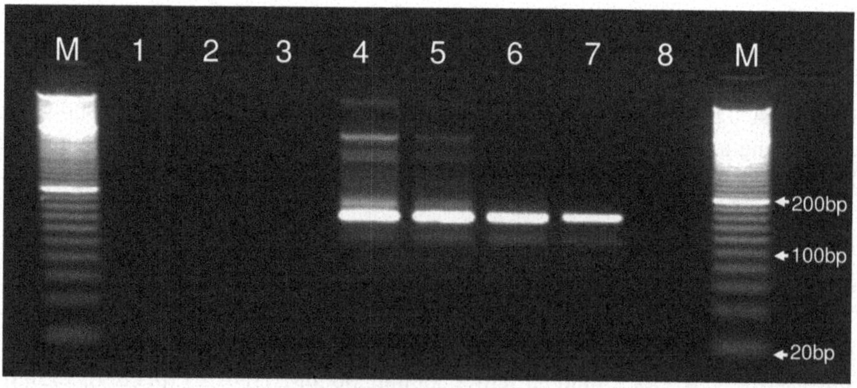

Fig. 32.2 Image of semi-nested endpoint PCR products in 3% agarose gel with 20 base pair marker – M. Lanes 1–3 are negative samples, Lanes 4–8 are serial dilutions of Sabin poliovirus serotype 1. Lane 4 = 3.95 log $CCID_{50}$, 5 = 2.95 log $CCID_{50}$, 6 = 1.95 log $CCID_{50}$, 7 = 0.95 log $CCID_{50}$, 8 = 0.05 log $CCID_{50}$

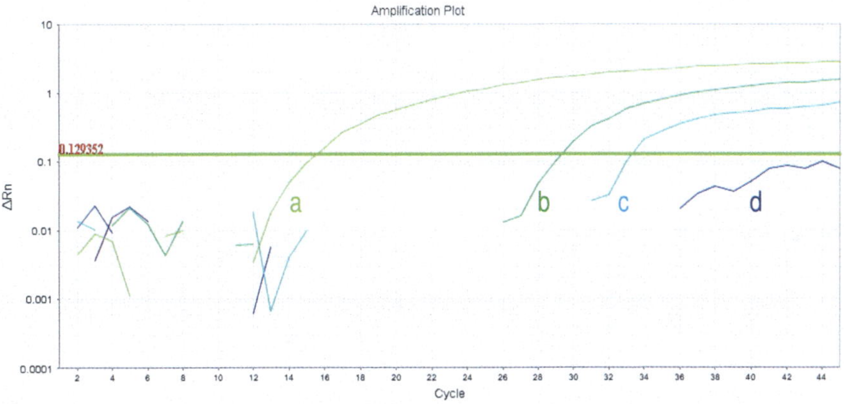

Fig. 32.3 Real-time Reverse Transcriptase PCR results for poliovirus control dilution series using ABI 2x Taqman® Universal Mastermix. (**a**) = 7.28 log $CCID_{50}$, (**b**) = 3.28 log $CCID_{50}$, (**c**) = 2.28 log $CCID_{50}$, (**d**) = 1.28 log $CCID_{50}$

References

1. Fowlkes AL, Honarmand S, Glaser C et al (2008) Enterovirus-associated encephalitis in the California Encephalitis Project, 1998–2005. J Infect Dis 198:1685–1691
2. Nix WA, Oberste MS, Pallansch MA (2006) Sensitive, semi-nested PCR amplification of VP1 sequences for direct identification of all enterovirus serotypes from original clinical specimens. J Clin Micro 44:2698–2704
3. Oberste MA, Maher K, Williams AJ et al (2006) Species-specific RT-PCR amplification of human enteroviruses: a tool for rapid species identification of uncharacterized enteroviruses. J Gen Virol 87:119–128
4. Verstrepen WA, Kuhn S, Kockx MM et al (2001) Rapid detection of enterovirus RNA in cerebrospinal fluid specimens with a novel single-tube real-time reverse transcription-PCR assay. J Clin Micro 39:4093–4096
5. World Health Organization (2003) Polio Laboratory Manual, 4th edn. Switzerland
6. Zoll GJ, Melchers WJG, Kopecka H et al (1992) General primer-mediated polymerase chain reaction for detection of enteroviruses: application for diagnostic routine and persistent infections. J Clin Micro 30:160–165

Chapter 33
Epstein Barr Virus

Jonathan Howard

33.1 Summary of Methods

Detection of Epstein Barr Virus (EBV) by the polymerase chain reaction (PCR) has been used to diagnose primary central nervous system lymphoma in AIDS, post-transplantation lymphoproliferative disease and infectious mononucleosis, although serology tests may be more useful here. The universal nature of PCR technology now makes it possible for many laboratories to diagnose infections caused by EBV.

33.2 Introduction

The catalyst for the discovery of Epstein-Barr virus (EBV) was Denis Burkitt, a surgeon working in East Africa in the 1950s, who first draw attention to a novel childhood tumour, now known as Burkitt's lymphoma (BL), which he found to be unusually common throughout equatorial Africa. The apparent influence of climatic factors in determining those regions of Africa where BL was "endemic" led Burkitt to postulate that an infectious agent, possibly arthropod-borne, was involved in the tumour's aetiology. This caught the attention of Tony Epstein who, with his Ph.D. student Yvonne Barr, succeeded in establishing a number of BL-derived cell lines in culture, of which the first to be examined in the electron microscope showed clear evidence of herpesvirus-like particles in a small proportion of the cells. The agent present in BL cells proved to be biologically and antigenically distinct from other members of the human herpesvirus (HHV) family, thereby identifying EBV as the first candidate human tumour virus. in the first decade of EBV research, a great deal was learned about the biology of EBV infection in vitro and

J. Howard (✉)
SEALS Virology Division, Prince of Wales Hospital, Randwick, NSW 2031, Australia
e-mail: jonathan.howard@sesiahs.health.nsw.gov.au

M. Schuller et al. (eds.), *PCR for Clinical Microbiology*,
DOI 10.1007/978-90-481-9039-3_33,

in vivo. The virus was found to cause infectious mononucleosis and to be latent in most adult humans. The tight linkage to BL and NPC, the rare occurrence of lymphoproliferative disease in young children and renal transplant recipients, and the ability of the virus to efficiently immortalise lympho- cytes in vitro and to induce lymphomas in marmosets provided a biologic framework indicating that EBV can be oncogenic under unusual circumstances. The discovery of cell lines that could be induced to replicate EBV led in the second decade to increasingly sophisticated biochemical analyses. The structure of the genome was determined, the genome cloned, the transcriptional program in latent and lytic infection worked out, and the entire genome sequenced. EBNA became a protein on an acrylamide gel, and then more than one protein. EBV mRNAs in latently infected and growth-transformed B lymphocytes were cloned, sequenced, and expressed in heterologous cells.

Detection of Epstein Barr Virus (EBV) by the polymerase chain reaction (PCR) has been used to diagnose primary central nervous system lymphoma in AIDS, post-transplantation lymphoproliferative disease and infectious mononucleosis, although serology tests may be more useful here. The universal nature of PCR technology now makes it possible for many laboratories to diagnose infections caused by EBV.

33.3 Specimen Collection

Specimen	Treatment
CSF	No preliminary processing required. Specimen may be processed for DNA extraction as specified in the Roche High Pure extraction kit
EDTA blood	Sample is spun and the plasma taken off for viral DNA/RNA extraction procedure
Biopsies	Sample is ground in PBS/VMM to a final volume of 200 μl

Processed specimens are stored at −20°C or used for DNA/RNA extraction immediately.

33.4 Specimen Preparation

Extract DNA using the Roche Diagnostics High Pure Kit. A positive control sample is also extracted as a control for extraction. Extracted specimens are stored at −20°C or used as template immediately. TE (Tris EDTA) is used in resuspension of DNA after extraction.

33.5 Primer and Probe Sequences

The nested EBV primers were taken from the publication of Meerbach et al. [1]. Sequences are:

1 Outer primer p23-1 5′ ATC AGA AAT TTG CAC TTT CTT TGC 3′
2 Outer primer p23-2 5′ CAG CTC CAC GCA AAG TCA GAT TG 3′
3 Inner primer p23-3 5′ TTG ACA TGA GCA TGG AAG AC 3′
4 Inner primer p23-4 5′ CTC GTG GTC GTG TTC CCT CAC 3′

The 5′ biotinylated probe sequence is: 5′biotin- AAC TAC CCG CAA TGA AAT GG 3′

These primers and the probe are purchased at a fixed concentration of 100 μM from Sigma Aldrich.

33.6 Equipment

Perkin Elmer GeneAmp PCR System 2400, Sanofi AIP4 incubator (Lab. No. V037), Biorad Immunowash model 1250, Sanofi Plate spectrophotometer (Lab No. TG115-18).

33.7 PCR Amplification and Product Detection

PCR reagents are diluted using nuclease free water, TE is Tris EDTA (Tris 10 mM, EDTA 1 mM) buffer pH 8, Roche Diagnostics High Pure Viral Nucleic Acid kit, Roche Diagnostics PCR ELISA kit.

Add the following to a sterile 0.2 ml thin walled tube

EBV NESTED AmpliTaq Gold PCR ROUND 1

Reagents		Volume/tube (μl) 1x	10 x
Nuclease-free water	–	12.5	125
AmplitaqGold	Applied Biosystems	25	250
EBV PRIMERS 1,2		2	20
Amperase	Applied Biosystems	0.5	5
Template		10	
TOTAL		50	400

Cycling is performed at 50°C × 5 min then 95°C × 5 min followed by 40 cycles of 94°C × 1 min, 60°C × 30 s, 72°C × 1 min followed by 72°C × 7 min and hold at 4°C until further analysis.

EBV NESTED AmpliTaq Gold PCR ROUND 2

Reagents		Volume/tube (μl) 1x	10 x
Nuclease-free water	-	20.8	208
AmplitaqGold	Applied Biosystems	25	250
EBV PRIMERS 3,4		2	20
Dig-11 dUTP	Applied Biosystems	0.2	2
Template		2	
TOTAL		50	478

Cycling is performed at 95°C × 5 min followed by 30 cycles of 94°C × 15 s, 60°C × 30 s, 72°C × 30 s, followed by 72°C × 7 min and hold at 4°C until further analysis.

No need to keep this mixture on ice as mispriming of Amplitaq gold will not occur without activation of the enzyme (this occurs in the first 95°C step). Note that at this stage 2 μl of water is added to the reagent (negative) control tube.

Add master mix followed by addition of 10 μl of extracted DNA into each tube. The 50 μl sample is then ready to be amplified. Program the thermal cycler (Perkin Elmer GeneAmp PCR system 2400) as detailed in the table above.

Prepare a 2.5% TBE agarose gel. Staining is performed using Ethidium Bromide (Sigma Cat. No. E1510). Prepare 1x TBE running buffer by adding 17.5 μl of ethidium bromide to 400 ml of 1x TBE and pour into submarine tank.

Add 2 μl loading dye to 10 μl PCR product and load onto the gel. Use 3 μl of a 1:10 dilution (300 ng) of pGEM (Promega Cat. No. G1741) as the marker, mixed with 2 μl of 6x loading dye. Run gel at 100 V for 40 min. View gel under ultraviolet light and photograph.

33.8 EIA Detection of PCR Products

Always Use Plugged pipette tips during latter parts of cycling, remove the Roche PCR ELISA kit from the cold room and warm to room temperature. Add 40 μl denaturation solution to each well (use multichannel pipette and change tips between strips), pipette up and down 6–10 times. Incubate at room temperature for 10 min to complete denaturation. Place appropriate number of Microwell strips from foil and insert into base ELISA plate. (seal foil bag after use). Add 165 μl Hybridisation solution (Solution IV) containing the probe (diluted 1:5 in distilled water) as per table:

33.9 Hybridisation Solution

No. of Strips	Volume of Probe (μl)	Volume of Buffer (μl)
1	10	1990
2	15	2985

Add 35 μl denatured amplified sample to a designated well of the microtitre test plate. Using plugged tips and gently mix without creating bubbles (pipette up and down 6–10 times). Cover and incubate at 40°C with shaking, for 70 min in microplate incubator/rocker. Wash plate six times, using the prepared wash solution, with 30 s soaks between each wash step. Tap plate dry. Add 200 μl of anti-DIG-POD conjugate diluted 1:100 in conjugate dilution buffer to each well and add plate sealer as per table below:

33.10 Conjugate

No. of Strips	Volume of conjugate (μl)	Volume of Buffer (μl)
1	20	1980
2	40	3960

Note: Reconstitute conjugate with 250 μl of distilled water

Cover and incubate at 40°C with shaking, for 30 min in AIP4 plate incubator/rocker. Wash plate six times, using the prepared wash solution, with 30 s soaks between each wash step. Tap plate dry. Add 200 μl of TMB substrate to each well. Cover with foil and incubate at 37°C without shaking, for 30 min in AIP4 plate incubator/rocker. Add stop solution and read microwell plate immediately at a wavelength of 405 nm (Measurement wavelength) and 620 nm (Reference wavelength).

Note: Wash Buffer needs to be prepared weekly and stored at 2–8°C as follows: Prepare working wash solution by adding 1 washing buffer tablet to 2000 ml of distilled water. Store in walk-in contaminated cold room with other EIA reagents for a maximum of 1 week only.

33.11 Quality Control

Development of PCR includes testing the primers with other viruses to test for cross-reactivity, to prevent false positives due to the presence of viruses other than EBV.

A number of control samples are included in each run to test both the extraction kit and the PCR amplification and EIA detection. A positive control is included from extraction through to the EIA. Negative controls are included, which are also extracted and run through the entire method. These include open tubes of water/buffer remain open throughout the extraction. These detect aerosol contamination arising during extraction. Closed tubes of water/buffer are included to detect contamination arising from sample handling during extraction. The final negative control is a negative reagent control for PCR, which has sterile water added in place of extracted DNA. Sensitivity of the assay is currently under investigation.

Reference

1. Meerbach A, Gruhn B, Egerer R et al (2001) Semiquantitative PCR analysis of Epstein-Barr virus DNA in clinical samples of patients with EBV-associated diseases. J Med Virol 65:348–357

Chapter 34
Flavivirus

Gerald B. Harnett and Julia A. Cattell

34.1 Summary of Methods

The Flaviviruses responsible for most cases of encephalitis in Australia and South East Asia include Murray Valley Encephalitis virus, Kunjin virus, Japanese B Encephalitis virus and West Nile virus. Flavivirus infection is usually diagnosed by the detection of virus-specific IgM antibody or by measuring a rising titre of IgG antibodies in the acute phase of illness. Local IgM antibody can also be detected in the CSF of patients with CNS disease. Because many infections with flaviviruses are inapparent, caution should be exercised in the interpretation of serological results. Patients may also have viral RNA in the blood or CSF and its presence constitutes a definite indicator of current infection.

This protocol describes a multiplex reverse transcriptase PCR that targets the NS5 region to detect Murray Valley Encephalitis virus, Kunjin virus and Japanese B Encephalitis virus and 3′ UTR to detect West Nile virus [4]. It is a rapid and economical method that only requires a small sample volume, of particular benefit in the case of CSF samples which are often only available in small volumes. The second-round individual real-time PCRs of the assay include the same primers as the first-round with the addition of TaqMan probes for each agent.

34.2 Acceptable Specimens

Arboviruses can be detected in CSF and CNS tissue specimens and in some cases in whole blood samples.

G.B. Harnett (✉)
Department of Microbiology and Infectious Diseases, PathWest Laboratory Medicine, QEII Medical Centre, Nedlands, WA 6009, Australia
e-mail: Gerry.Harnett@health.wa.gov.au

M. Schuller et al. (eds.), *PCR for Clinical Microbiology*,
DOI 10.1007/978-90-481-9039-3_34,

34.3 Sample Extraction

RNA is extracted from clinical specimens using the QIAmp Viral RNA Mini Kit (Qiagen Pty Ltd) according to the manufacturer's instructions. Tissue samples are digested with proteinase K before extraction.

34.4 Primer and Probe Sequences

The first-round reverse transcriptase PCR reaction is multiplex, incorporating primer pairs that target the four flaviviruses and the internal control, MS2 RNA coliphage. The second-round PCRs are individual real-time PCR reactions which use the same primer pairs as the first-round, with the addition of TaqMan probes for each target. Details of primer and probe sequences are shown in Table 34.1.

34.5 PCR Amplification and Product Detection

If sample volumes permit it is good practice to perform assays in duplicate. The initial multiplex RT-PCR is performed in a 20 μl volume consisting of 1X RT-PCR buffer (Invitrogen Life Technologies), 10 U RNAsin (Invitrogen Life Technologies), 0.3 μl One-step SS RT enzyme (Invitrogen Life Technologies), 0.5 U iSTAR Taq (Scientifix Australia), 0.2 μM of each primer (MVE-F, MVE-R, WN-10533, WN-10606, KUN-F, KUN-R, JE-F, JE-R, MS2-F and MS2-R), 2.5% DMSO and 8 μl of RNA sample. Amplification is performed in an ABI 2700 (Applied Biosystems) thermal cycler under the following conditions: 50°C for 30 min, 95°C for 5 min, followed by 20 cycles of 94°C for 30 s, 50°C for 30 s and 68°C for 45 s. Following amplification, the first-round PCR products are diluted 1:10 with molecular biology grade water to reduce the transfer of possible non-specific products into the second round mixes [2].

The second-round individual real-time TaqMan PCRs are performed in 20 μl volumes consisting of 1X PCR buffer (Applied Biosystems), 4 mM $MgCl_2$, 0.2 mM dNTPs (Fisher Biotec Australia), 0.75 U DNA polymerase (Applied Biosystems), 0.2 μM of forward primer (MVE-F or WN-10533 or KUN-F or JE-F, and MS2-F), 0.2 μM of reverse primer (MVE-R or WN-10606 or KUN-R or JE-R, and MS2-R), 0.2 μM of TaqMan probe (MVE-Probe or WN-10560-Probe or KUN-Probe or JE-Probe, and MS2-Probe), 0.01% BSA, and 1 μl of diluted first-round PCR product. Amplification is performed in a RotorGene 6000 (Corbett Life Science, Australia) thermal cycler under the following conditions: 95°C for 10 min, followed by 35 cycles of 94°C for 10 s, 55°C for 90 s and 72°C for 15 s. Fluorescence is measured at the end of the 72°C extension step (expressed as a cycle threshold (Ct) value).

Table 34.1 Primer and probe sequences for the flavivirus multiplex PCR assay

Target virus	Primer and probe sequences (5′ to 3′)		Size bp
Murray Valley Encephalitis virus	MVE-F	AGC CGA AGC GGT CAT AGG T	72
	MVE-R	CAT GTG CGG ACT GCA AAT TT	
	MVE-Probe	6FAM-TTT CAA TGC TTT CAA TGT CA-MGBNFQ	
West Nile virus	WN-10533	AAG TTG AGT AGA CGG TGC TG	92
	WN-10606	AGA CGG TTC TGA GGG CTT AC	
	WN-10560-Probe	6FAM-CTC AAC CCC AGG AGG ACT GG-MGBNFQ	
Kunjin virus	KUN-F	CAT TGG TTG AGG ATA CAG TAT TG	132
	KUN-R	CCT GAC TTC CTC ACT AAA ATT TT	
	KUN-Probe	6FAM-AAG CTA AAC TTC TAC ACA TAA TAA CA-MGBNFQ	
Japanese Encephalitis virus	JE-F	CAG AAC TGA TTT AGA AAA TGA AGC TAA GG	99
	JE-R	TTG TGC CTG TAA GTC AGT TCA ATT ATG	
	JE-Probe	6FAM-ACC GCA TGC TCG CCC GAG-MGBNFQ	
MS-2 RNA Coliphage	MS2-F	GTC GAC AAT GGC GGA ACT G	66
	MS2-R	TTC AGC GAC CCC GTT AGC	
	MS2-Probe	CalO-ACG TGA CTG TCG CCC CAA GCA ACT T-BHQ2	

34.6 Quality Control and Validation Data

A standardized amount of MS-2 phage is incorporated into the lysis buffer of the extraction kit to serve as an internal control and provide an assessment of the efficiency of extraction, reverse transcription and PCR amplification. The MS-2 phage sample inhibitor controls must give similar Ct values to the negative controls. Higher Ct values indicate a degree of inhibition and the assay must be repeated with diluted RNA extract or re-extracted sample.

Positive control RNA is included in each test batch at 10X the expected titre and their Ct values are plotted in Shewhart charts to monitor variation in assay performance. Mean values are calculated from every 100 test runs and those with positive control Ct values falling within two standard deviations (2SD) from the means give 95% confidence limits for the assays. Test runs with control Ct values above 2SD should be repeated.

Negative control water blanks are extracted between every five samples to detect any cross contamination events.

To determine the limit of detection, replicate PCRs were performed on a three-fold dilution series of reference materials from each of the flaviviruses included in the multiplex assay. The results from these experiments were subjected to Probit Analysis using StatsDirect software [1, 3]. The assay limits of detection at a 95% confidence level ranged from 390 to 440 gen/equ. per ml. To date the assay has only been validated against reference strains of the relevant arboviruses because few positive clinical samples are available in Australia.

34.7 Assay Limitations

The West Nile virus assay [4] also detects Kunjin virus while the Kunjin assay is virus-specific. Kunjin virus is now recognised as a clade of West Nile virus.

Acknowledgments We would like to thank Eileen J. Pratt for valuable technical assistance provided.

References

1. Jalal H, Stephen H, Curran MD et al (2006) Development and validation of a RotorGene real-time PCR assay for detection, identification and quantification of *Chlamydia trachomatis* in a single reaction. J Clin Microbiol 44:206–213
2. Stanley KK, Szewezuk SA (2005) Multiplexed tandem PCR: gene profiling from small amounts of RNA using SYBR Green detection. Nucleic Acid Res 33:e180
3. Sykes PJ, Neoh MJ, Hughes E et al (1992) Quantitation of targets for PCR by use of limiting dilution. Biotechniques 13:444–449
4. Tanaka M (1993) Rapid identification of flaviviruses using the polymerase chain reaction. J Virol Methods 41:311–322

Chapter 35
Hepatitis A Virus

Scott Bowden

35.1 Clinical Background

Hepatitis A infection is common throughout many parts of the world and occurs most frequently in developing nations whenever there may be a lower standard of sanitation. In such countries, most of the population can be infected but paradoxically, this usually presents no significant public health problem as infection early in life is asymptomatic and provides the individual with life-long immunity. Transmission occurs mainly by the faecal-oral route and poor sanitation practices can lead to contaminated water which also allows concentration in shellfish. Infection with hepatitis A virus (HAV) is self-limiting and can produce symptoms ranging from subclinical to fulminant hepatitis. The chances of developing clinical disease can be correlated with the age at which infection occurs. In young children, most infections are asymptomatic but in older children and adults, symptoms occur commonly, most notably jaundice, fever, malaise and dark urine.

Historically, hepatitis A was known as "short term incubation" hepatitis to distinguish it from hepatitis B, known as "long term incubation" hepatitis. The average incubation period is around one month and infection is not associated with chronic or persistent infection. HAV is excreted in bile and is shed in feces in large numbers, peaking before the onset of symptoms and declining soon after their appearance. Viremia can be detected one to two weeks after exposure and may persist for up to two months, although virus titres are lower than that found in the feces. Serological detection of hepatitis A IgM is a reliable assay and is the mainstay of diagnosis. IgM antibodies appear around the time of symptom development and persist for several months. Their decline is mirrored by an increase in HAV total antibodies, predominantly IgG which provides life-long immunity. A vaccine is available for hepatitis A.

S. Bowden (✉)
Victorian Infectious Diseases Laboratory, VIDRL, Wreckyn St, Melbourne, VIC, Australia
e-mail: scott.bowden@mh.org.au

M. Schuller et al. (eds.), *PCR for Clinical Microbiology*,
DOI 10.1007/978-90-481-9039-3_35, © Springer Science+Business Media B.V. 2010

HAV is a non-enveloped single-stranded plus-sense RNA virus, classified in the genus *Hepatovirus* in the *Picornaviridae*. Unusually for an RNA virus it shows a relatively high degree of conservation and exists as a single serotype.

There is sufficient genetic diversity to allow HAV to be divided into seven genotypes, of which genotypes I–IV represent human isolates and of these genotypes I and III are the most prevalent. A real-time HAV reverse transcription PCR assay is commercially available (QIAGEN Artus HAV RT-PCR kit). The assay described below allows qualitative detection of HAV RNA and the PCR product generated is suitable for sequence analysis and subsequent genotyping.

35.2 Acceptable Specimens

Appropriate collection and handling of blood samples is an important part of the testing procedure. While it appears the nucleic acid of the virus is relatively stable, inappropriate collection and storage can impact on the accuracy of the final result. Serum separator tubes (SST), whole blood, ACD and EDTA tubes are the preferred collection tubes.

35.3 Unacceptable Specimens

Heparin tubes are not acceptable, as heparin has been shown to be a potent inhibitor of PCR. Similarly, severely haemolysed samples may cause inhibition of the PCR assay.

35.4 Sample Extraction

Extract 140 μl serum according to the QIAamp Viral RNA Mini Kit instructions (QIAGEN, Melbourne, Australia). Store the eluted material on ice until ready to continue the cDNA synthesis step. If the cDNA synthesis is not to be performed on the same day, store at –70°C. For each sample incubate the extracted RNA at 65°C for 10 min to relax the secondary structure of the RNA. Cool on ice.

35.5 Primer Sequences

The in-house HAV RT-PCR is a nested PCR assay. The primers used are modified from those described in the review of Nainan et al. [1], which have been designed to the region encoding the HAV VP1/P2A junction. Sequence analysis of this region can allow samples to be genotyped.

Primer name	Sequence (5′–3′)	No. of bases	Function	Product size (bp)
HAV-OU1	TTG TCT GTC ACA GAA CAA TCA G	22	1st R	
HAV-OD1	AGT CAC ACC TCT CCA GGA AAA CTT	24	RT and 1st R	358
HAV-IU2	TCY CAG AGC TCC ATT GAA	18	2nd R	
HAV-ID2	AGG RGG TGG AAG YAC TTC ATT TGA	24	2nd R	233

35.6 PCR Amplification and Product Detection

Reverse transcription and first round PCR is carried out using SuperScript III One-Step RT-PCR System with Platinum *Taq* Polymerase (Invitrogen).

The master mix (25 μl total volume) consists of 12.5 μl 2x Reaction Mix, 10 μl Template RNA, 1.25 μl HAV-OU1 (10 μM) and HAV-OD1 (10 μM), 0.25 μl RNasin (40 units/μl), 1 μl RT/Taq Mix.

Reverse transcription and first round PCR amplification conditions consist of 50°C for 30 min (Reverse Transcription), 94°C for 2 min, 35 cycles of 94°C for 15 s, 55°C for 30 s, 72°C for 30 s followed by 72°C for 5 min then 4°C indefinitely.

The second round PCR is carried out using QIAGEN *Taq* polymerase and buffers. The master mix (40 μl total volume) consists of 25 μl Nuclease free water, 8 μl 5x Q Buffer, 4 μl 10x Taq buffer (+Mg), 2 μl Template (1st round PCR product), 0.4 μl HAV-IU2 + HAV-ID2 (25 μM), 0.4 μl 20 mM dNTP, 0.3 μl Taq DNA polymerase (5 units/μl).

Second round PCR amplification conditions consist of 94°C for 2 min, 25 cycles of 94°C for 30 s, 55°C for 30 s, 72°C for 30 s followed by 72°C for 5 min and 4°C indefinitely.

Both first round and second round PCR products are electrophoresed through a 1.5% TAE agarose gel, stained with ethidium bromide and photographed using the BIO-RAD Gel Doc 2000 system.

35.7 Quality Control and Validation Data

Negative serum and low positive control serum samples are extracted along with test samples and included in the nested RT-PCR. An additional control containing the reaction mix with water added in place of sample is also included in the RT-PCR to monitor for contamination of reaction mix reagents.

The specificity of the assay was validated with 50 clinical samples sent to VIDRL for testing unrelated to hepatitis A and which had no detectable HAV antibodies. No detectable PCR product was generated. Cross-reactivity was tested by assaying samples in triplicate known to be positive for HBV DNA, HCV RNA, HDV RNA and HEV RNA and no amplification product was generated. By comparison with the WHO International Standard for Hepatitis A Virus RNA (NIBSC Code 00/560), the lower limit of detection of the assay was estimated to be approximately 500 IU/ml.

Reference

1. Nainan OV, Xia G, Vaughan G, Margolis HS (2006) Diagnosis of hepatitis A virus infection: a molecular approach. Clin Micro Rev 19: 63–79

Chapter 36
Hepatitis B Virus

Scott Bowden

36.1 Clinical Background

Hepatitis B is the most common form of viral hepatitis with more than 400 million people chronically infected. Of these chronic carriers, around 25% will develop serious liver disease, including cirrhosis and hepatocellular carcinoma. In Australia, it is estimated that 1% of the population is chronically infected with hepatitis B; around 50% have a South East Asian background.

The diagnosis of hepatitis B is nearly always based on interpretation of serological assays. During the course of infection, the viral surface antigens (HBsAg), core protein (HBcAg) and non-structural secreted "e" antigen (HBeAg) stimulate the immune system, which responds by producing the corresponding antibodies (anti-HBc, anti-HBs and anti-HBe). The persistence of HBsAg for longer than 6 months defines chronic infection and the presence of its corresponding antibody (anti-HBs) is a marker of immunity and disease resolution. Usually, the loss of HBeAg is associated with a significant clinical improvement and normalization of serum transaminase levels despite the lasting presence of HBsAg. However, sometimes levels of HBV DNA remain high and disease progression continues. This is caused by HBV variants that limit the expression of HBeAg due to mutations in the precore region of the HBV core gene. The most common mutation is a single base substitution of G>A at nucleotide 1896 (G1896A) which gives rise to a stop codon (TAG) in place of the usual amino acid tryptophan (TGG). Further common mutations are seen in the core promoter region (also known as the basal core promoter) at nucleotide positions 1762 and 1764 (A1762T and G1764A), which lead to a reduction in HBeAg synthesis.

HBV is an enveloped, partially double-stranded DNA virus that is the prototype member of the *Hepadnaviridae*. The virus replicates its DNA genome via an RNA intermediate, an unusual strategy that generates a heterogeneous population of genetic variants during the normal course of infection and provides enormous

S. Bowden (✉)
Victorian Infectious Diseases Laboratory, VIDRL, Wreckyn St, Melbourne, VIC, Australia
e-mail: scott.bowden@mh.org.au

M. Schuller et al. (eds.), *PCR for Clinical Microbiology*,
DOI 10.1007/978-90-481-9039-3_36, © Springer Science+Business Media B.V. 2010

potential for adaptation to changing environments and response to particular selection pressures. The HBV genome is a circular, partially double-stranded relaxed circular (RC) DNA molecule of around 3.2 kb in length. Eight HBV genotypes (A–H) have been defined based on nucleotide sequence divergence of >8% over the genome. Major genotypes A-D are found worldwide; genotypes B and C are the predominant forms in Asia and Oceania. Genotype E is largely restricted to Central and West Africa, and genotype F to Alaska, Central America and Polynesia. Genotype G is limited to a small number of people in Europe and the USA with HBV Genotype H largely confined to Central and South America.

Several commercial kits are available for measuring HBV load, using both real-time PCR and signal amplification. The assay described below has been designed to characterize the mutations associated with HBeAg-negative chronic hepatitis B by sequence analysis of the PCR product.

36.2 Acceptable Specimens

Appropriate collection and handling of blood samples is an important part of the testing procedure. While it appears the nucleic acid of the virus is relatively stable, inappropriate collection and storage can impact on the accuracy of the final result. Serum separator tubes (SST), whole blood, ACD and EDTA tubes are the preferred collection tubes.

36.3 Unacceptable Specimens

Heparin tubes are not acceptable, as heparin has been shown to be a potent inhibitor of PCR. Similarly, severely haemolysed samples may cause inhibition of the PCR assay.

36.4 Sample Extraction

The viral DNA is extracted using the Corbett X-tractor Gene System (QIAGEN, Melbourne, Australia) or alternatively, the viral DNA can be extracted using the QIAamp DNA Mini Kit (QIAGEN).

36.5 Primer Sequences

The in-house HBV DNA assay is a hemi-nested PCR assay. The primers used are modified from those described by Omata et al. [1] which have been designed to the region encoding the precore/core protein.

Primer name	Sequence 5′ to 3′	No. of bases	Function	Product size (bp)
PC2	GGC AAA AAC GAG AGT AAC TC	20	1st R primers PC5 + PC2	356
PC5	TCG CAT GGA GAC CAC CGT GA	20		
527	GTA ACT CCA CAG WAG CTC C	19	2nd R primers PC5 + 527	343

36.6 PCR Amplification and Product Detection

The first round PCR in the nested PCR protocol is carried out using QIAGEN *Taq* polymerase and buffers. The 40 μl master mix consists of 23 μl nuclease free water, 8 μl 5X Q buffer (QIAGEN), 4 μl 10x Taq buffer (QIAGEN), 4 μl template, 0.4 μl 20 mM dNTP, 0.4 μl HBV primers PC5 and PC2 (25 μM), 0.3 μl Taq DNA polymerase (5 units/μl). First round PCR amplification conditions consist of 94°C for 3 min, 35 cycles of 94°C for 30 s, 50°C for 30 s, 72°C for 30 s followed by 72°C for 5 min and 4°C for 5 min, then indefinitely.

The second round PCR is carried out using QIAGEN *Taq* polymerase and buffers. The 40 μl master mix consists of 25 μl nuclease free water, 8 μl 5x Q buffer, 4 μl 10x Taq buffer (+Mg), 2 μl template (1st round PCR product), 0.4 μl primers mix PC5 and 527 (25 μM), 0.4 μl 20 mM dNTP, 0.3 μl Taq DNA polymerase (5 units/μl). Second round PCR amplification conditions consist of 94°C for 3 min, 28 cycles of 94°C for 30 s, 50°C for 30 s, 72°C for 30 s followed by 72°C for 5 min and 4°C for 5 min, then indefinitely.

Both first round and second round PCR product are separated by electrophoresis through a 1.5% TAE agarose gel, stained with ethidium bromide and photographed using the BIO-RAD Gel Doc 2000 system.

36.7 Quality Control and Validation Data

Negative and positive control serum samples are extracted along with test samples and included in the nested RT-PCR. An additional control containing the reaction mix with water added in place of sample is also included in the RT-PCR to monitor for contamination of reaction mix reagents.

The specificity of the assay was validated with 50 clinical samples shown to be HBsAg negative and sent to VIDRL for testing unrelated to hepatitis B. No detectable PCR product was generated. Cross-reactivity was tested by assaying samples in triplicate known to be positive for HAV RNA, HCV RNA, HDV RNA and HEV RNA and no amplification product was generated. By comparison with the

WHO International Standard for Hepatitis B Virus (NIBSC Code 97/746), the lower limit of detection of the assay was estimated to be approximately 200 IU/ml.

Reference

1. Omata M, Ehata T, Yokosuka O et al (1991) Mutations in the precore region of hepatitis B virus DNA in patients with fulminant and severe hepatitis. N Engl J Med 324:1699–1704

Chapter 37
Hepatitis C Virus

Scott Bowden

37.1 Clinical Background

Infection with hepatitis C virus (HCV) has become a global health problem with a prevalence estimated by the World Health Organization of 2%, representing 123 million people. In developed countries, HCV infection is the most common form of chronic viral hepatitis and up to 20% of those chronically infected will develop cirrhosis and 1–4% of these will develop hepatocellular carcinoma. The burden of current HCV related complications is considerable and modeling predicts that the prevalence of serious sequelae due to chronic hepatitis C will increase substantially during the next few decades.

The diagnosis of HCV infection relies on both serological assays for detection of specific antibodies and on molecular assays for the detection of HCV RNA. Serological assays have been developed for the detection of virus-specific antibodies (anti-HCV) to recombinant HCV antigens and synthetic peptides by enzyme immunoassay (EIA) and since their introduction they have undergone frequent modification by the manufacturers to improve sensitivity and specificity. They represent the most cost effective screening method for determining exposure to HCV but they have limitations in that they cannot be used to diagnose active or chronic HCV infection. This role can be provided by the molecular assays, which detect the presence of HCV RNA either qualitatively or quantitatively. A number of commercially available kits are available for the diagnostic detection of HCV RNA in qualitative or viral load assays and the Therapeutic Goods Administration regulates their use in Australia (Therapeutic Goods Regulations 1990 Schedule 3 Part 2 Item 7).

Since the discovery of HCV, nucleotide sequence data has accumulated and has shown that HCV can be divided up into six phylogenetically distinct groups. Each phylogenetic group has been assigned a numerical genotype, HCV genotype 1 to 6, designated in the order of their discovery [1]. Each genotype in turn can be further

S. Bowden (✉)
Victorian Infectious Diseases Laboratory, VIDRL, Wreckyn St, Melbourne, VIC, Australia
e-mail: scott.bowden@mh.org.au

M. Schuller et al. (eds.), *PCR for Clinical Microbiology*,
DOI 10.1007/978-90-481-9039-3_37, © Springer Science+Business Media B.V. 2010

divided into a number of more closely related subtypes, denoted by a lower case letter (e.g. 1a).

Interferon-based therapy has been the mainstay of chronic hepatitis C treatment and improved outcomes have been achieved as knowledge is gained about the predictors of therapy response. Virus genotype and viral load have been shown to be the key viral characteristics to guide treatment and clinical management of patients with chronic hepatitis C.

Genotyping of HCV is usually carried out by first amplifying 5′ untranslated region (5′ UTR), and this is the most common target in diagnostic HCV RNA assays, it has also become the substrate used for most genotyping assays. Although the 5′ UTR is relatively conserved, there are generally sufficient nucleotide differences to discriminate between most genotypes. In Australia, the most commonly used genotyping assay is the commercial Line Probe Assay (LiPA) assay, originally developed by Innogenetics (Belgium). The LiPA is a reverse-phase hybridization assay in which denatured PCR product is hybridized to genotype-specific oligonucleotides bound to a nitrocellulose strip. The assay, the VERSANT HCV Genotype Assay, is marketed in Australia by Siemens Healthcare Diagnostics. The disadvantage of this method is the reliance on a single targeted region of the genome, the 5′UTR, which lacks discrimination for some of the HCV genotype 6 isolates which are common in South East Asia. Furthermore, the banding pattern observed with the LiPA was sometimes atypical and not conforming to the supplied template. To overcome some of these problems, a core region RT-PCR was designed and with subsequent sequencing the HCV genotype and/or subtype could often be determined after comparison to the HCV public database at Los Alamos.

More recently, a new generation LiPA (VERSANT HCV Genotype Assay 2.0), which simultaneously detects sequences in the 5′ UTR and the HCV core region, has been developed which may overcome some of the deficiencies of the previous assay.

37.2 Acceptable Specimens

Appropriate collection and handling of blood samples is an important part of the testing procedure. While it appears the nucleic acid of the virus is relatively stable, inappropriate collection and storage can impact on the accuracy of the final result. Serum separator tubes (SST), whole blood, ACD and EDTA tubes are the preferred collection tubes.

37.3 Unacceptable Specimens

Heparin tubes are not acceptable, as heparin has been shown to be a potent inhibitor of PCR. Similarly, severely haemolysed samples may cause inhibition of the PCR assay.

37.4 Sample Extraction

Extract 140 μl serum according to the QIAamp Viral RNA Mini Kit instructions (QIAGEN, Melbourne, Australia). Store the eluted material on ice until ready to continue the cDNA synthesis step. If the cDNA synthesis is not to be performed on the same day, store at −70°C. For each sample incubate the extracted RNA at 65°C for 10 min to relax the secondary structure of the RNA. Cool on ice.

37.5 Primer Sequences

Primer name	Sequence 5′ to 3′	No. of Bases	Function	Product size (bp)
HCV C1R	ATG TAC CCC ATG AGG TCG GC	20	RT and 1st R	
HCV C1F	ACT GCC TGA TAG GGT GCT TGC GAG	24	1st R	463
HCV C2F	TAG GGT GCT TGC GAG TGC CCC G	22	2nd R	
HCV C2R	AGG GTA TCG ATG ACC TTA C	19	2nd R	403

37.6 PCR Amplification and Product Detection

Reverse transcription and first round PCR is carried out using SuperScript III One-Step RT-PCR System with Platinum *Taq* Polymerase (Invitrogen). The 25 μl master mix consists of 12.5 μl 2x reaction mix, 10 μl template RNA, 1.25 μl R176 (10 μM) and HCV NC1 (10 μM), 0.25 μl RNasin (40 Units/μl), 1 μl RT/Taq Mix.

Reverse transcription and first round PCR amplification conditions consist of 50°C for 30 min (Reverse Transcription), 94°C for 2 min, 35 cycles of 94°C for 15 s, 55°C for 30 s, 72°C for 30 s followed by 72°C for 5 min, then 4°C indefinitely.

Second round PCR is carried out using QIAGEN *Taq* polymerase and buffers. The 40 μl master mix consist of 25 μl nuclease free water, 8 μl 5x Q Buffer, 4 μl 10x Taq buffer (+Mg), 2 μl template (1st round PCR product), 0.4 μl HCV NC3 and HCV NC4 (25 μM), 0.4 μl 20 mM dNTP, 0.3 μl Taq DNA polymerase (5 units/μl).

Second round PCR amplification conditions consist of 94°C for 2 min, 25 cycles of 94°C for 30 s, 55°C for 30 s, 72°C for 30 s followed by 72°C for 5 min, then 4°C indefinitely.

Both first round and second round PCR products are separated by electrophoresis through a 1.5% TAE agarose gel, stained with ethidium bromide and photographed using the BIO-RAD Gel Doc 2000 system.

37.7 Quality Control and Validation Data

Negative and positive control serum samples are extracted along with test samples and included in the nested RT-PCR. An additional control containing the reaction mix with water added in place of sample is also included in the RT-PCR to monitor for contamination of reaction mix reagents.

The specificity of the assay was validated with 50 clinical samples sent to VIDRL for testing unrelated to hepatitis C and which had no detectable HCV antibodies. No detectable PCR product was generated. Cross-reactivity was tested by assaying in triplicate samples known to be positive for HAV RNA, HBV DNA, HDV RNA and HEV RNA and no amplification product was generated. By titration against samples tested in a commercial viral load assay the sensitivity of the assay is approximately 200 IU/ml.

Reference

1. Simmonds P, Bukh J, Combet C et al (2005). Consensus proposals for a unified system of nomenclature of hepatitis C virus genotypes. Hepatology 42:962–973

Chapter 38
Hepatitis D Virus

Scott Bowden

38.1 Clinical Background

The hepatitis D virus (HDV) or delta agent is a defective RNA virus that requires HBV infection to complete its replication. Thus, HDV infection only occurs with simultaneous HBV co-infection or by super-infection of someone chronically infected with HBV. Acute HDV infection acquired by co-infection with HBV can range from mild to severe hepatitis and is clinically indistinguishable from acute hepatitis B. With super-infection, there is clinically a more severe acute hepatitis and because of the underlying persistent HBV infection, HDV infection generally is associated with more rapidly progressive liver disease and an increased risk of hepatocellular carcinoma. HDV infection is endemic worldwide but with the control of HBV infection by vaccination, the prevalence of HDV infection has declined substantially over the last twenty years. Despite this, it is estimated that around 5% of those chronically infected with HBV are also co-infected with HDV and these are at risk of severe and progressive liver disease.

Diagnosis of HDV infection can be done by serology based on detection of anti-HDV and while the serology assay for HDV antigen can be helpful, it has limitations because of sequestration of antigen in immune complexes. Thus, the detection of HDV RNA by PCR is an important tool for diagnosis of HDV infection [1].

HDV is unlike other animal viruses and has been classified in a so-called floating genus *Deltavirus*. The HDV genome is a single piece of circular RNA, approximately 1700 nucleotides long and shows considerable heterogeneity. Sequence data has allowed HDV to be classified into three major genotypes, designated I–III and these differ in their global distribution with genotype I being the most widespread.

S. Bowden (✉)
Victorian Infectious Diseases Laboratory, VIDRL, Wreckyn St, Melbourne, VIC, Australia
e-mail: scott.bowden@mh.org.au

M. Schuller et al. (eds.), *PCR for Clinical Microbiology*,
DOI 10.1007/978-90-481-9039-3_38,

38.2 Acceptable Specimens

Appropriate collection and handling of blood samples is an important part of the testing procedure. While it appears the nucleic acid of the virus is relatively stable, inappropriate collection and storage can impact on the accuracy of the final result. Serum separator tubes (SST), whole blood, ACD and EDTA tubes are the preferred collection tubes.

38.3 Unacceptable Specimens

Heparin tubes are not acceptable, as heparin has been shown to be a potent inhibitor of PCR. Similarly, severely haemolysed samples may cause inhibition of the PCR assay.

38.4 Sample Extraction

Extract 140 μl serum according to the QIAamp Viral RNA Mini Kit instructions (QIAGEN, Melbourne, Australia). Store the eluted material on ice until ready to continue the cDNA synthesis step. If the cDNA synthesis is not to be performed on the same day, store at −70°C. For each sample incubate the extracted RNA at 65°C for 10 min to relax the secondary structure of the RNA. Cool on ice.

38.5 Primer Sequences

Primer name	Sequence 5′ to 3′	No. of bases	Function	Product size (bp)
HDV 01	GATGCCATGCCGACCCGAAGAGG	23	1st R	
HDV 01a	AGAGGCAGGATCACCGACGAAGG	23	1st R	421
HDV N1	CCCGCTTTATTCACTGG	17	2nd R	
HDV N2	GAAGGCCCTCGAGAAC	16	2nd R	339

38.6 PCR Amplification and Product Detection

Reverse transcription and first round PCR is carried out using SuperScript III One-Step RT-PCR System with Platinum Taq Polymerase (Invitrogen). The 20 μl master mix consists of 10 μl 2x reaction mix, 8 μl template RNA, 0.25 μl primer O1 (75 ng/μl), 0.25 μl primer O1A (75 ng/μl), 0.25 μl primer N1 (200 ng/μl),

0.25 μl primer N2 (200 ng/μl), 0.25 μl RNasin (40 U/μl), 0.8 μl RT/Taq Mix (SuperScript III).

Reverse transcription and PCR amplification conditions consist of 50°C for 30 min, 94°C for 2 min, 20 cycles of 94°C for 30 s and 72°C for 60 s, followed by 30 cycles of 94°C for 30 s, 52°C for 30 s, 72°C for 30 s and finally 72°C for 5 min then 4°C indefinitely.

PCR product is separated by electrophoresis through a 1.5% TAE agarose gel, stained with ethidium bromide and photographed using the BIO-RAD Gel Doc 2000 system.

38.7 Quality Control and Validation Data

Negative and positive control serum samples are extracted along with test samples and included in the RT-PCR. An additional control containing the reaction mix with water added in place of sample is also included in the RT-PCR to monitor for contamination of reaction mix reagents.

The specificity of the assay was validated with 50 clinical samples shown to be HBsAg negative and sent to VIDRL for testing unrelated to hepatitis B or hepatitis D (note: HDV is defective and requires HBV as a helper virus to supply HBsAg, hence samples negative for HBV cannot be positive for HDV). No detectable PCR product was generated. Cross-reactivity was tested by testing samples, known to be positive for HAV RNA, HCV RNA and HEV RNA, in triplicate and no amplification product was generated.

Reference

1. Simpson LH, Battegay M, Hoofnagle JH (1994) Hepatitis delta virus RNA in serum of patients with chronic delta hepatitis. Digestive Diseases and Sciences 39:2650–2655

Chapter 39
Hepatitis E Virus

Scott Bowden

39.1 Clinical Background

Hepatitis E presents with the typical clinical and morphologic features of acute hepatitis. Discovered in 1990, hepatitis E virus (HEV) is now recognised as the agent responsible for nearly all enterically transmitted non-A, non-B hepatitis. Hepatitis E is principally the result of water borne infection and while large-scale outbreaks have been reported in several developing countries, sporadic cases can occur in developed countries where the seroprevalence is around 1%. Acute hepatitis E infection, like hepatitis A, is an acute, self-limiting infection that may vary in severity from subclinical (more likely in children) to fulminant hepatitis. The incubation period is approximately 40 days (range 15–60 days). The death rate of patients infected with HEV is <1%, however, the mortality from HEV in pregnant patients is high (around 20%) for reasons which remain unclear. The diagnosis of acute hepatitis E relies on serology and molecular testing for HEV RNA. Hepatitis E can be diagnosed by the demonstration of anti-HEV IgM and/or IgG in the serum or by detection of a viral RNA in the serum or faeces. Anti-HEV IgM testing is not performed due to poor sensitivity so anti-HEV IgG testing is the routine first test for the investigation of possible acute hepatitis E infection in most developed countries. Genomic HEV RNA is detected in the acute phase of HEV infection and is found in the serum from 2 to 7 weeks after the appearance of symptoms.

HEV is the only member of the Hepevirus genus in the *Hepeviridae*. The viral RNA genome is approximately 7.2 kb in length and contains three partially overlapping open reading frames. Phylogenetic analysis shows that HEV can be classified into four genotypes, 1–4 and a large number of subtypes. Like hepatitis A virus, there is only one serotype of HEV.

S. Bowden (✉)
Victorian Infectious Diseases Laboratory, VIDRL, Wreckyn St, Melbourne, VIC, Australia
e-mail: scott.bowden@mh.org.au

M. Schuller et al. (eds.), *PCR for Clinical Microbiology*,
DOI 10.1007/978-90-481-9039-3_39, © Springer Science+Business Media B.V. 2010

39.2 Acceptable Specimens

Appropriate collection and handling of blood samples is an important part of the testing procedure. While it appears the nucleic acid of the virus is relatively stable, inappropriate collection and storage can impact on the accuracy of the final result. Serum separator tubes (SST), whole blood, ACD and EDTA tubes are the preferred collection tubes.

39.3 Unacceptable Specimens

Heparin tubes are not acceptable, as heparin has been shown to be a potent inhibitor of PCR. Similarly, severely haemolysed samples may cause inhibition of the PCR assay.

39.4 Sample Extraction

Extract 140 μl serum according to the QIAamp Viral RNA Mini Kit instructions (QIAGEN, Melbourne, Australia). Store the eluted material on ice until ready to continue the cDNA synthesis step. If the cDNA synthesis is not to be performed on the same day, store at −70°C. For each sample incubate the extracted RNA at 65°C for 10 min to relax the secondary structure of the RNA. Cool on ice.

39.5 Primer Sequences

The in-house HEV RT-PCR comprises two nested PCR assays, with primers designed to regions of identity between the open reading frame 1 (ORF1) and open reading frame two (ORF2) respectively, of Burmese and Mexican isolates [1].

39.6 PCR Amplification and Product Detection

Reverse transcription and first round PCR amplification is carried out using SuperScript III One-Step RT-PCR System with Platinum *Taq* Polymerase (Invitrogen). The 25 μl ORF1 1st R reaction mix – (use ORF2 1st R primers for ORF2 1st R reaction mix) consists of 12.5 μl 2x reaction mix, 10 μl template RNA, 1.25 μl HEVORF1 – outerF (10 μM) and HEVORF1 – outerR (10 μM), 0.25 μl RNasin (40 Units/μl), 1 μl RT/Taq Mix.

Reverse transcription and first round PCR amplification conditions are 50°C for 30 min (reverse transcription), 94°C for 2 min, 40 cycles of 94°C for 30 s, 55°C for 30 s, 72°C for 30 s followed by 72°C for 5 min then 4°C indefinitely.

Second round PCR amplification is carried out using QIAGEN *Taq* polymerase and buffers. The 40 μl ORF1 2nd R reaction mix (use ORF2 primers for ORF2 2nd reaction mix) consists of 25 μl nuclease free water, 8 μl 5x Q Buffer, 4 μl 10x Taq buffer (+Mg), 2 μl template (1st round PCR product), 0.4 μl HEVORF1-InnerF +

Primer name	Sequence 5′ to 3′	No. of bases	Function	Product size (bp)
HEVORF1-OuterF	CTG GCG ATY ACT ACT GCY ATT GAG C	25	1st R	
HEVORF1-OuterR	CCA TCR ARR CAG TAA GTG CGG TC	23	RT and 1st R	336
HEVORF1-InnerF	CTG CCY TKG CGA ATG CTG TGG	21	2nd R	
HEVORF1-InnerR	GCG AGW RTA CCA RCG CTG AAC ATC	24	2nd R	288
HEVORF2-OuterF	GAC AGA ATT RAT TTC GTC GGC TGG	24	1st R	
HEVORF2-OuterR	CTT GTT CRT GYT GGT TRT CAT AAT C	25	RT and 1st R	198
HEVORF2-InnerF	GTY GTC TCR GCG AAT GGC GAG C	22	2nd R	
HEVORF2-InnerR	GTT CRT GYT GGT TRT CAT AAT CCT G	25	2nd R	146

HEVORF1-OuterR (25 μM), 0.4 μl 20 mM dNTP, 0.3 μl Taq DNA polymerase (5 units/μl).

Second round PCR amplification conditions are 94°C for 2 min, 25 cycles of 94°C for 30 s, 55°C for 30 s, 72°C for 30 s followed by 72°C for 5 min and 4°C indefinitely.

Both first round and second round PCR products are separated by electrophoresis through a 1.5% TAE agarose gel, stained with ethidium bromide and photographed using the BIO-RAD Gel Doc 2000 system.

39.7 Quality Control and Validation Data

Negative and positive control serum samples are extracted along with test samples and included in the nested RT-PCR. An additional control containing the reaction mix with water added in place of sample is also included in the RT-PCR to monitor for contamination of reaction mix reagents.

The specificity of the assay was validated with 50 clinical samples sent to VIDRL for testing unrelated to hepatitis E and which had no detectable HEV antibodies. No detectable PCR product was generated. Cross-reactivity was tested by assaying samples in triplicate known to be positive for HAV RNA, HBV DNA, HCV RNA and HDV RNA and no amplification product was generated.

Reference

1. Erker JC, Desai SM, Mushahwar I (1999) Rapid detection of hepatitis E virus RNA by reverse transcription-polymerase chain reaction using universal oligonucleotide primers. J Virol Methods 81:109–113

Chapter 40
Herpes Simplex Virus Type 1 and 2

Neisha Jeoffreys and David M. Whiley

40.1 HSV Conventional PCR Assay

Neisha Jeoffreys

40.1.1 Summary of Methods

Traditionally, herpes simplex virus (HSV) central nervous system infections have been diagnosed by inoculation of cell culture with brain biopsy material. In more recent times however, PCR detection of HSV in CSF has become the 'gold standard' for diagnosis of HSV encephalitis [3]. Occasionally, in cases of retrospective or neonate infections, serology and culture are used as an adjunct to PCR [1].

The most common PCR targets to detect HSV are the DNA polymerase, thymidine kinase and glycoprotein B, C, D and G genes [4]. This protocol describes a conventional 'in-house' PCR assay to detect both HSV 1 and 2 by targetting the DNA polymerase gene (*polA*).

40.1.2 Acceptable Specimens

CSF ($\geq$200 μl) is the most appropriate specimen for identification of HSV infection in the CNS. Brain biopsy material, including fresh, formalin fixed or paraffin embedded (PE), is also acceptable, although fresh tissue is preferable.

N. Jeoffreys (✉)
Molecular Biology, Centre for Infectious Diseases and Microbiology, Institute for Clinical Pathology and Medical Research, Westmead Hospital, Westmead, NSW 2145, Australia
e-mail: Neisha.Jeoffreys@swahs.health.nsw.gov.au

M. Schuller et al. (eds.), *PCR for Clinical Microbiology*,
DOI 10.1007/978-90-481-9039-3_40,

40.1.3 Sample Extraction

DNA is extracted from specimens using the GeneElute Genomic DNA extraction kit (Sigma-Aldrich Co.) according to the manufacturer's instructions. In brief, for CSF, 200 μl of sample is combined with 200 μl of Lysis C solution and 20 μl of proteinase K (20 mg/ml) and incubated for 10 min at 55°C. Two hundred microlitres of 100% ethanol is then added and the solution centrifuged through a prepared silica spin column. The membrane is washed twice with Wash solution and the DNA is eluted with 200 μl of Elution solution.

For fresh tissue, the sample is initially ground in phosphate buffered saline (PBS). Two hundred microlitres of ground tissue is then combined with 180 μl of Lysis T solution and 20 μl proteinase K (20 mg/ml) and incubated at 55°C overnight. Two hundred microlitres of Lysis C solution is then added and the mixture incubated for a further 10 min at 70°C before addition of 200 μl of 100% ethanol and transfer to a prepared silica spin column as described above.

For PE tissues, 10–20 slices (10 μm thickness) are combined with 5 ml of histolene (Sigma-Aldrich Co.) to dissolve the paraffin. The tissue is then collected by centrifugation at 4800 rpm and washed twice with 100% ethanol, followed by a third wash with 70% ethanol. The ethanol is then removed and the tissue pellet air dried before addition of Lysis T and proteinase K as described for fresh tissues.

40.1.4 Primer and Probe Sequences

The primers PolA1 (5′-ATC ATC TAC GGG GAC ACG GAC T-3′) and PolA2 (5′-TCC ACG CCC TTG ATG A GC ATC T-3′) are used to amplify a 224 bp region of the *polA* gene [10].

40.1.5 PCR Amplification and Product Detection

PCR is performed in a 25 μl volume containing 800 nM of each primer (polA1 and polA2), 400 μM dNTPs (Roche Applied Science), 4 mM $MgCl_2$,, 20 mM Tris-HCl (pH 8.4), 50 mM KCl, 0.75 U of Platinum Taq Polymerase (Invitrogen Life Technologies) and 10 μl of DNA. Amplification is performed in a MasterCycler® gradient (Eppendorf) thermal cycler under the following cycling conditions: 95°C for 10 min, followed by 45 cycles of 95°C for 15 s, 65°C for 30 s, and 72°C for 90 s, and a final extension of 72°C for 5 min.

PCR amplicons are separated by agarose gel electrophoresis and visualized using SYBR Safe DNA stain (Invitrogen Life Technologies). PCR amplicons with an identical length to the positive control are verified and differentiated into HSV

1 or HSV 2 using restriction endonuclease digestion with *AluI* (New England Biolabs).

For restriction endonuclease digestion, 6 μl of PCR product is added to 6 U of *AluI* and 1X Buffer 2 to give a final volume of 10 μl and incubated for 1 h at 37°C. Digested products are separated using polyacrylamide gel electrophoresis (PAGE) and visualized using SYBR Safe DNA stain.

HSV 1 positive samples produce two fragments; 153 and 71 bp in size. HSV 2 positive samples produce three fragments; 126, 71 and 27 bp in size. Negative samples are not digested or produce fragments of the incorrect size and/or number.

40.1.6 Quality Control and Validation Data

All samples are tested both neat and diluted (1:100) to ensure both strong and weak positive specimens are detected. Each sample is also tested in parallel with an inhibition control and two negative water blanks which act as contamination controls during DNA extraction and PCR set-up. The inhibition control contains both dilute sample and positive control material (1:1). A negative result in this tube indicates inhibition in the sample.

A serial dilution of both the HSV 1 and HSV 2 positive controls is also included with every run to ensure the limit of detection of the assay for both HSV types remains constant over time. These positive controls are also digested using the restriction endonuclease to ensure the accuracy of the verification process.

The assay has a limit of detection of 1–10 virus copies/μl. The sensitivity and specificity of the assay against brain biopsy tissue for the diagnosis of encephalitis has not been determined due to the small number of brain biopsy samples available. In the literature, HSV PCR on CSF samples is generally considered to have a sensitivity of 96% and specificity of 99% [5].

40.1.7 Assay Limitations

The sensitivity of the assay is most limited by the quality and type of specimens received and the duration of infection. PE and formalin fixed tissue samples or samples with insufficient volume have a significant impact on assay sensitivity and false negative results are not uncommon. Additionally, the sensitivity of the assay may be reduced in samples collected within the first 72 h of infection [9]; in patients who have been symptomatic for longer than a week [2] or in neonates [6]. In these cases, it is recommended that repeat samples be requested for testing if negative PCR results are obtained from patients with a high clinical suspicion of disease.

40.2 HSV Real-Time PCR Assay

David M. Whiley

40.2.1 Summary of Methods

This is an alternative HSV PCR protocol to the one described above. Although either method may be used for the detection of HSV-1 and -2 in STI, our laboratory prefers to use an alternative method to prevent cross contamination of the CSF testing environment with amplicon from the high volume swab (STI) testing protocol.

This protocol describes a duplex 5′ nuclease-based real-time PCR assay, utilising two type-specific reactions in the one mix, for simultaneous detection of HSV-1 and -2. Each reaction comprises two primers and one probe, and targets type-specific sequences of the glycoprotein D genes of HSV-1 and -2. Amplification of either virus is distinguished through the use of different fluorophores (FAM and JOE) on each probe.

40.2.2 Acceptable Specimens

Acceptable specimens are dry swabs, including genital, oral, and rectal swabs.

40.2.3 Sample Extraction

Swabs are rotated 15 times in 400 μl of phosphate buffered saline (PBS) to dislodge cells. 200 μl of PBS or urine are then extracted manually using the High Pure Viral Nucleic Acid kit (Roche Applied Science, Australia), or mechanically using the MagNAPure instrument (Roche Applied Science, Australia), according to the manufacturer's instructions.

40.2.4 Primers and Probe Sequences

Two primer and probe sets are used in this assay one each for HSV-1 and HSV-2 respectively. These are both directed to the gene coding for glycoprotein D. The

D.M. Whiley (✉)
Queensland Paediatric Infectious Diseases Laboratory, Sir Albert Sakzewski Virus Research Centre, Queensland Children's Medical Research Institute, Children's Health Service District, Brisbane, QLD 4029, Australia
e-mail: d.whiley@uq.edu.au

primer and probe sequences for HSV-1 are: HSV1UP: 5′-CGG CCG TGT GAC ACT ATC G-3′; and HSV1DP: 5′-CTC GTA AAA TGG CCC CTC C-3′ with probe sequence HSV1P: 5′-FAM- CCA TAC CGA CCA CAC CGA CGA ACC -BHQ1-3′ [8]. The primer and probe sequences for HSV-2 are: HSV2-F: 5′-CGC CAA ATA CGC CTT AGC A-3′ and HSV2-R: 5′- GAA GGT TCT TCC CGC GAA AT-3′. The HSV-2 specific probe sequence is: HSV2-TM: 5′- JOE-CTC GCT TAA GAT GGC CGA TCC CAA TC-BHQ1-3′ [7].

40.2.5 PCR Amplification and Product Detection

Note: Both the HSV-1 and -2 PCR reactions are performed in the same reaction mix.

The PCR reactions are performed in 0.2 ml or 0.1 ml reaction tubes. Each reaction mix contains 2.5 pmol of each forward and reverse primer (HSV1UP, HSV1DP, HSV2-F and HSV2-R), 4 pmol of each probe (HSV1P and HSV2-TM), 1 × QIAGEN QuantiTect Probe PCR Master Mix (QIAGEN, Australia), and 5.0 μl of DNA, in a final reaction volume of 25 μl.

Thermocycling is performed on a RotorGene 3000 or 6000 instrument (QIAGEN, Australia) with an initial 15 min at 95°C followed by 45 cycles of 95°C 15 s, 60°C 60 s (fluorescence acquisition). HSV-1 results are analysed in the Rotorgene FAM channel and HSV-2 results in the Rotorgene JOE channel. An amplification curve observed in each channel is indicative of a positive reaction for the relevant virus.

40.2.6 Quality Control and Validation Data

Extracted DNA from HSV-1 and -2 isolates are used as a positive control. Positive control should be diluted to a concentration providing a cycle-threshold (Ct) value of approximately 30 cycles. Extract from PCR-grade water is used as a negative control.

The clinical sensitivity and specificity of these assays are over 95%. Although such measurements are dependent on the gold standard, sensitivity and specificity is achieved when compared with an expanded gold standard composed of consensus results from two to three different nucleic acid amplification assays.

The clinical PPV and NPV, similar to sensitivity and specificity are dependent on the gold standard. PPV and NPV of over 95% are generally achieved using these assays.

References

1. Boivin G (2003) Diagnosis of herpesvirus infections of the central nervous system. Herpes (Suppl 2):48A–56A
2. Cinque P, Cleator GM, Weber T et al (1996) The role of laboratory investigation in the diagnosis and management of patients with suspected herpes simplex encephalitis: a consensus

report. The EU concerted action on virus meningitis and encephalitis. J Neurol Neurosurg Psychiatry 61:339–345

3. DeBiasi RL, Kleinschmidt-DeMasters BK, Weinberg A et al (2002) Use of PCR for the diagnosis of herpesvirus infections of the central nervous system. J Clin Vir 25(Suppl 1): S5–S11
4. Mitchell PS, Espy MJ, Smith TF et al (1997) Laboratory diagnosis of central nervous system infections with herpes simplex virus by PCR performed with cerebrospinal fluid specimens. J Clin Microbiol 35:2873–2877
5. Tebas P, Nease RF, Storch GA (1998) Use of the polymerase chain reaction in the diagnosis of herpes simplex encephalitis: a decision analysis model. Am J Med 105:287–295
6. Troendle-Atkins J, Demmler GJ, Buffone GJ (1993) Rapid diagnosis of herpes simplex virus encephalitis by using the polymerase chain reaction. J Pediatr 123:376–380
7. Watzinger F, Suda M, Preuner S et al (2004) Real-time quantitative PCR assays for detection and monitoring of pathogenic human viruses in immunosuppressed pediatric patients. J Clin Microbiol 42:5189–5198
8. Weidmann M, Meyer-König U, Hufert FT (2003) Rapid detection of herpes simplex virus and varicella-zoster virus infections by real-time PCR. J Clin Microbiol 41:1565–1568
9. Weil AA, Glaser CA, Amad Z et al (2002) Patients with suspected herpes simplex virus encephalitis: rethinking an initial negative polymerase chain reaction result. CID 34: 1154–1157
10. Yamamoto LJ, Tedder DG, Ashley R et al (1991) Herpes simplex virus type 1 DNA in cerebrospinal fluid of a patient with Mollaret's meningitis. N Eng J Med 325:1082–1085

Chapter 41
Human Bocavirus

Sarah J. Tozer

41.1 Summary of Methods

Human bocavirus (HboV) has been detected worldwide in respiratory samples. This protocol describes two real-time PCR assays, targeting the non-structural protein (NP-1) and viral protein (VP-1) genes, and was validated to detect HboV in patients with respiratory disease, gastroenteritis, or systemic illness [2].

41.2 Acceptable Specimens

Respiratory secretions, nose/throat swabs, stool samples, whole blood.

41.3 Sample Extraction

Two hundred microlitres of respiratory specimens were extracted using the QIAGEN X-tractor Gene (QIAGEN, Australia); the stool samples using the Roche MagNA Pure LC Instrument (Roche Applied Science, Australia); and the blood samples using the Roche High Pure Viral Nucleic Acid kit (Roche Applied Science, Australia).

41.4 Primers and Probe Sequences

The primers and Taqman probes for HboV real-time PCR were designed using Bioedit 7.0.5.3 and Primer Express 2.0 software. Two sets of primers and probe sequences were used, one targeting the HboV viral structural protein VP-1 and the

S.J. Tozer (✉)
Queensland Paediatric Infectious Diseases Laboratory, Sir Albert Sakzewski Virus Research Centre, Queensland Children's Medical Research Institute, Children's Health Service District, Brisbane, QLD 4029, Australia
e-mail: s.tozer@uq.edu.au

M. Schuller et al. (eds.), *PCR for Clinical Microbiology*,
DOI 10.1007/978-90-481-9039-3_41, © Springer Science+Business Media B.V. 2010

second the gene coding for the nuclear protein (NP). The primers targeting the VP-1 region were STBoVP-1f: GGC AGA ATT CAG CCA TAC TCA AA; and STBoVP-1r: TCT GGG TTA GTG CAA ACC ATG A; and targeting the NP-1 were STBoNP-1f: AGC ATC GCT CCT ACA AAA GAA AAG; and STBoNP-1r: TCT TCA TCA CTT GGT CTG AGG TCT.

The respective probes were STBoVP-1pr: JOE-AGA GTA GGA CCA CAG TCA TCA GAC ACT GCT CC-BHQ; and STBoNP-1pr: FAM-AGG CTC GGG CTC ATA TCA TCA GGA ACA-BHQ. (BHQ = black hole quencher).

41.5 PCR Amplification and Product Detection

Both assays used the same PCR mix and cycling conditions. Briefly 12.5 μl of QIAGEN Quantitect Probe Master Mix (QIAGEN, Australia), 10 pmol of each primer, 4 pmol of the corresponding probe, and 2 μl of sample nucleic acid extract in a final reaction volume of 25 μl. Amplification was performed on a RotorGene 6000 (QIAGEN, Australia) using the following parameters; initial incubation of 15 min at 95°C, followed by 50 cycles of 95°C for 15 s, and 60°C for 1 min, with fluorescence acquired at the end of each 60°C step.

41.6 Quality Control and Validation Data

An HboV positive control (previously confirmed positive clinical specimens in our laboratory) is included in every assay. Saline is extracted as per the procedure applied to clinical specimens, and used as a negative control. Using a conventional assay as the reference standard [1], the sensitivity of both real-time NP-1 and VP-1 assays was 100%, with specificities of 94 and 93%, respectively. The reliable detection limit of the VP-1 and NP-1 assays was 10 copies per reaction. The conventional PCR assay had a reliable detection limit of 100 copies per reaction. Analytical specificity was assessed by testing the assays against unrelated viruses ($n = 25$) and bacteria ($n = 160$). No cross-reactions were identified with these organisms or human genomic DNA in either real-time assay.

References

1. Allander T, Tammi MT, Eriksson M et al (2005) Cloning of a human parvovirus by molecular screening of respiratory tract samples. Proc Natl Acad Sci USA 102:12891–12896
2. Tozer SJ, Lambert SB, Whiley DM et al (2009) Detection of human bocavirus in respiratory, faecal and blood samples by real time PCR. J Med Virol 81:488–493

Chapter 42
Human Coronaviruses

Rebecca Rockett

42.1 Summary of Methods

Human Coronaviruses (HCoVs) are recognised to be an important cause of the common cold. In 1962 HCoV-229E and HCoV-OC43 where first recognised, more recently HCoV-NL63 and HCoV-HKU-1 have been discovered in respiratory specimens from children and adults [3]. This protocol describes two real-time reverse-transcriptase polymerase chain reaction (RT-PCR) methods, a single target and triplex RT-PCR that identifies and differentiates HCoV infection.

42.2 Acceptable Specimens

Respiratory specimens (Nasopharyngeal aspirates, nose and throat swabs and bronchial samples).

42.3 Sample Extraction

Respiratory specimens (200 μl) were extracted using the QIAGEN QIAxtractor, as per manufacturer's instructions giving a final volume of 50 μl.

42.4 Primer and Probe Sequences

Primers and probe targets were obtained from previously published articles [1–3] (Table 42.1).

R. Rockett (✉)
Queensland Paediatric Infectious Diseases Laboratory, Sir Albert Sakzewski Virus Research Centre, Queensland Children's Medical Research Institute, Children's Health Service District, Brisbane, QLD 4029, Australia
e-mail: r.rockett@uq.edu.au

M. Schuller et al. (eds.), *PCR for Clinical Microbiology*,
DOI 10.1007/978-90-481-9039-3_42,

Table 42.1 Sequence details of primers and probes, used for the detection of human coronaviruses by real-time PCR

Oligonucleotides	Sequence (5′–3′)	Target gene	Refs.
Single Target		Replicase 1b	(2)
HCoV-HKU-1-F	CCTTGCGAATGAATGTGCT		
HCoV-HKU-1-R	TTGCATCACCACTGCTAGTACCAC		
HCoV-HKU-1-R	FAM–TGTGTGGCGGTTGCTATTATGTTAAGCCTG-BHQ-1		
Triplex RT-PCR			
HCoV-229E-F	CAGTCAAATGGGCTGATGCA		
HCoV-229E-R	AAAGGGCTATAAAGAGAATAAGGTATTCT	N gene	(1)
HCoV-229E-Pr	FAM-CCCTGACGACCACGTTGTGGTTCA-BHQ		
HCoV-OC43-F	CGATGAGGCTATTCCGACTAGGT		
HCoV-OC43-R	CCTTCCTGAGCCTTCAATATAGTAACC	N gene	(1)
HCoV-OC43-Pr	QUASAR-TCCGCCTGGCACGGTACTCCCT-BHQ		
HCoV-NL63-F	ACGTACTTCTATTATGAAGCATGATATTAA		
HCoV-NL63-R	AGCAGATCTAATGTTATACTTAAAACTACG	1a gene	(3)
HCoV-NL63-Pr	VIC-ATTGCCAAGGCTCCTAAACGTACAGGTGTT-TAMRA (300)		

42.5 PCR Amplification and Product Detection

The first assay contains a single target for HCoV-HKU1, the conserved primer and probe sequences are found in the replicase 1b gene. The second assay detects three HCoV strains, HCoV-229E, HCoV-OC43 both contain target sequences in the N gene, and HCoV-NL63 which target sequences are in the 1a gene (Table 42.1). Both assays are performed using the QIAGEN One-step RT-PCR kit (QIAGEN, Australia) comprising 0.8 μM of forward and reverse primers (HCoV-HKU-1-F and HCoV-HKU1-R for the single target PCR and the triplex HCoV-229E-F, HCoV-229E-R, HCoV-OC43-F, HCoV-OC43-R, HCoV-NL63-F and HCoV-NL63-R) (Table 42.1), 0.2 μM of TaqMan probe (single target PCR, HCoV-HKU1-Pr and in the triplex HCoV-229E-Pr, HCoV-OC43-Pr and HCoV-NL63-Pr) (Table 42.1), in a total reaction volume of 25 μl including 5 μl of sample RNA. Both amplification reactions were performed on the RotorGene 3000 or 6000 (QIAGEN, Australia) using the following parameters; initial RT incubation of 20 min at 50°C, followed by 50 cycles of 95°C for 15 s, and 60°C for 1 min, with fluorescence acquired at the end of each 60°C step.

42.6 Quality Control and Validation Data

The single target HCoV-HKU1 assay was validated using 31 clinical specimens positive for HCoV, detected using various different techniques. No cross reaction was detected to samples positive for other respiratory pathogens. Sensitivity was measured by serial titration, 15 replicates where tested with 100% detection at 50 copies and 33.3% at 5 copies [1].

The HCoV triplex was validated by ATCC positive specimens for each strain. Each positive was tested as an individually target and in the triplex reaction with no loss of sensitivity or cross reaction between strains [2].

References

1. Dare RK, Fry AM, Chittaganpitch M et al (2007) Human coronavirus infections in rural Thailand: a comprehensive study using real-time reverse-transcription polymerase chain reaction assays. J Infect Dis 196:1321–1328
2. Gunson RN, Collins TC, Carman WF (2005) Real-time RT-PCR detection of 12 respiratory viral infections in four triplex reactions. J Clin Virol 33:341–344
3. van Elden LJR, van Loon AM, van Alphen F et al (2004) Frequent Detection of human coronaviruses in clinical specimens from patients with respiratory tract infection by use of a novel real-time reverse-transcriptase polymerase chain reaction. J Infect Dis 189:652–657

Chapter 43
Human Herpes Viruses -6, -7 and -8

Jonathan Howard

This multiplex PCR is a nested PCR requiring two rounds of amplifications.

43.1 Acceptable Specimens

EDTA blood, bone marrow and body fluids (amniotic fluid, CSF), biopsies, newborn screening cards.

43.2 Sample Extraction

For all samples except new-born screening cards, total nucleic acid is extracted using MagNA Pure LC System (Roche, Australia) using MagNA Pure Total Nucleic Acid Extraction kit following the manufacturer's protocol. The extraction of DNA from newborn screening cards (NBSC), is similar to the CMV DNA extraction [1, 2]. Briefly, three disks of 3 mm in diameter is prepared from NBSC using a Wallac 1296-071 DBS Puncher (PerkinElmer, Finland). Carryover DNA contamination is excluded by punching 30 disks from a clean, blank NBSC after punching each sample card. Three blank disks (as negative controls) are processed as test samples to detect any carryover DNA between cards. Blood is eluted from the NBSC by incubating in 45 μl minimal essential media at 55°C for 60 min prior to boiling at 100°C for 7 min. Samples were rapidly cooled, centrifuged at $10{,}000\times g$ for 3 min and frozen at −80°C for at least 1 h before addition to the PCR mix.

43.3 Primer Sequences

HHV-6, -7 and -8 primer sequences are shown in Table 43.1.

J. Howard (✉)
SEALS Virology Division, Prince of Wales Hospital, Randwick, NSW 2031, Australia
e-mail: jonathan.howard@sesiahs.health.nsw.gov.au

M. Schuller et al. (eds.), *PCR for Clinical Microbiology*,
DOI 10.1007/978-90-481-9039-3_43, © Springer Science+Business Media B.V. 2010

Table 43.1 HHV-6, -7 and -8 primer sequences

Primer	Sequences	Product size (bp)	Target gene	Limit of detection (copies per reaction)
HHV6-OF	5′ GCA CTG CTG TTC TCC AAA GT 3′	344	large tegument	10^2
HHV6-OR	5′ CTA CAT TGT CAT CAT CTT CAT GGA T 3′		protein gene	
HHV6-IF	5′ ATG ATG ATA CTG GTT TGA TTA 3′	119	(LTP)	
HHV6-IR	5′ TTG ATT CTG AAG GGT CTA CTA TGT 3′			
HHV7-OF	5′ – ATA GTT GGG TTA GGC ATC AC - 3′	296	U57 region	10^2
HHV7-OR	5′ – CAC AAA AGC GTC GCT ATC AA - 3′		(major capsid protein)	
HHV7-IF	5′ – ATC CTT CCG AAA CCG ATG CTT - 3′	174		
HHV7-IR	5′ – ATA CAC GAA GCC CTT CCT CT - 3′			
HHV8-EF	5′ – CCA GCT AGC AGT GCT ACC CCC ATT - 3′	419	Glycoprotein M	10^2
HHV8-ER	5′ - ATG GAC AGA TCG TCA AGC ACT CGC - 3′		(DNA rep. protein)	
KS1	5′ – AGC CGA AAG GAT TCC ACC AT -3′	223		
KS2	5′ – TCC GTG TTG TCT AGT CCA G -3′			
Reference [3]				

43.4 Composition of Primer Mixes

Outer forward (OF) mix = 3.75 μM HHV6-OF + 2.5 μM HHV7-OF + 2 μM HHV8-EF
Outer reverse (OR) mix = 3.75 μM HHV6-OR + 2.5 μM HHV7-OR + 2 μM HHV8-ER
Inner forward (IF) mix = 3.75 μM HHV6-IF + 2.5 μM HHV7-IF + 2 μM KS1
Inner reverse (IR) mix = 3.75 μM HHV6-IR + 2.5 μM HHV7-IR + 2 μM KS2

43.5 PCR Amplification and Product Detection

Nuclease-free water	4.2 μl
5X GoTaq Hot Start buffer	5.0 μl
$MgCl_2$ (25 mM)	3.5 μl
dNTPs (1 mM)	5.0 μl
Outer Forward Primer Mix	1 μl
Outer Reverse Primer Mix	1 μl
GoTaq Hot Start polymerase	0.3 μl
Template	5.0 μl

The first and second rounds of amplifications consist of initial denaturation step at 94°C for 2 min followed by 30 cycles of 94°C for 30 s, 58°C for 40 s and 72°C for 50 s with a final extension at 72°C for 5 min and a 4°C hold cycle.

The reaction mixture for the second round PCR is the same as the first round reaction mixture except replacing the outer forward and reverse primers with inner forward and reverse primers.

Products are subjected to electrophoresis on 2% agarose gel containing ethidium bromide (0.5 μg/ml). Human herpes virus types 6, 7 and 8 positive DNA and negative water controls are also included with every reaction.

References

1. Barbi, M. et al (2000) Cytomegalovirus DNA detection in Guthrie cards: A powerful tool for diagnosing congenital infection. J Clin Virol 17: 159–165
2. Howard J et al (2009) Utility of newborn screening cards for detecting CMV infection in cases of stillbirth. J Clin Virol 44:215–218
3. McIver CJ et al (2005) Development of multiplex PCRs for detection of common viral pathogens and agents of congenital infections. J Clin Microbiol 43:5102–5110

Chapter 44
Human Papillomavirus

Sepehr N. Tabrizi

44.1 Summary of Methods

This protocol utilises a PCR-ELISA system. Multiple primers are used to amplify HPV L1 sequences by PCR in a conventional thermocycler. Detection of amplification product is achieved by biotinylated probes in a microtitre plate using an ELISA format [3].

44.2 Acceptable Specimens

A wide range of mucosal specimens can be utilised for HPV testing. Acceptable specimens are dry swab (including self-collected vaginal), tampon [2], cells collected in liquid based cytology medium PreservCyt™ (Hologic, Massachusetts, USA) and SurePath™ (BD, New Jersey, USA) and biopsy (including paraffin embedded)

44.3 Unacceptable Specimens

Samples other than those indicated above, would need to be validated prior to their use.

44.4 Sample Extraction

Perform the following ***pre-extraction*** steps prior to extraction depending on the specimen type tested:

S.N. Tabrizi (✉)
Microbiology and Infectious Diseases Department, Royal Women's Hospital,
Parkville, VIC 3052, Australia; Department of Obstetrics and Gynecology, University of Melbourne, Parkville, VIC 3052, Australia
e-mail: Sepehr.Tabrizi@thewomens.org.au

M. Schuller et al. (eds.), *PCR for Clinical Microbiology*,
DOI 10.1007/978-90-481-9039-3_44,

- *Dry swab* – rotate 15 times in 400 μl of phosphate buffered saline (PBS).
- *Tampon samples* – This sample should be transported into 10 ml PBS. Using a sterile glove, twist tampon and collect the resultant cell suspension. Pellet at 4000×*g* 10 min and resuspend cell pellet in 500 μl of PBS.
- *Liquid cytology medium* – Aliquot 1 ml of PreservCyt and SurePath media and pellet at 15,000×*g* for 10 min and resuspend cell pellet in 200 μl of PBS [6].
- *Biopsy* – Incubate 1–2 mm [2] of fresh tissue with 400 μl of ATL buffer (QIAGEN, Australia) and 4 μl (20 mg/ml) of proteinase K (Sigma, Missouri, USA) at 55°C until the tissue is completely digested.
- *Paraffin embedded tissue* – Mix 5–10 μm paraffin section with 800 μl Histolene (Grale Scientific, Victoria Australia). After gentle mixing, add 400 μl absolute ethanol. Centrifuge at 15,000×*g* and discard supernatant. Wash tissue pellet with 1 ml of absolute ethanol followed by centrifugation as above. Allow tissue pellet to dry briefly and subsequently add 190 μl ATL (QIAGEN, Australia) and 10 μl (20 mg/ml) proteinase K (Sigma) and incubate at 55°C until the tissue is completely digested.

Extraction of 200 μl of processed sample can be performed by the manual QIAamp DNA Blood Mini Kit (QIAGEN, Australia) or automated MagNA Pure LC (Roche Applied Science, Australia) using the DNA Isolation Kit 1 (Roche Applied Science, Australia) according to the manufacturer's instructions. DNA should be eluted in 100 μl final volume and a 10 μl aliquot used for PCR.

44.5 Primers and Probe Sequences

The primers and probe sequences used for HPV PCR testing are shown in Table 44.1.

44.6 PCR Amplification and Product Detection

Each PCR reaction mix contains 2.5 units of AmpliTaq Gold® DNA Polymerase (Applied Biosystems, Australia) 1x reaction buffer (50 mM KCl, 10 mM Tris pH 8.3), 200 μM of each of dATP, dGTP and dCTP and 190 mM dTTP, 10 mM digoxigenin-dUTP, 4 mM $MgCl_2$, 10 pmol of each of HPV primers and 5 pmol of beta globin primers GH20-PCO4 in a total of 50 μl. The addition of digoxigenin-dUTP allows the incorporation of digoxigenin dUTP in amplicons.

Consensus probe is synthesised using diluted HPV 16, 18 and 31 amplicons. Briefly, a 50 μl reaction contains 1 μl of the appropriate diluted (1:1000) Ll product, 2.5 unit of AmpliTaq Gold® DNA Polymerase (Applied Biosystems, Australia), 1x reaction buffer (50 mM KCl, 10 mM Tris pH 8.3), 200 μM of each of dATP, dGTP and dCTP and 100 μM dTTP, 100 μM Bio-16-dUTP.

Table 44.1 Primers and probes used for HPV PCR testing

Primers targeting 450 bp region of L1 gene [3]:	
PGMY11-A	5′-GCACAGGGACATAACAATGG-3′
PGMY11-B	5′-GCGCAGGGCCACAATAATGG-3′
PGMY11-C	5′-GCACAGGGACATAATAATGG-3′
PGMY11-D	5′-GCCCAGGGCCACAACAATGG-3′
PGMY11-E	5′-GCTCAGGGTTTAAACAATGG-3′
PGMY09-F	5′-CGTCCCAAAGGAAACTGATC-3′
PGMY09-G	5′-CGAGGTAAAGGAAACTGATC-3′
PGMY09-H	5′-CGTCCAAAAGGAAACTGATC-3′
PGMY09-I	5′-GCCAAGGGGAAACTGATC-3′
PGMY09-J	5′-CGTCCCAAAGGATACTGATC-3′
PGMY09-K	5′-CGTCCAAGGGGATACTGATC-3′
PGMY09-L	5′-CGACCTAAAGGGAATTGATC-3′
PGMY09-M	5′-CGACCTAGTGGAAATTGATC-3′
PGMY09-N	5′-CGACCAAGGGGATATTGATC-3′
PGMY09-P	5′-GCCCAACGGAAACTGATC-3′
PGMY09-Q	5′-CGACCCAAGGGAAACTGGTC-3′
PGMY09-R	5′-CGTCCTAAAGGAAACTGGTC-3′
HMB01	5′-GCGACCCAATGCAAATTGGT-3′

The following primers targeting 260 bp region of human beta-globin gene should also be included as internal control:

PC04 5′-CAACTTCATCCACGTTCACC-3′
GH20 5′-GAAGAGCCAAGGACAGGTAC-3′

The generic Ll probes are synthesized from dilute solutions of the Ll PCR fragments of HPV 16, 18, and 31 with nested type-specific primers [1, 5] as listed.

HPV 16F 5′-CATTTGTTGGGGTAACCAAC-3′
HPV 16R 5′-TAGGTCTGCAGAAAACTTTTC-3′
HPV 18F 5′-TGTTTGCTGGCATAATCAAT-3′
HPV 18R 5′-TAAGTCTAAAGAAAACTTTTC-3′
HPV 31F 5′-TATTTGTTGGGGCAATCAG-3′
HPV 31R 5′-CTAAATCTGCAGAAAACTTTT-3′

Set up three reactions with the following changes in primer and magnesium concentrations.

HPV16 reaction: 10 pmol each primer (HPV 16F and 16R), 4 mM $MgCl_2$;
HPV18 reaction: 50 pmol each primer (HPV 18F and 18R), 8 mM $MgCl_2$;
HPV31 reaction: 10 pmol each primer (HPV 31R and 31F, 6 mM $MgCl_2$.

Amplify as per parameters below and purify the resultant 400 bp probe using MinElute PCR Purification Kit (QIAGEN, Australia).

Cycling parameters include an initial 9 min at 94°C followed by 40 cycles of 94°C 1 min, 55°C 1 min, 72°C 1 min, with a 10 min 72°C step at cycle 40.

PCR products are detected using the PCR ELISA (Dig Detection Kit (Roche Applied Science, Australia). Briefly, add 100 ng/ml of each of three probes synthesised and 5 μl of amplicon to 25 μl of 50 mM NaOH and incubate for 10 minutes. Add this denatured amplicon/probe mixtures to streptavidin coated microtiter wells containing 175 μl hybridisation buffer (Roche PCR ELISA kit). Seal the plates well and place in a hybridisation oven for at least 6 h at 55°C. Wash and detect colour in the microtiter wells according to the manufacturer's instructions.

44.7 Quality Control and Validation Data

A positive and negative control is included in each PCR run. The positive control consists of extracted DNA from SiHA cells (containing 1 copy/cell of integrated HPV 16) equivalent to 50 HPV 16 copies. Extracted DNA from HPV negative cell line such as A549 (carcinomic human alveolar basal epithelial cells) equivalent to 200 copies is used as a negative control.

A limit of detection of 1–10 copies per PCR reaction should be routinely achievable. This assay will detect presence of all mucosal HPV types present in the clinical sample at sensitivity of 1–10 copies/reaction. Analytical specificity of this assay is close to 100%.

The clinical sensitivity and specificity of the assay are dependent on the gold standard. Generally, detection of high risk-HPV DNA has over 90% sensitivity and 50% specificity for detection of histologically diagnosed high grade cervical abnormality. A clinical PPV and NPV of over 70 and 80% respectively are achieved using this assay for the detection of histologically diagnosed high grade cervical abnormality.

44.8 Assay Limitations

A positive result using this assay will only indicate presence of HPV DNA and does not necessarily infer disease in patients. Further genotyping with type specific probes allows for detection of high-risk HPV type (HPV 16, 18, 31, 33, 35, 39, 45, 51, 52, 56, 58, 59 and 68). This can be done using biotinylated type specific oligonucleotide probes with the same amplicon generated above as described by Resnick et al. [5]

In addition, this method does not detect all HPV types associated with non-mucosal lesions. Alternative assays for detection of HPV in keratinized tissue should be followed [4].

Commercial assays, such as Hybrid Capture II (QIAGEN, Australia) and Amplicor HPV (Roche Applied Science, Australia), detect the 13 high-risk types highlighted above. HPV typing can be performed using Linear Array HPV Genotyping Test (Roche Applied Science, Australia). Archival samples that have undergone extensive nucleic acid degradation require amplification of a shorter target sequence. INNO-LIPA HPV Genotyping Extra (Innogenetics, Belgium) utilises the SPF10 primer sets which allow amplification and detection of a 65 bp region. Utility of this assay is more likely to produce assessable results with older archival samples.

References

1. Bauer HM, Ting Y, Greer CE et al (1991) Genital human papillomavirus infection in female university students as determined by a PCR-based method. JAMA 265:472–477
2. Fairley CK, Chen S, Tabrizi SN et al (1992) Tampons: a novel patient-administered method for the assessment of genital human papillomavirus infection. J Infect Dis 165:1103–1106
3. Gravitt, P, Peyton CL, Alessi TQ et al (2000) Improved amplification of genital human papillomaviruses. J Clin Microbiol 38:357–361
4. Nindl I, Köhler A, Gottschling M et al (2007) Extension of the typing in a general-primer-PCR reverse-line-blotting system to detect all 25 cutaneous beta human papillomaviruses. J Virol Methods 146:1–4
5. Resnick RM, Cornelissen MT, Wright DK et al (1990) Detection and typing of human papillomavirus in archival cervical cancer specimens by DNA amplification with consensus primers. J Natl Cancer Inst 82:1477–1484
6. Stevens MP, Rudland E, Garland SM et al (2006) Assessment of MagNA pure LC extraction system for detection of human papillomavirus (HPV) DNA in PreservCyt samples by the Roche AMPLICOR and LINEAR ARRAY HPV tests. J Clin Microbiol 44:2428–2433

Chapter 45
Human Polyomaviruses – JCV and BKV

Ian Carter and David M. Whiley

45.1 Summary

Detection of the human polyomaviruses JC and BK by the polymerase chain reaction (PCR) has been used to diagnose the neurological disease, progressive multifocal leukoencephalopathy, and to study viruria in immunocompromised and nonimmunosuppressed individuals. The universal nature of PCR technology now makes it possible for many laboratories to diagnose infections produced by these viruses.

45.2 Introduction

The human polyomaviruses JC virus (JCV) and BK virus (BKV) are distributed worldwide and have been shown to infect a high proportion of the population. JC is the etiologic agent of progressive multifocal leukoencephalopathy (PML). Once a rare disease, the incidence of PML increased substantially because it is a complication of AIDS and nearly all cases of PML occur in subjects with conditions known to impair T-cell function and are the result of reactivation of latent infections, presumably in the kidney [9]. The virus reaches the brain via the blood stream and infects the myelin-producing oligodendrocytes. Patients with this progressively fatal disease commonly experience limb weakness and cognitive abnormalities [4].

Haemorrhagic cystitis in recipients of bone marrow transplants is the most frequent pathological condition associated with BKV [2]. Unlike JCV, BKV does not appear to be neurotropic, and infection and pathology are limited to the urinary tract. Viruria with JC and BK occurs frequently in patients with conditions that impair immunity, e.g., pregnancy, malignancy, immunosuppressive therapy, tissue and organ transplantation, but it also develops in individuals without any evidence of

I. Carter (✉)
SEALS Microbiology Division, Prince of Wales Hospital, Randwick, NSW 2031, Australia
e-mail: ian.carter@sesiahs.health.nsw.gov.au

M. Schuller et al. (eds.), *PCR for Clinical Microbiology*,
DOI 10.1007/978-90-481-9039-3_45,

disease [2]. The primary infections with JCV and BKV occur during childhood and adolescence and produce viraemia that results in infection of the kidneys. Symptoms accompanying primary infections, if present, are mild and nonspecific. Persistence in renal tissue of both JCV and BKV has been documented by examining autopsy tissue from immunocompetent subjects [6, 6].

The detection of JCV and BKV in clinical specimens once primarily depended on nonculture techniques. Diagnostic techniques have included cytologic and histologic examination of specimens for characteristic cytoplasmic viral inclusions, electron microscopy study, immunologic techniques to detect viral antigens, and nucleic acid hybridization techniques for the detection of viral genomes. Serologic techniques are of little diagnostic value since high proportions of the general population are infected and disease is usually the result of viral reactivation.

The application of the polymerase chain reaction (PCR) for the detection of polyomaviruses in urine and brain tissue was described in 1989 [1]. This method took advantage of the 75% nucleotide sequence homology between JCV and BKV to select a single pair of primers capable of amplifying both viruses. The region flanked by these primers was unique for each virus, and the amplification products could be identified by *BamHI* cleavage patterns and by hybridisation with JCV- and BKV-specific oligonucleotide probes.

The advent of real-time PCR methodology [10] has now further enhanced the rapid diagnostic utility for recognition of infection by BKV or JCV. This sensitive, specific, and rapid assay represents a notable improvement over previously described techniques for diagnosing polyomavirus infection. Both PCR methods for JCV and BKV are outlined below.

45.3 JCV and BKV Block PCR Assay

Ian Carter

45.3.1 Principle of Procedure

This assay is based on the amplification of a specific nucleic acid sequence found in both BK and JC virus using PCR [1]. The final product is detected by agarose gel electrophoresis, which is stained with SYBR green and visualised under ultraviolet light. An enzyme immunoassay (EIA) detection (using oligonucleotide probes specific for BK and JC virus) can be performed as well as electrophoresis.

I. Carter (✉)
SEALS Microbiology, Prince of Wales Hospital, Randwick, NSW 2031, Australia
e-mail: ian.carter@sesiahs.health.nsw.gov.au

45.3.2 Sample Extraction

Extract DNA using the Magna Pure LC Roche Total Nucleic Acid Isolation Kit or the Magna Pure LC DNA Isolation Kit and the Magna Pure LC Instrument (refer to Magna Pure LC Extraction Methods)

Specimen	Treatment
Blood treated with EDTA	Spin EDTA tube at 2500 rpm for 8 min and remove plasma and store at –20°C
CSF, body fluids	No preliminary processing required. Specimen may be processed for DNA extraction
Urine	Aliquot and store at −20°C prior to extraction
Biopsies	Sample is ground in VMM to a final volume of 500 μl. Store at −20°C prior to extraction

Probes used in the EIA detection are 5′ biotinylated.

BK Virus	**BEP**	Biotin – 5′ – GAG AAT CTG CTG TTG CTT CT
JC Virus	**JEP**	Biotin – 5′ – TTG GAT CCT GTG TTT TCA TC

45.3.3 Equipment

Magna Pure LC, Roche, Perkin Elmer GeneAmp PCR System 2400 or Biorad MyCycler, Biorad Gel Doc, Sanofi AIP4 incubator, Microtitre Plate Washer, Spectrophotometer Expert Plus.

45.3.4 PCR Amplification

PCR reagents are diluted using Sigma nuclease-free water (Cat No. W4502), and stored at –20°C; TBE is Tris Borate EDTA (Tris 10 mM, Borate 10 mM, EDTA 1 mM) buffer pH 8.I (Cat No. 816202); Magna Pure LC Total Nucleic Acid Isolation Kit or Magna Pure LC DNA Isolation Kit; Roche PCR ELISA (DIG detection) kit is stored at 4°C when not in use; Size marker for electrophoresis: 50 bp Promega (Cat. No. G4521); Amplitaq Gold Master Mix (Applied Biosystems P/L); TMB substrate (Sigma Cat No T-0440).

Add 40 μl of the master mix into a 0.2 ml sterile PCR tubes. Make up the master mix in a sterile nuclease free 0.2 ml thin-walled tube: 37.3 μl nuclease-free water, 50 μl 2x Amplitaq Gold, 3 μl Pep1 primer (0.1 pmol/μl), 4.65 μl Pep2 primer (0.1 pmol/μl), 0.2 μl DIG-11-dUTP (1 nmol/μl), 0.5 μl AmpErase.

Add 10 μl of extracted DNA into each tube containing the master mix for the first round of PCR.

The 50 μl reaction mix is then ready to be amplified using the following protocol. 50°C × 2 min, 95°C × 5 min, 40 cycles of 94°C × 30 s, 42°C × 30 s, 72°C × 1 min followed by 72°C × 7 min then hold at 4°C.

45.3.5 Product Detection by Agarose Gel Electrophoresis

Prepare a 2% agarose gel using the following: 200 ml 1x TBE, 4.0 g Agarose (Promega Cat. No. V312A), SYBR Safe DNA gel stain (10 μl/100 ml).

Add 2.0 μl loading dye to 10.0 μl PCR product and load onto the gel. Use 10.0 μl of a 1:10 dilution (300 ng) of 50 bp Promega (Cat.No.G4521) as the marker mixed with 2.0 μl of 6x loading dye. Run gel at 110 V for 35 min. View gel under ultraviolet light and print and save image.

45.3.6 Product Detection Using PCR-ELISA

Allow PCR ELISA (DIG Detection Kit-Roche) to stand at room temperature for 1 h before using. Add a 40 μl volume of denaturation solution (protocol A solution in kit) to each tube containing amplicon, aspirate several times, and incubate at room temperature for 10 min. Add JEP (JCV) or BEP (BKV) probe to the following volume of Hybridisation solution.

Strips	Hybridisation solution (μl)	JCV probe (μl)	BKV probe (μl)
1	2000	5	5
2	3000	7.5	7.5

NB: Always use plugged pipette tips (art tips)

Place appropriate number of Micro-well strips from foil and insert into the base ELISA plate (seal foil bag after use). Add 85.0 μl hybridisation solution (Solution IV) containing the probe. Add 15.0 μl denatured amplified sample to the designated well of the microtitre test plate probes Using plugged tips and gently mix without creating bubbles (pipette up and down 6–10 times). Cover and incubate at 40°C with shaking for 70 min (plate incubator/rocker). Wash strips/plate 6 times, using the prepared wash solution*, with 30 s soaks between each wash step. Tap plate dry. Add 100 μl of anti-DIG-POD conjugate diluted 1:100 in conjugate dilution buffer to each well and add plate sealer as per table below:

Conjugate dilution (100 μl per well = 0.8 ml per strip)

No. of strips	Volume of conjugate (μl)	Volume of buffer (μl)
2	20	2000
4	40	4000

Note: Reconstitute conjugate with 250 μl of distilled water

Cover and incubate at 40°C with shaking for 30 min. Wash strips/plate six times, using the prepared wash solution*, with 30 s soaks between each wash step. Tap plate dry. Add 100 μl of TMB substrate to each well. Incubate at 37°C in the dark without shaking for 10 min. Add 100 μl of STOP reagent. Using a spectrophotometer, read the Microwell strips/plate immediately at a wavelength of 450 nm. Check correlation between Gel results.

**Note: Wash Buffer needs to be prepared weekly and stored at 2–8° C for a maximum of one week only. Prepare working wash solution by adding 1 washing buffer tablet to 2000 ml of distilled water.*

Negative cut-off and equivocal ranges may vary with spectrophotometer available but calculation may be as follows:

Negative	Average absorbance value of negative controls + 0.100 <0.20
Equivocal	Negative cut-off to 3x (Negative Cut-off) 0.20–0.80

45.3.7 Quality Control

Development of this PCR included testing the primers with other viruses to test for cross-reactivity, to prevent false positives due to the presence of viruses other than BK/JC. No cross-reaction with other viruses was encountered.

A number of control samples are included in each run to test both the extraction kit and the PCR amplification and EIA detection. A positive control is included from extraction through to the EIA. Many negative controls are included, which are also extracted and run through the entire method. These include open tubes of water/buffer remain open throughout the extraction. These detect aerosol contamination arising during extraction. Closed tubes of water/buffer are included to detect contamination arising from sample handling during extraction. The final negative control is a negative reagent control for PCR, which has sterile water added in place of extracted DNA. Sensitivity of the assay is currently under investigation.

45.4 JCV and BKV Real-Time PCR Assay

David M. Whiley

45.4.1 Summary of Methods

Human polyomaviruses JCV and BKV can cause several clinical manifestations in immunocompromised hosts, including progressive multifocal leukoencephalopathy (PML) and haemorrhagic cystitis. Molecular detection by PCR is recognised

D.M. Whiley (✉)
Queensland Paediatric Infectious Diseases Laboratory, Sir Albert Sakzewski Virus Research Centre, Queensland Children's Medical Research Institute, Children's Health Service District, Brisbane, QLD 4029, Australia
e-mail: d.whiley@uq.edu.au

as a sensitive and specific method for detecting human polyomaviruses in clinical samples [10]. The assay described here uses two sets of specific primers and TaqMan probes for the detection of JC and BK polyomaviruses in a duplex real-time PCR. The primers and probes have previously been described by others [7, 8]. The reaction conditions were modified for optimal performance in a duplex reaction.

45.4.2 Acceptable Specimens

Whole blood, urine and respiratory secretions have been validated for the method described. CSF specimens may also be used for the diagnosis of PML.

45.4.3 Sample Extraction

Nucleic acids from NPA, CSF, blood and urine samples were extracted from 0.2 ml of each specimen using the MagNA Pure automated extractor (Roche Diagnostics, Australia), according to the manufacturer's instructions. Alternatively, nucleic acids may be extracted using the High Pure viral nucleic acid kit (Roche Diagnostics, Australia). Stool samples may also be analysed but require pre-treatment with STAR Buffer (Roche Diagnostics, Australia) before extraction.

45.4.4 Primer and Probe Sequences

The primers and probes used for the detection of JCV were previously described by Pal et al. [8] and target the VP1 gene. The primers and TaqMan probe for BKV have been published by Hirsch et al. [7]:

Primers for JCV were JC-AF: 5′-AAG GGA GGG AAC CTA TAT TTC TTT TG-3′; JC-AR: 5′-TCT AGC CTT TGG GTA ACT TCT TGA A-3′; and the JCV-specific TaqMan probe JC-Aprobe: 5′-FAM-CTC ATA CAC CCA AAG TAT AGA TGA TGC AGA CAG CA-BHQ-3′.

Primers for BKV detection were BK-HF: 5′-GAA GCA ACA GCA GAT TCT CAA CA-3′; BK-HR: 5′-AGC AGG CAA GGG TTC TAT TAC TAA AT-3′; with TaqMan probe: BK-Hprobe: 5′-YAK-AAG ACC CTA AAG ACT TTC CCT CTG ATC TAC ACC AGT TT-BHQ1-3′ (*YAK = Yakima Yellow).

**Note: that JOE can be used in place of YAK.*

45.4.5 PCR Amplification and Product Detection

PCR was performed using 25 μl reaction mixtures containing 5 μl of nucleic acid extract, 10 pmol of each primer, 4 pmol of each probe, and 12.5 μl of QIAGEN QuantiTect Probe PCR master mix (Qiagen, Australia). PCR cycling was performed on a RotorGene 3000 cycler (Qiagen, Australia), with an initial activation at 95°C for 15 min and 50 cycles of 95°C for 15 s and 60°C for 60 s. NOTE: JCV is detected by the FAM channel; BK is detected by the JOE channel.

45.4.6 Quality Control and Validation Data

Blood and urine samples from previously proven positive patients were used as positive controls. Water for injection (Baxter, Australia) was subjected to the extraction process and used as a negative control. Commercial positive controls for JCV and BKV may be purchased from Quality Control for Molecular Diagnostics (QCMD – http://www.qcmd.org/).

45.4.7 Assay Limitations

JCV and BKV may occur in the same patient at the same time. The accurate detection of two targets of different concentrations in a single sample may be compromised due to competitive inhibition [3].

References

1. Arthur RR, Dagostin S, Shah KV (1989) Detection of BKV and JCV in urine and brain tissue by the polymerase chain reaction. J Clin Microbiol 27:1174–1179
2. Arthur RR, Shah KV (1989) Occurrence and significance of papoviruses BK and JC in the urine. In: Melnick J (ed) Progress in Medical Virology, vol 36. Karger, Basel, pp 42–61
3. Bialasiewicz S, Whiley DM, Nissen MD et al (2007) The impact of competitive inhibition and sequence variation upon the sensitivity of malaria PCR. J Clin Microbiol 45:1621–1623
4. Berger JR, Kaszovitz B, Donovan Post MJ, Dickinson G (1987) Clinical review: progressive multifocal leukoencephalopathy associated with human immunodeficiency virus infection. Arm Intern Med 107:78–87
5. Chesters PM, Heritage J, McCance DJ (1983) Persistence of DNA sequences of BK virus and JC virus in normal human tissues and in diseased tissues. J Infect Dis 147:676–684
6. Heritage J, Chesters PM, McCance DJ (1981) The persistence of papovavirus BK DNA sequences in normal human renal tissue. J Med Virol 8:143–150
7. Hirsch HH, Mohaupt M, Klimkait T (2001) Prospective monitoring of BK virus load after discontinuing sirolimus treatment in a renal transplant patient with BK virus nephropathy. J Infect Dis 184:1494–1495
8. Pal A, Sirota L, Maudru T et al (2006) Real-time, quantitative PCR assays for the detection of virus-specific DNA in samples with mixed populations of polyomaviruses. J Virol Methods 135:32–42
9. Walker D, Padgett BL (1983) Progressive multifocal leukoencephalopathy, In: Fraenkel Conrat H, Wagner RR (eds) Comprehensive Virology, vol 18. Plenum, New York, pp 161–193
10. Whiley DM, Arden KE, Mackay IM et al (2004) Simultaneous detection and differentiation of human polyomaviruses JC and BK by a Rapid and sensitive PCR-ELAHA assay and a survey of the JCV subtypes within an Australian population. J Med Virol 72:467–472

Chapter 46
Human Polyomaviruses – KIV and WUV

Seweryn Bialasiewicz

46.1 Summary of Methods

Human polyomaviruses WU (WUV) and KI (KIV) were both originally discovered in respiratory samples of paediatric patients suffering from respiratory tract disease [1, 3], and the respiratory tract is still the predominant body area of detection for both of these viruses. The real-time PCR assays described below can be run individually (singleplex) or in the case of the WUV assays, as a duplex in order to guard against unforseen sequence variation and act as complementary confirmatory assays. The WUV assays target the regulatory region and the C-terminal end of the Large T Antigen (LTAg), while the KIV assays target the regulatory region and the unique portion of the Small T Antigen (STAg). The assays were originally validated for respiratory tract samples [2], but have subsequently been successfully used on blood and faecal samples. Assay KI-B may have limited sensitivity in respiratory samples, but can be used as a secondary assay, and recent data has suggested it may be more sensitive in detecting KIV in faecal samples.

46.2 Acceptable Specimens

Nasopharyngeal aspirates, bronchial washes and other respiratory secretions (excluding KI-B), nose/throat swabs, cerebrospinal fluid, urine, whole blood, and stool samples (excluding KI-A).

46.3 Sample Extraction

The respiratory specimens were extracted using the Roche High Pure Viral Nucleic Acid kit (Roche Applied Science, Australia).

S. Bialasiewicz (✉)
Queensland Paediatric Infectious Diseases Laboratory, Sir Albert Sakzewski Virus Research Centre, Queensland Children's Medical Research Institute, Children's Health Service District, Brisbane, QLD 4029, Australia
e-mail: seweryn@uq.edu.au

M. Schuller et al. (eds.), *PCR for Clinical Microbiology*,
DOI 10.1007/978-90-481-9039-3_46,

46.4 Primers and Probe Sequences

Published genomes of either WUV or KIV were aligned and conserved regions identified using Bioedit 7.0.5.3. Conserved regions were input into Primer Express 2.0 software (Applied Biosystems, Australia) to generate candidate primer and hydrolysis probe sets which were then cross-referenced with BLAST searches to exclude any assays with non-specific sequence homology. Final primer and probe sequences are shown in Table 46.1.

Table 46.1 Primer and probe sequences for the detection of the novel polyomaviruses, KIV and WUV

Oligonucleotides	Sequence (5′–3′)	Target	Reference
WUV:			
WU-B-2729-F	CTACTGTAAATTGATCTATTGCAACTCCTA	LTAg	[2]
WU-B-2808-R	GGGCCTATAAACAGTGGTAAAACAACT		
WU-B-2797-TM	FAM-CCTTTCCTCCACAAAGGTCAAGTAAA-BHQ1	Regulatory	[2]
WU-C-4824-F	GGCACGGCGCCAACT	Region	
WU-C-4898-R	CCTGTTGTAGGCCTTACTTACCTGTA		
WU-C-4861-TM	HEX-TGCCATACCAACACAGCTGCTGAGC-BHQ1		
KIV:			
KI-A-141-F	ACCTGATACCGGCGGAACT	Regulatory	[2]
KI-A-200-R	CGCAGGAAGCTGGCTCAC	Region	
KI-A-182-TM	FAM-CCACACAATAGCTTTCACTCTTGGCGTGA-BHQ1		
KI-B-4603-F	GAATGCATTGGCATTCGTGA	STAg	[2]
KI-B-4668-R	GCTGCAATAAGTTTAGATTAGTTGGTGC		
KI-B-4632-TM	FAM-TGTAGCCATGAATGCATACATCCCACTGC-BHQ1		

46.5 PCR Amplification and Product Detection

In singleplex format, all assays used the same PCR mix and cycling conditions; 12.5 μl of QIAGEN Quantitect Probe Master Mix (QIAGEN, Australia), 10 pmol of each primer, 4 pmol of the corresponding probe, and 2 μl of sample nucleic acid extract in a final reaction volume of 25 μl, which was then cycled under the following parameters: incubation of 15 min at 95°C, followed by 45 cycles of 95°C for 15 s, and 60°C for 1 min. Signal acquisition was obtained at the latter end of extension on each cycle on the corresponding channel (FAM or HEX). Amplification was performed on a RotorGene 3000. (QIAGEN, Australia) The duplex WU-B&C

method involved the addition of 10 pmol of each of the other assay's primers and 4 pmol of the respective probe to the existing reaction mix, with an adjustment of water to maintain the final 25 μl reaction volume. Cycling parameters for the duplex were identical to the singleplex with the exception of acquiring signals on both the FAM and HEX channels.

46.6 Quality Control and Validation Data

Analytical sensitivity and amplification dynamics were obtained with the use of serially-diluted genomic standards, which were quantified by parallel analyses with known assay-specific plasmid preparations. All real-time assays could reliably detect down to 10 copies per reaction, in comparison to the conventional assays' 100 copies per reaction [1, 3]. Duplexing the WUV assays did not impact on either assays' limit of detection. A panel of human genomic DNA, 22 unrelated respiratory viruses and 12 bacterial species was used to establish the analytical specificity of the assays; all four exhibited exclusive detection of their specified target. A panel of 200 respiratory samples from paediatric patients was used to determine clinical sensitivity and specificity of the real-time PCR assays, with the original conventional PCR detection assays [1, 3] serving as the reference comparators. The assays WU-B, C, and KI-A all showed 100% sensitivity, and 97.7, 97.2, and 94.6% specificity, respectively. In contrast, KI-B produced a lower clinical sensitivity of 86.7% and specificity of 97.3%. The lower specificity rates of all four assays were due to additional WUV or KIV detections, which indicates that the lower figures were an artifact of the reference assays' inferior sensitivity.

References

1. Allander T, Andreasson K, Gupta S et al (2007) Identification of a third human polyomavirus. J Virol 81:4130–4136
2. Bialasiewicz S, Whiley DM, Lambert SB et al (2007) Development and evaluation of real-time PCR assays for the detection of the newly identified KI and WU polyomaviruses. J Clin Virol 40:9–14
3. Gaynor AM, Nissen MD, Whiley DM et al (2007) Identification of a novel polyomavirus from patients with acute respiratory tract infections PLoS Pathog 3:595–604

Chapter 47
Human Rhinoviruses

Ian M. Mackay

47.1 Background

Human rhinoviruses (HRVs) are a pervasive cause of upper respiratory tract infections that occur throughout life. The impracticality and insensitivity of early culture-based diagnostic methods resulted in the misconception that compared to influenzavirus and RSV, HRVs had a relatively minor role in illness. HRVs are the most common cause of acute respiratory tract illness globally, infecting both upper and lower respiratory tract tissues. In addition to the majority of cold and flu-like illness, HRVs also cause more asthma and chronic obstructive pulmonary disease exacerbations than any other agent identified to date.

47.2 Acceptable Specimens

Suitable respiratory tract specimens include nasopharyngeal aspirates, nose and throat swabs, nasal or bronchial washings, bronchoalveolar lavage or middle ear fluids.

47.3 Sample Extraction

Nucleic acids are prepared by manual, column-based method e.g. High Pure Viral Nucleic Acid Kit (Roche Applied Science, Australia) or automated extractions resulting in a final extraction volume of 50 μl.

I.M. Mackay (✉)
Queensland Paediatric Infectious Diseases Laboratory, Sir Albert Sakzewski Virus Research Centre, Queensland Children's Medical Research Institute, Children's Health Service District, Brisbane, QLD 4029, Australia
e-mail: ian.mackay@uq.edu.au

M. Schuller et al. (eds.), *PCR for Clinical Microbiology*,
DOI 10.1007/978-90-481-9039-3_47,

47.4 Primers and Probe Sequences

Our favoured rapid screening RT-PCR assay, derived from Lu et al. [1], employs a dual-labelled nuclease oligoprobe (5′–3′; FAM-TCC TCC GGC CCC TGA ATG YGG C-BHQ1) to specifically detect amplicon generated by two oligonucleotides (FORWARD – CYX GCC ZGC GTG GY; REVERSE – GAA ACA CGG ACA CCC AAA GTA where X=LNA-dA and Z=LNA-dT [1]) which hybridise to the 5′ untranslated region and amplify a 207 nt region common to all known HRV strains.

47.5 PCR Amplification and Product Detection

The 20 μl reaction mix contains 2 μl of purified nucleic acids, 8 μl of the two primers mixed at 2 μM (0.8 μM final concentration), 1 μl of oligoprobe at 2 μM (0.1 μM), 0.4 μl of 25 mM $MgCl_2$, 0.8 μl of 10 mM dNTP mix (QIAGEN, Australia) and 0.4 μl enzyme mix (OneStep RT-PCR kit, QIAGEN, Australia).

To reverse transcribe and amplify the viral genome, a 20 min 50°C reverse transcription is followed by a 15 min denaturing "Hot-Start" at 95°C in a RotorGene 3000 or 6000 (QIAGEN, Australia). The reactions are then exposed to 45 cycles of 94°C for 10 s and 60°C for 60 s; fluorescent emission data are collected at 60°C.

PCR products can be identified by a suitable reporter signal generated by the destruction of the nuclease oligoprobe after its specific hybridisation to target. Electrophoresis can be used to confirm the presence of the 207 nt amplicon on 1.5–2.0% agarose gels.

47.6 Quality Control and Validation Data

A mid-range (C_T25 to C_T30) positive control consisting of HRV-positive clinical material previously characterised by nucleotide sequencing (>99% nucleotide identity to a GenBank sequence is considered a variant of the expected strain) is included. Two no-template controls are included along with another for each 10 clinical samples being tested.

The assay can detect 50 synthetically created RNA transcripts per reaction, has a 7-$\log_{10}$ dynamic range, does not cross-react with other common respiratory viruses and is capable of detecting all 100 serotypes of HRV and members of the putative HRV C species.

47.7 Assay Limitations

The assay also detects some human enterovirus strains, although C_T values are commonly greater than 33 cycles, requiring at least 10^5 transcripts per reaction for detection.

Although extensively tested against a wide range of HRV subtypes, the assay may miss some subtypes not commonly circulating or not previously identified in the population.

Reference

1. Lu X, Holloway B, Dare RK et al (2008) Real-time reverse transcription-PCR assay for comprehensive detection of human rhinoviruses. J Clin Microbiol 46:533–539

Chapter 48
Influenza Virus A H5N1 ("Avian Influenza")

Greg Smith and Ina Smith

48.1 Acceptable Specimens

A nasal pharyngeal aspirate (NPA) is the specimen of choice for exclusion of influenza H5 strains in young children while a nasopharyngeal swab may be easier to obtain for older children and adults. Swabs should be collected and transported in the primary collection container containing viral transport medium (VTM). A range of other respiratory specimens are recommended by the World Health Organisation and include a throat swab, nasal swab or nasal wash.

48.2 Sample Extraction

All samples for H5 exclusion should be handled in a Class II Biosafety cabinet. This assay was validated for both automated magnetic bead based kits (QIAGEN EZ1 Virus Mini Kit V2.0, QIAGEN, Australia) and silica based kits for automated platforms (QIAGEN RNeasy 96 BioRobot 9604 Kit, QIAGEN, Australia) and manual centrifugation based approaches (QIAamp Viral RNA Mini Kit, QIAGEN, Australia). The extractions are performed according to manufactures instructions.

NPAs are transferred to a labelled 2.0 ml tube and vortexed briefly (15 s) to break up any thick mucus. The final volume is adjusted to 1.0 ml with sterile water or VTM. The sample is stored at 4°C until extraction if processing will occur within 48 h otherwise sample should be stored at –80°C.

Nasopharyngeal swabs are agitated (by repeated squeezing of collection tube) or vortexed briefly. The swab is removed and the VTM transferred into a sterile pre-labelled 2.0 ml collection tube. The final volume is adjusted to 1.0 ml by addition of sterile water or VTM for subsequent processing. Dry swabs are diluted in 1.0 ml sterile water or VTM.

I. Smith (✉)
Australian Animal Health Laboratory, CSIRO Livestock Industries, Geelong, VIC 3220, Australia
e-mail: Ina.Smith@csiro.au

M. Schuller et al. (eds.), *PCR for Clinical Microbiology*,
DOI 10.1007/978-90-481-9039-3_48, © Springer Science+Business Media B.V. 2010

48.3 Primer and Probe Sequences

The assay is a real-time PCR assay (TaqMan) employing a minor groove binding (MGB) FAM-labelled probe. The assay targets the influenza haemagglutinin gene and is based on the sequence of A/Vietnam/1196/2004 H5N1. Redundancies (R) have been incorporated into the primer design to account for observed sequence divergence. Forward Primer: 5′-GAG GAT GGC AGG GMA TGG TA-3′; Reverse Primer 5′-TGA CCT TAT TGG TRA CTC CAT CTA TT-3′ and MGB Probe: 5′- ATC CAC TCA AAA GGC-3′.

48.4 PCR Amplification and Product Detection

The Superscript III Platinum One-Step Quantitative RT-PCR System (Invitrogen, Australia) is used according to manufactures instructions with a final concentration of forward and reverse primers of 300 nM and a final probe concentration of 150 nM. Five microlitres of recovered RNA is added to each real-time PCR reaction with a final reaction volume of 25 μl.

Following a 5 min 50°C reverse transcription step, samples are denatured at 94°C for 2 min followed by 40 cycles of 95°C for 3 s and 60°C for 30 s. Samples with Ct values of <38 and compliant curves are deemed positive.

48.5 Quality Control and Validation Data

This laboratory uses a synthetic probe control and a synthetic primer control to eliminate the possibility of generating a false positive result due to inadvertent contamination [1]. The current, modified approach employs synthetic DNA probe and primer controls (without a T7 RNA polymerase binding site) in conjunction with an in house developed real-time PCR assay for bovine viral diarrheal virus. The latter serves as both a reverse transcriptase control and an extraction control. The probe control is comprised of a single synthetic oligonucleotide

5′-TGC ACC ACC AAC TGC TTA GAA TCC ACT CAA AAG GCA GAA C AT CAT CCC TGC ATC C-3′. The primer control sequence is 5′-GGA TGG CAG GGM ATG GTA ACA GAA GAC TGT GGA TGG CCC CTC AAA TAG ATG GAG TYA CCA ATA AGG TCA-3′.

Due to the absence of clinical material and the scarcity of virus isolates, validation of this assay is incomplete. When inactivated virus has been available no titre of the virus stocks has been known to allow determination of sensitivity against intact virus. The assay has consistently detected clade 1 and clade 2 strains which have been generously provided by Paul Selleck from the CSIRO Australian Animal Health Laboratory in Geelong, Victoria. No cross reactivity has been observed when tested against an extensive panel of clinical samples and culture supernatants including influenza A/H1 and H3 isolates, influenza B, respiratory syncytial virus, parainfluenza virus types 1, 2 and 3 and adenovirus.

48.6 Assay Limitations

Due to the constant evolution of influenza viruses and the emergence of a number of different clades and sub-clades of influenza A/H5 virus strains it is critical that laboratories periodically review influenza A/H5N1 sequences to ensure that the primer and probe sequences are still able to identify contemporary strains.

Samples submitted for influenza H5 exclusion are tested using a panel of real-time assays targeting the influenza A matrix gene (presented elsewhere in this chapter) as well as the influenza A H1, influenza A/H2 [2] and the influenza A/H5 genes. The laboratory also has real-time PCR assays for the influenza A/H7 and influenza A/H9 genes. It is critical that the haemagglutinin type of any influenza A matrix gene positive is identified to ensure that the failure to detect H5 is not due to assay failure as a result of sequence divergence.

References

1. Smith G, Smith I, Harrower B et al (2006) A simple method for preparing synthetic controls for conventional and real-time PCR for the identification of endemic and exotic disease agents. J Virol Methods 135:229–234
2. Stone B, Burrows J, Schepetiuk S et al (2004) Rapid detection and simultaneous subtype differentiation of influenza A viruses by real-time PCR. J Virol Methods 117:103–112

Chapter 49
Influenza Virus A H1N1 (2009) ("Human Swine Influenza")

David M. Whiley and Theo P. Sloots

49.1 Summary of Methods

Novel influenza A (H1N1) virus ("human swine flu") continues to expand globally, and its accurate and rapid diagnosis is important to minimise further spread and to rapidly respond through the administration of appropriate antiviral treatment. Real-time PCR has been widely used to diagnose influenza viruses [1], but a serious limitation of PCR is that false-negative results may occur due to sequence variation in primer and probe targets binding sites. This is particularly relevant for the detection of influenza viruses which undergo continuous mutation during their evolution. Our previous experience has shown that the use of multiple targets can reduce such limitations [2] and may serve as a means of confirming positive results. For this reason two fully validated TaqMan-based reverse transcription PCR (RT-PCR) methods for the detection of novel influenza A (H1N1) virus are described [3].

49.2 Acceptable Specimens

Nasopharyngeal aspirates, bronchial washings, throat swab, nose swab.

49.3 Sample Extraction

Respiratory samples and controls were extracted using the QIAGEN X-tractor Gene (QIAGEN, Australia) according to manufacturer's instructions.

D.M. Whiley (✉)
Queensland Paediatric Infectious Diseases Laboratory, Sir Albert Sakzewski Virus Research Centre, Queensland Children's Medical Research Institute, Children's Health Service District, Brisbane, QLD 4029, Australia
e-mail: d.whiley@uq.edu.au

M. Schuller et al. (eds.), *PCR for Clinical Microbiology*,
DOI 10.1007/978-90-481-9039-3_49,

49.4 Primers and Probe Sequences

The primers and TaqMan probes used in the RT-PCR methods were designed using influenza A sequences available on the GenBank database (accessed 29 April 2009) and Primer Express 2.0 software (Applied Biosystems, Australia). Two TaqMan assays were designed; H1-PCR and N1-PCR, targeting the novel influenza A (H1N1) virus haemagglutinin and neuraminidase genes, respectively. Probe and primer sequences for the haemagglutinin target were: H1-F: GGT TTG AGA TAT TCC CCA AGA CA; H1-R: GAG GAC ATG CTG CCG TTA CA; and a probe H1-TM: FAM-TCA TGG CCC AAT CAT GAC TCG AAC A-BHQ. Primers and probe specific for the novel influenza A (H1N1) virus neuraminidase gene were: N1-F: CAG AGG GCG ACC CAA AGA GA; N1-R: GGC CAA GAC CAA CCC ACA with a probe N1-TM: FAM-CAC AAT CTG GAC TAG CGG GAG CAG CAT-BHQ.

49.5 PCR Amplification and Product Detection

The assays were performed using the QIAGEN One-Step RT-PCR Kit (QIAGEN, Australia) comprising 0.8 μM of forward and reverse primers (H1-F and H1-R for the H1-PCR; N1-F and N1-R for the N1-PCR) and 0.2 μM of TaqMan probe (H1-TM for the H1-PCR; N1-TM for the N1- PCR) in a total reaction volume of 25 μl. 5 μl of nucleic acid extract or control was added to the mix. Amplification and detection were performed on a RotorGene 6000 (QIAGEN, Australia) with the following conditions: initial holds at 50°C for 20 min and 95°C for 15 min followed by 45 cycles at 95°C for 15 s and 60°C for 1 min.

49.6 Quality Control and Validation Data

An influenza A (H1N1) virus isolate (Auckland, 2009) provided by the Australian World Health Organization Collaborating Centre for Reference and Research on Influenza (Melbourne, Australia), was used as a positive control. Extracted distilled water was used as negative controls, and were included in each test run.

A total of 152 clinical respiratory samples (127 nasopharyngeal aspirates, 16 bronchial specimens and 9 swabs) collected in Queensland, during 2008–2009 were retrospectively tested to validate the assays. All 152 clinical samples and 12 seasonal H1N1 and H3N2 isolates from 2000–2002 which were positive in a PCR targeting the influenza A matrix gene [1], gave negative results in both the H1-PCR and N1-PCR assays. This indicates the assays are specific and do not cross-react with seasonal H1 and H3 influenza A strains affecting humans.

The positive control (influenza A (H1N1) virus; Auckland, 2009) showed similar cycle threshold (Ct) values (23 cycles) in both assays.

49.7 Assay Limitations

The assays were validated against as many influenza isolates as were readily available. However, the number of novel influenza A (H1N1) virus isolates available at the time of assay development was extremely limited. It is possible that as this virus becomes endemic, mutations may occur that will affect assay results. It is important therefore that sequencing of the primer and probes target sites in the novel influenza A (H1N1) virus is performed to monitor potential mutation over time.

Also, some wild-type swine H1N1 isolates gave a positive result in the N1-PCR but a negative result in the H1-PCR assay, indicating that the N1-PCR assay may potentially cross-react with influenza A strains infecting swine. As it is unlikely that humans would be infected with wild type swine influenza virus, this may not be an issue. Also it is unlikely that a wild-type swine H1N1 would produce positive results in both methods.

References

1. Whiley DM, Sloots TP (2005) A 5′-nuclease real-time reverse transcriptase-polymerase chain reaction assay for the detection of a broad range of influenza A subtypes, including H5N1. Diagn Microbiol Infect Dis 53:335–337
2. Whiley DM, Lambert SB, Bialasiewicz S et al (2008) False negative results in nucleic acid amplification tests-do we need to routinely use two genetic targets in all assays to overcome problems caused by sequence variation? Crit Rev Microbiol 34:71–76
3. Whiley DM, Bialasiewicz S, Bletchly C et al (2009) Detection of novel influenza A (H1N1) virus by real-time RT-PCR. J Clin Virol 45:203–204

Chapter 50
Influenza Type C

Cassandra Faux

50.1 Summary of Methods

Serological and molecular methods indicate influenza C virus is globally distributed and that the majority of the population has antibodies to the virus [1–3]. This method is a real time RT-PCR duplex assay targeting the matrix protein and non-structural genes of influenza C virus.

50.2 Acceptable Specimens

Respiratory secretions, bronchial lavage and washings, nose/throat swabs.

50.3 Unacceptable Specimens

All specimen types have not been evaluated.

50.4 Sample Extraction

Nucleic acid was extracted using High Pure Viral Nucleic Acid kit (Roche Applied Science, Australia) according to manufacturer's instructions giving a final volume of 50 μl.

C. Faux (✉)
Queensland Paediatric Infectious Diseases Laboratory, Sir Albert Sakzewski Virus Research Centre, Queensland Children's Medical Research Institute, Children's Health Service District, Brisbane, QLD 4029, Australia
e-mail: cassyfaux@uq.edu.au

M. Schuller et al. (eds.), *PCR for Clinical Microbiology*,
DOI 10.1007/978-90-481-9039-3_50,

50.5 Primers and Probe Sequences

Primers and probes were designed and optimised using Bioedit 7.0.9.0 and Primer Express 2.0. Primers and probe for the non-structural (NS) target sequence were: FluC_NS-F: GAG ATT GAG CTA TGC ATT TGT TTT G; and FluC_NS-R: TGC ATC TGA AAC ATT TTC TAA TTC CT; with probe FluC_NS-FAM: FAM-TGC AGA AAT ACT AAG AAG ATC TGT GGA TAC CTC ATC G-BHQ1.

Primers and probe sequences for the matrix gene were FluC_M-F: CAT AAT TGA ACT TGT CAA TGG TTT TGT; and FluC_M-R: TTC AGG CAT AAT TGT GGT CTT TAT ATC T; with probe FluC_M-YAK: YAK-CTC GGC AGA TGG GAG AGA TGG TGT G-BHQ1.

50.6 PCR Amplification and Product Detection

OneStep RT-PCR kit (QIAGEN, Australia) was used as per manufacturer's instructions. The reaction mix included 0.8 μM of forward and reverse primers and 0.2 μM of probe. The assays have been performed on the Roche LightCycler 480, QIAGEN Rotor-Gene 3000 and Applied Biosystems 7500 using the following conditions: RT step 50°C for 20 min, Incubation 95°C for 15 min; cycling 95°C for 15 s and 60°C for 1 min for 45 cycles.

50.7 Quality Control and Validation Data

Positive control material (sequenced patient positive material) and RNAse free water as negative controls are included in all runs. All positive results were sequenced and deemed to be influenza C virus, no cross reactions with other viruses or bacteria present in the sample occurred.

PPV and NPV are not available for this assay.

References

1. Gouarin S, Vabret A, Dina J et al (2008) Study of influenza C virus infection in France. J Med Virol 80:1441–1446
2. Jennings R (1968) Respiratory viruses in Jamaica: a virological and serological study Am J Epidemiol 87:440–446
3. Matsuzaki Y, Sugawara K, Mizuta K et al (2002) Antigenic and genetic characterization of influenza C viruses which caused two outbreaks in Yamagata city, Japan, in 1996 and 1998. J Clin Microbiol 40:422–429

Chapter 51
Measles Virus

Greg Smith

51.1 Summary of Methods

Measles virus (MV) is a highly contagious acute viral disease, that can result in serious complications and death. As a number of viral infections can mimic measles and cause measles-like rashes (e.g. rubella, parvovirus B19, enteroviruses), confirmation by laboratory testing of clinically suspected cases is recommended. A recent publication utilising multiplex detection of rubella and measles virus [1] and a real-time method have been previously published [2]. This protocol describes a real-time RT-PCR assay targeting the F gene which codes for the fusion protein. The classical symptoms of measles include a fever for at least 3 days duration, and the three C's – cough, coryza (runny nose) and conjunctivitis (red eyes). The fever may reach up to 40°C (105°F). Koplik's spots seen inside the mouth are pathognomic (diagnostic) for measles but are not often seen, even in real cases of measles because they are transient and may disappear within a day of arising.

51.2 Acceptable Specimens

Urine – first passed morning specimen preferred, combined throat and nasopharyngeal swab, nasopharyngeal aspirates and blood in EDTA or ACD tube.

A further optional step in the processing of urine is ultracentrifugation (UC). UC may increase the sensitivity of the assay by concentrating viral particles.

51.3 Ultracentrifugation

For urine samples less than 2 ml in volume, no concentration is necessary. For samples greater than 2 ml, complete the following:

G. Smith (✉)
Australian Animal Health Laboratory, CSIRO Livestock Industries, Geelong, VIC 3220, Australia
e-mail: Greg.A.Smith@csiro.au

M. Schuller et al. (eds.), *PCR for Clinical Microbiology*,
DOI 10.1007/978-90-481-9039-3_51, © Springer Science+Business Media B.V. 2010

Inspect sample for presence of red cells or possible bacterial contamination (check for cloudy or turbid appearance). If present, centrifuge the sample for 10 min at 1,000×*g*. Remove the urine and transfer to 5 ml polyallomer tubes. Discard the pellet. Balance the tubes and place into centrifuge rotor buckets. Pellet the virus by ultracentrifugation at 40,000 rpm (152,280×*g*) for 1 h at 4°C. Draw off the urine, transferring it back into the original specimen container. Store at 4°C. Re-suspend the pellet in 1 ml of sterile PBS or 1 ml of the urine supernatant. Transfer to a 2 ml tube. The sample is now ready for extraction.

51.4 Sample Extraction

The specimens were extracted using the Qiagen Biorobot Universal System, Qiagen EZ1, or Qiagen EZ1 AdXL (Qiagen Hilden Germany).

51.5 Primers and Probe Sequences

The primers and Taqman probes for MV real-time PCR were designed using Primer Express 2.0 software (Applied Biosystems Pty. Ltd., Scoresby, Australia) and are listed in Table 51.1.

Table 51.1 Primer and probe sequences for measles virus PCR assay

Oligonucleotides	Sequence (5′–3′)	Target gene
Measles MGB FP	GCTCAAATTGCTCAGATACTATACAGAAA	F
Measles MGB Probe	CCTGTCATTATTTGGCC	F
Measles MGB RP	GCAGATATGGGGTCCCGTAA	F

51.6 PCR Amplification and Product Detection

This assay uses Invitrogen SuperScript III Platinum one-step Quantitative RT-PCR System Cat. No. 11732-020. Components of the reaction mix are outlined in Table 51.2.

Fifteen microliters of master mix is combined with 5 μl of nucleic acid extract giving a final volume of 20 μl.

Amplification is performed on a Roto-Gene Q (Qiagen Hilden Germany) and Applied Biosystems 7500 Fast Real Time PCR System (Applied Biosystems Pty. Ltd., Scoresby, Australia).

Both instruments are programmed with the following parameters; initial incubation of 5 min at 50°C, followed by 1 cycle of 2 min at 95°C then 40 cycles of 95°C for 3 s, and 60°C for 30 min, with fluorescence acquired at the end of each 60°C step.

Table 51.2 Master mix preparation for measles virus PCR assay

Reagent	Volume (μl)	Concentration
Primer Measles MGB FP	0.04	150 pmol/μl
Primer Measles MGB RP	0.04	150 pmol/μl
Probe Measles MGB Probe	0.031	100 pmol/μl
2X Reaction Mix (Superscript) P/N551777	10.00	
Rox Reference Dye [P/N54881]	0.04	25 μM
Superscript/Platinum Taq Mix [P/N55176]	0.4	25 μM
RNAse DNAse free water	4.449	
Total	15.00	

51.7 Quality Control and Validation Data

Specific probe and primer controls are included in each run as well as no template control (NTC). A negative extraction control is extracted and processed in a manner similar to that applied to clinical specimens, and used as a negative extraction control. NTC and negative extraction control are RNAse/DNAse free water.

The sensitivity of this assay is 100%, with specificities yet to be determined. The reliable detection limit of this assay is less than 1 Cell Culture Infective Dose $(CCID)_{50}$. Analytical specificity was assessed by testing the assays against viruses that cause measles-like rashes. No cross-reactions were identified with these organisms or human genomic DNA in either real-time assay.

References

1. Hubschen JM, Kremer JR, De Landtsheer S et al (2008) A multiplex Taqman assay for the detection of measles and rubella virus. J Virol Methods 149:246–250
2. El Mubarak HS, De Swart RL, Osterhaus AD et al (2005) Development of a semi-quantitative real-time RT-PCR for the detection of measles virus. J Clin Virol 32:313–317

Chapter 52
Norovirus

Carl Kirkwood

52.1 Summary of Methods

Tests such as ELISA that use antibodies against a mixture of norovirus strains are available commercially but lack specificity and sensitivity. RT-PCR assays are the most commonly used methods to detect noroviruses. PCR assays are targeted at conserved regions on either the polymerase or capsid genes. A variety of methods are described in the literature, however, no single assay is able to detect all known norovirus genogroups because of the high genetic variation. Here we describe two different RT-PCR assays [1–4].

52.2 Organism

Norovirus (formerly Norwalk virus) is an RNA virus (taxonomic family *Caliciviridae*) which causes approximately 90% of epidemic non-bacterial outbreaks of gastroenteritis around the world, and may be responsible for 50% of all food-borne outbreaks of gastroenteritis in the US. Norovirus affects people of all ages. The viruses are transmitted by faecally contaminated food or water and by person-to-person contact.

Outbreaks of norovirus infection often occur in long-term care facilities, overnight camps, hospitals, prisons, dormitories, and cruise ships where the infection spreads very rapidly by either person-to-person transmission or through contaminated food.

When a person becomes infected with norovirus, the virus begins to multiply within the small intestine. After approximately 24 and 48 h after exposure, norovirus

C. Kirkwood (✉)
Enteric Virus Research Group, Murdoch Childrens Research Institute, Royal Children's Hospital, Parkville, VIC 3052, Australia
e-mail: carl.kirkwood@mcri.edu.au

M. Schuller et al. (eds.), *PCR for Clinical Microbiology*,
DOI 10.1007/978-90-481-9039-3_52, © Springer Science+Business Media B.V. 2010

symptoms can appear with the principal symptom being acute gastroenteritis that lasts for 24–60 h. The disease is usually self-limiting, and characterised by nausea, vomiting, diarrhoea, and abdominal pain; and in some cases, loss of taste. General lethargy, weakness, muscle aches, headache, and low-grade fever may occur.

Noroviruses are classified into five different genogroups (GI, GII, GIII, GIV, and GV) which can be further divided into different genetic clusters or genotypes. For example genogroup II, the most prevalent human genogroup, presently contains 19 genotypes. Genogroups I, II and IV infect humans, whereas genogroup III infects bovine species and genogroup V has recently been isolated in mice.

Most noroviruses that infect humans belong to genogroup GI and GII. Noroviruses from Genogroup II, genotype 4 (abbreviated as GII.4) account for the majority of adult outbreaks of gastroenteritis and often sweep across the globe. Recent examples include US95/96-US strain, associated with global outbreaks in the mid- to late-1990s, Farmington Hills virus associated with outbreaks in Europe and the United States in 2002 and in 2004 Hunter virus was associated with outbreaks in Europe, Japan and Australasia. In 2006 there was another large increase in norovirus infection around the globe with two new GII.4 variants causing around 80% of norovirus associated outbreaks.

52.3 Sample Preparation and RNA Extraction

A 20% (w/v) stool suspension of 1 ml is made in 0.01 M Tris buffer containing 0.01 M $CaCl_2$ and 0.15 M NaCl (pH 7.5). The sample is vortexed for 30 s and centrifuged for 3 min at 13,000×*g*. The clarified supernatant is collect and stored at –70°C.

Viral RNA is extracted using the QIAmp Viral RNA minikit according to manufacturer's instructions (Qiagen), and RNA resuspended in 80 μl sterile RNase-free water before storage at –70°C.

52.4 Primer Sequences

Primer	Sequence (5′-3′)	Rd	Region	Size (bp)	Nucleotide position[a]
NV2oF2	G GAG GGC GAT CGC AAT C	1	ORF1	379	5050–5066
NV2oR	GTR AAC GCR TTY CCM GC		ORF2		5412–5428
G2F3	TTG TGA ATG AAG ATG GCG TCG A	2	ORF1/2	311	5079–5100
G2SKR	CCR CCN GCA TRH CCR TTR TAC AT		ORF2		5367–5389

[a]position based on Lordsdale strain (GenBank accession number X86557)

52.5 PCR Amplification and Product Detection

52.5.1 Nested RT-PCR Assay

This method is used for the detection of commonly circulating noroviruses using primers directed to conserved regions of the 3′ end of the polymerase gene and the 5′ end of the capsid gene [1, 2, 4].

A reverse transcription reaction is performed in a final volume of 20 μl consisting of 5.4 μl dH_2O, 4 μl 5x RT buffer, 2 μl DTT (0.1 M), 2 μl dNTP (10 mM), 2 μl random primers (150 μM), 0.6 μl AMV-RT (9 U/μl) and 4 μl extracted RNA. Reverse transcription proceeds at 42°C for 60 min, and then is stopped by increasing the temperature to 72°C for 15 min.

The first round PCR is undertaken in a 20 μl final volume consisting of 10.52 μl dH_2O, 2 μl 10x buffer, 1.2 μl $MgCl_2$(25 mM), 0.4 μl dNTP (10 mM), 0.24 μl primer NV2oF (50 pmol/μl), 0.24 μl primer NV2oR (50 pmol/μl), 0.4 μl Taq (5 U/μl) and 5 μl cDNA produced from the reverse transcription. After a brief vortex, the mixture is placed into a thermal cycler. The tubes are heated at 95°C for 5 min, then by 25 cycles of 95°C for 30 s, 55°C for 30 s and 72°C for 1 min. All samples undergo a final extension at 72°C for 7 min prior to storage at 4°C.

The nested second round PCR is prepared using a similar procedure to the first round PCR. The second round PCR is undertaken in a 20 μl final volume consisting of 10.52 μl dH_2O, 2 μl 10x buffer, 1.2 μl $MgCl_2$(25 mM), 0.4 μl dNTP (10 mM), 0.24 μl primer G2F3 (50 pmol/μl), 0.24 μl primer G2SKR (50 pmol/μl), 0.4 μl Taq (5 U/μl) and 5 μl Template-cDNA produced from the first round PCR. After a brief vortex, the mixture is placed into a thermal cycler. The tubes are heated at 95°C for 5 min, then by 25 cycles of 95°C for 30 s, 45°C for 30 s and 72°C for 1 min. All samples undergo a final extension at 72°C for 7 min prior to storage at 4°C.

PCR products are separated on a 1.5% (w/v) agarose gel using 0.5% TBE, and stained with ethidium bromide prior to visualisation under UV light. The 1st round PCR products have an expected size of 379 bp, while the 2nd round PCR products are 311 bp. All specimens producing DNA products of appropriate size are confirmed as norovirus by sequence analysis of the second round PCR products.

52.5.2 Single Step RT-PCR Assay

This assay is used for the detection of both norovirus and sapovirus genera targeting conserved regions of the polymerase gene, and amplifies either a 319- or 321-base region [3].

The reverse transcription step is performed in a 50 μl volume consisting of 33.45 μl dH_2O, 5 μl 10x buffer, 4 μl $MgCl_2$(25 mM), 2.5 μl 2%BSA, 1 μl dNTP

(20 mM), 2 μl primer P289 (100 ng/μl), 0.25 μl RNase inhibitor (40 U/μl), 0.3 μl reverse transcriptase (40 U/μl) and 1.5 μl template-RNA. Reverse transcription proceeds at 42°C for 45 min.

The PCR reaction is undertaken in a 100 μl final volume consisting of 38.7 μl dH_2O, 5 μl 10x PCR buffer, 4 μl $MgCl_2$(25 mM), 2 μl P290 (100 ng/μl), 0.3 μl Taq (5 U/μl) and 50 μl RT reaction mix.

After a brief vortex and spin, the tubes are placed in a thermal cycler. Samples are heated to 94°C for 3 min, then 40 cycles of 94°C for 45 s, 49°C for 90 s and 72°C for 90 s. A final extension of 72°C for 10 min is performed.

The PCR products are separated on a 1.5% (w/v) TBE agarose gel, stained with ethidium bromide and visualized under UV light. Samples of appropriate sizes are confirmed as norovirus/sapovirus by sequence analysis of the specific products.

References

1. Bull RA, Tu ETV, McIver CJ et al (2006) Emergence of a new norovirus genotype II.4 variant associated with global outbreaks of gastroenteritis. J Clin Microbiol 44:327–333
2. Hansman GS, Katayama K, Maneekarn N et al (2004). Genetic diversity of norovirus and sapovirus in hospitalized infants with sporadic cases of acute gastroenteritis in Chiang Mai, Thailand. J Clin Microbiol 42:1305–1307
3. Jiang X, Huang PW, Zhong WM et al (1999). Design and evaluation of a primer pair that detected both Norwalk-like and sapporo-like caliciviruses by RT-PCR. J Virol Methods 83: 145–154
4. Kojima S, Kageyama T, Fukushi S et al (2002) Genogroup-specific PCR primers for detection of Norwalk-like viruses. J Virol Methods 100:107–114

Chapter 53
Respiratory Syncytial Virus Types A and B

Gerald B. Harnett and Glenys R. Chidlow

53.1 Summary of Methods

This protocol describes three singleplex real-time RT-PCR assays to detect the nucleoprotein genes of respiratory syncytial viruses types A and B (RSVA and RSVB) and the coat protein gene of MS2 RNA coliphage. MS2 is added to the lysis buffer of the extraction kit to serve as a control of RNA extraction, reverse transcription, the PCR reaction and removal of PCR inhibitors.

53.2 Acceptable Specimens

Pernasal aspirate is the preferred sample for RSV detection but respiratory swabs, sputa or respiratory tissues may also be tested.

53.3 Sample Extraction

Many RNA extraction kits may be used but we have found the QIAGEN Viral RNA Mini kit suitable for individual samples and the Applied Biosystems MagMax kit, used in the MagMax Express 96 robot, suitable for large batches of samples. Water blanks are positioned between every 5th test sample, extracted, and further processed with the test samples.

G.B. Harnett (✉)
Department of Microbiology and Infectious Diseases, Pathwest Laboratory Medicine, QEII Medical Centre, Nedlands, WA 6009, Australia
e-mail: Gerry.Harnett@health.wa.gov.au

M. Schuller et al. (eds.), *PCR for Clinical Microbiology*,
DOI 10.1007/978-90-481-9039-3_53,

53.4 Primers and Probe Sequences

Primer and probe sequences for RSV type A were: RSV-AF: 5′-CAA CTT CTG TCA TCC AGC AAA -3′; RSV-AR: 5′- TGC ACA TCA TAA TTA GGA GTA TCA AT-3′; and the RSV-A MGB probe sequence was: RSV-Aprobe 5′-6FAM-CAC CAT CCA ACG GAG C-MGB-NFQ-3′. Primers for RSV type B were: RSV-BF: 5′-ATT CAA CGT AGT ACA GGA GAT AAT A-3′; RSV-BR: 5′-CCA CAT AGT TTG TTT AGG TGT TT-3′; with a RSV-B specific MGB probe: RSV-Bprobe: 5′-6FAM-TGA CAC TCC CAA TTA T-MGB-NFQ-3′.

The MS2 RNA coliphage control was detected with primers MS2-F: 5′-GTC GAC AAT GGC GGA ACT G-3′ and MS2-R: 5′-TTC AGC GAC CCC GTT AGC-3′ and probe MS2-probe: 5′-Quasar 670-ACG TGA CTG TCG CCC CAA GCA ACT T-BHQ2-3′. MGB probes were manufactured by Applied Biosystems, Australia and the BHQ2 probe by Biosearch Technologies, USA.

53.5 PCR Amplification and Product Detection

After extraction, 8 μl of sample extract or control is added to each of the PCR tubes which contain 0.3 μM of the primers and 0.2 μM of the TaqMan probes, 1x Superscript III One Step buffer, 0.3 μl of Superscript III (Invitrogen, Australia), 10 units of RNaseOUT (Invitrogen, Australia) and 0.5 units of iSTAR DNA polymerase (Intron) in a total reaction volume of 20 μl. The tubes are cycled in a real-time RotorGene 6000 instrument (QIAGEN, Australia) and probe emissions acquired on the appropriate wavelength channel. The cycling programme consists of an RT reaction at 50°C for 30 min and a DNA denaturation period of 5 min at 95°C, followed by 50 cycles of 12 s at 94°C, 15 s at 55°C and 20 s at 72°C.

53.6 Quality Control and Validation Data

The MS2 assay serves as a control of RNA extraction, reverse transcription and removal of PCR inhibitors and should show a Ct value consistent with the copy number of the MS2 target introduced. High Ct values indicate assay problems which call for repeat testing. Positive controls are included with each batch of tests at a dilution 10x the expected end-point titre and all controls must show the expected results for test batches to be accepted. The CT values of any positive control need to be determined on every test batch as to calibrate the signal strength against that of the internal control used: MS2 RNA coliphage in this particular test. The CT values for both the positive controls and the internal control need to be consistent.

The one-step TaqMan assays were compared against a nested RT-PCR for RSV [1] which had been in use for several years in this laboratory and which itself had been shown to be more sensitive than cell culture. Stored RSV positive samples (n=116) and 131 negative samples were retested by the nested method and also by the TaqMan assays. The samples comprised various swabs and aspirates. The TaqMan assays detected seven extra positive samples compared with the nested method and only one false negative. Probit analysis was performed using three-fold dilutions of target with 8 replicates per dilution and showed that the 95% confidence limit of detection was 300 copies/ml of extract for RSVA and 660 copies/ml for RSVB. It is recommended that Ct values for control positive materials be plotted on a Shewhart chart. When about 100 test batches have been processed, the mean Ct value ± two standard deviations can be determined and utilised to monitor subsequent test batches on a daily basis. When control Ct values fall within these limits they provide a 95% confidence limit for the assay and can be used for Uncertainty of Measurement calculations.

Acknowledgments We would like to thank Natasha Ryan and Eileen Pratt who provided valuable technical assistance for this study.

Reference

1. Cubie HA, Inglis JM, Leslie EE et al (1992) Detection of respiratory syncytial virus in acute bronchiolitis in infants. J Med Virol 38:283–287

Chapter 54
Rotavirus

Carl Kirkwood

54.1 Summary of Methods

Diagnosis of infection with rotavirus A is made by identification of the virus in the patient's stool by enzyme immunoassay. There are several licensed test kits on the market which are sensitive, specific and detect all serotypes of rotavirus A. Other methods such as electron microscopy and polyacrylamide gel electrophoresis are used in research laboratories.

Reverse transcription-polymerase chain reaction (RT-PCR) can detect and identify all species and serotypes of human rotavirus. A variety of sensitive conventional or real-time reverse-transcription polymerase chain reaction (RT-PCR) methods have been developed based on primers specific for rotavirus genes [2, 3, 5]. These methods have been useful in detecting rotavirus in extra-intestinal tissues, in studies of the duration of viral shedding in stool and the correlation between disease severity and virus load [1, 2, 7].

In molecular epidemiology studies, genotyping of rotavirus strains based on the 2 outer capsid proteins, VP7 (G typing) and VP4 (P typing) is the method of choice for many laboratories [4].

54.2 Organism

Rotavirus is a *genus* of *double-stranded RNA virus* in the family *Reoviridae*. It is the leading single cause of *severe diarrhoea* among infants and young children. By the age of five, nearly every child in the world has been infected with at least one rotavirus strain. However, with each infection, *immunity* develops with subsequent infections less severe. There are seven *species* of the virus, referred to as A, B, C, D, E, F and G. rotavirus A is the most common and causes more than 90% of infections in humans. It was 'discovered' in 1973 by Ruth Bishop and colleagues

C. Kirkwood (✉)
Enteric Virus Research Group, Murdoch Childrens Research Institute, Royal Children's Hospital, Parkville, VIC 3052, Australia
e-mail: carl.kirkwood@mcri.edu.au

M. Schuller et al. (eds.), *PCR for Clinical Microbiology*,
DOI 10.1007/978-90-481-9039-3_54, © Springer Science+Business Media B.V. 2010

using electron microscopy. They observed a 70-nm virus in the duodenal epithelium of children with diarrhoea and was subsequently termed rotavirus (Latin, *rota* = wheel) because of its appearance.

Rotavirus is transmitted by the *faecal-oral route*. It infects *cells* that line the *small intestine* and produces an *enterotoxin*, which induces *gastroenteritis*, leading to severe diarrhoea and sometimes death through *dehydration*. In addition to its impact on human health, rotavirus also infects other animals, and is a *pathogen* of livestock.

In 2006, two vaccines against rotavirus A infection were shown to be safe and effective in children: Rotarix by GlaxoSmithKline http://en.wikipedia.org/wiki/Rotavirus - cite_note-81 and RotaTeq by Merck. Both are taken orally and contain live attenuated rotavirus strains. Rotavirus vaccines are now available in Australia and on June 5, 2009 the World Health Organization (WHO) recommended that rotavirus vaccine be included in all national immunization programs.

54.3 Sample Preparation and RNA Extraction

A 20% (w/v) stool suspension of 1 ml is made in 0.01 M Tris buffer containing 0.01 M $CaCl_2$ and 0.15 M NaCl (pH 7.5). The sample is vortexed for 30 s and centrifuged for 1 min at 13,000×*g*. The clarified supernatant is collected and stored at –70°C.

Viral RNA is extracted using the QIAmp Viral RNA kit according to manufacturer's instructions, and RNA resuspended in 50 μl sterile Rnase-free water before storage at −70°C.

54.3.1 VP7 Genotyping Assay (G Typing)

VP7 genotyping is conducted using a 2 step hemi-nested multiplex RT-PCR assay using gene specific consensus primers in the RT-PCR, and genotype specific primers in the second round PCR [3, 5, 6].

54.3.1.1 VP7 Primer Sequences

Primer	Sequence (5′–3′)	Rd	Gene/ genotype	Size (bp)	Nucleotide position
VP7F	ATGTATGGTATTGAATATACCAC	1	VP7	881	51–71
VP7R	AACTTGCCACCATTTTTTCC		VP7		932–914
VP7R	AACTTGCCACCATTTTTTCC	2	Consensus		932–914
aBT-1	CAAGTACTCAAATCAATGATGG		G1		314–335
aCT-2	CAATGATATTAACACATTTTCTGTG		G2		411–435
G3	ACGAACTCAACACGAGAGG		G3		250–269
ADT-4	CGTTTCTGGTGAGGAGTTG		G4		480–499
AAT-8	GTCACACCATTTGTAAATTCG		G8		178–198
G9	CTTGATGTGACTAYAAATAC		G9		757–776

54.3.1.2 VP7 PCR Amplification and Product Detection

The first round RT-PCR is performed in a 100 μl volume consisting of 77.5 μl dH_2O, 1 μl 1 M Tris-HCl pH 8.3, 1.5 μl 100 mM $MgCl_2$, 4.0 μl 1 M KCl, 7 μl DMSO, 1.0 μl dNTP (20 mM each), 1.0 μl VP7F primer (100 ng/μl), 1.0 μl VP7R primer (100 ng/μl) and 5 μl Template- dsRNA. The mixture is heated at 97°C for 3 min to denature the dsRNA.

Following denaturation of dsRNA, 0.25 μl of AMV reverse transcriptase (22 U/μl) and 0.5 μl *Amplitaq* DNA polymerase (5 U/μl) are added to make the total volume 100 μl. Reverse transcription proceeds for 60 min at 42°C, followed by 30 cycles of 94°C for 1 min, 42°C for 2 min and 68°C for 1 min. Samples are then held at 4°C.

To determine the G genotype of a specimen 1 μl of the first round RT-PCR product is added to 99 μl of genotyping PCR reaction mix, prepared with genotype specific primers for G1, G2, G3, G4, G8 and G9 types. The VP7 G genotyping reaction mix consists of 79 μl dH_2O, 1 μl 1 M Tris-HCl pH 8.3, 1.5 μl 100 mM $MgCl_2$, 4 μl 1 M KCl, 7 μl DMSO, 1 μl dNTP (20 mM each), 1 μl VP7R primer (100 ng/μl), 4 μl Genotype primer mix (100 ng/μl of each primer), 0.5 μl AmpliTaq (5 U/μl) and 5 μl Template – cDNA.

The cycling conditions are 2 min at 94°C, then 30 cycles of 94°C for 1 min, 42°C for 2 min for 68°C 1 min. The samples are held at 4°C.

PCR products are separated by gel electrophoresis using a 1.5% (w/v) agarose gel, in 0.5% TBE buffer with ethidium bromide, and visualized by UV light. Each G genotype has a specific cDNA size.

The expected cDNA sizes are; Genotype G1 = 618 bp, Genotype G2 = 521 bp, Genotype G4 = 452 bp, Genotype G3 = 682 bp, Genotype G8 = 754 bp, Genotype G9 = 179 bp.

54.3.2 VP4 Genotyping (P Typing)

54.3.2.1 VP4 Primer Sequences

Primer	Sequence (5′–3′)	Rd	Gene/ genotype	Size (bp)	Nucleotide position
VP4F	TATGCTCCAGTNAATTGG	1	VP4	663	132–149
VP4R	ATTGCATTTCTTTCCATAATG		VP4		795–775
VP4F	TATGCTCCAGTNAATTGG	2	Consensus		132–149
2T-1	CTATTGTTAGAGGTTAGAGTC		P4	362	492–474
3T-1	TGTTGATTAGTTGGATTCAA		P6	146	278–259
1T-1D	TCTACTGGATAACGTGC		P8	224	356–339
4T-1	TGAGACATGCAATTGGAC		P9	270	402–385
5T-1	ATCATAGTTAGTAGTCGG		P10	462	594–575
P[11]	GTAAACATCCAGAATGTG		P11	191	323–305

54.4 PCR Amplification and Product Detection

The first round consensus RT-PCR for VP4 genotyping is performed in a 100 μl volume consisting of 77.5 μl dH_2O, 1 μl 1 M Tris-HCl (pH 8.3), 1.5 μl 100 mM $MgCl_2$,4 μl 1 M KCl, 7 μl DMSO, 1 μl dNTP (20 mM each), 1 μl VP4F primer (100 ng/μl), 1 μl VP4R primer (100 ng/μl) and 5 μl Template-dsRNA.

After a brief vortex, the mixture is heated at 97°C for 3 min, to denature the dsRNA, then transferred immediately to ice. Subsequently 0.25 μl of reverse transcriptase AMV (22 U/μl) and 0.5 μl *Amplitaq* DNA polymerase (5 U/ul) are added to each tube for a total of 100 μl, briefly vortexed, then placed into a thermal cycler. Reverse transcription proceeds for 60 min at 42°C, followed by 30 cycles of 94°C for 1 min, 50°C for 2 min and 68°C for 1 min. Samples are then held at 4°C.

To determine the P genotype of a specimen, a volume of 5 μl of the first round RT-PCR product is added to 95 μl of genotyping PCR reaction mix prepared with genotype specific primers for P[4], P[6], P[8], P[9] and P[10] types. The VP4 genotyping reaction mix consists of 79.75 μl dH_2O, 1 μl 1 M Tris-HCl pH 8.3, 1.75 μl 100 mM $MgCl_2$,4 μl 1 M KCl, 1 μl dNTP (20 mM each), 1 μl VP4F Primer (100 ng/μl), 6 μl P genotyping primer mix (each at 100 ng/μl), 0.5 μl Amplitaq (5 Units/μl) and 5 μl Template-cDNA.

The cycling conditions are 2 min at 94°C, then 40 cycles of 94°C for 1 min, 50°C for 2 min for 68°C 1 min. The samples are held at 4°C.

PCR products are separated by gel electrophoresis using a 2% (w/v) agarose gel, in 0.5% TBE buffer with ethidium bromide, and visualized by ultraviolet light. Each P genotype has a specific cDNA size (Fig. 54.1).

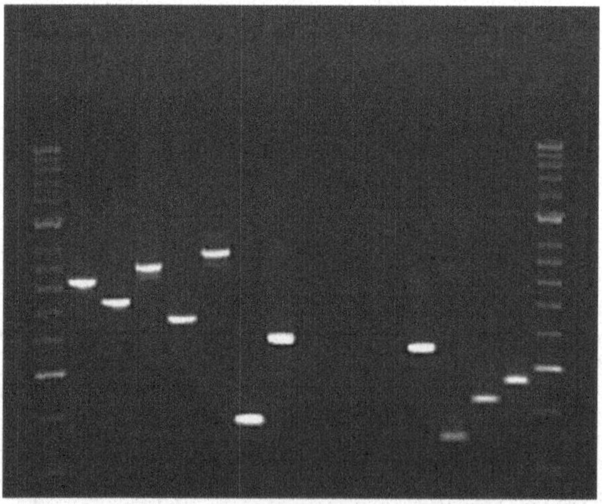

Fig. 54.1 Ethidium bromide stained agarose gel showing products G- and P-Genotyping DNA products generated by hemi-nested multiplex RT-PCR assays. (**a**) VP7 G genotyping products generated with VP7R and G- specific-typing primers (lanes 2–8). (**b**) VP4 P genotyping products generated with VP4F and P-specific typing primers (lanes 12–15). The expected cDNA sizes are; P10 = 462 bp, P4 = 362 bp, P9 = 270 bp, P8 = 224 bp, P6 = 146 bp, P11 = 191 bp

54.5 Assay Limitations with Strain Genotyping

A small proportion of strains cannot be typed for P and/or G genotype. These nontypeable strains may result from genetic variation of common strains or because of the presence of novel strains. Characterisation of these strains is achieved by sequence analysis of first round PCR products and subsequent alignment of sequence with that from standard strains.

Samples positive for rotavirus antigen might fail to yield any PCR products. These nontypeable samples might be the result of a false-positive EIA, insufficient or degraded RNA, or the presence of residual stool inhibitors in the RNA extract.

References

1. Chiappini E, Azzari C, Moriondo M et al (2005) Viraemia is a common finding in immunocompetent children with rotavirus infection. J Med Virol 76:265–267
2. Elschner M, Prudlo J, Hotzel H et al (2002) Nested reverse transcriptase-polymerase chain reaction for the detection of group A rotaviruses. J Vet Med B Infect Dis Vet Public Health 49:77–81
3. Gentsch JR, Glass RI, Woods P et al (1992) Identification of group A rotavirus gene 4 types by polymerase chain reaction. J Clin Microbiol 30:1365–1373.
4. Gentsch JR, Laird AR, Bielfelt B et al (2005) Serotype diversity and reassortment between human and animal rotavirus strains: implications for rotavirus vaccine programs. J Infect Dis 192(Suppl 1):S146–S159
5. Gouvea V, Glass RI, Woods P et al (1990) Polymerase chain reaction amplification and typing of rotavirus nucleic acid from stool specimens. J Clin Microbiol 28:276–282
6. Simmonds MK, Armah G, Asmah R et al (2008) New oligonucleotide primers for P-typing of rotavirus strains: strategies for typing previously untypeable strains. J Clin Virol 42:368–373
7. Zeng SQ, Halkosalo A, Salminen M et al (2008) One-step quantitative RT-PCR for the detection of rotavirus in acute gastroenteritis. J Virol Methods 153:238–240

Chapter 55
Varicella Zoster virus

Mala Ratnamohan

55.1 Summary of Methods

Varicella Zoster virus (VZV) is an alpha herpesvirus that causes chicken pox (varicella), usually in childhood. Following primary infection, VZV establishes a latent infection in the sensory nerve ganglia. Reactivation of VZV from the dorsal root ganglia, known as shingles or herpes zoster, may be seen decades later, usually in older people and immunocompromised patients. Shingles presents with or without rash and may lead to neurological disease [5], however CNS involvement is a rare manifestation after primary varicella infection. Detection of VZV in CSF by conventional virological methods is difficult as the virus is highly cell associated and has a narrow host range. DNA amplification methods have improved sensitivity over viral culture for the detection of viruses in CSF samples. Although enteroviruses are the major cause of aseptic meningitis in immunocompetent adults, VZV and herpes simplex virus are the next two most frequently detected aetiological agents [3, 4]. In recent years some leading laboratories have incorporated the practice of routinely testing for these three viruses in all CSF samples from patients presenting with suspected neurological symptoms [4]. As symptoms may not easily differentiate the diagnosis of viral meningitis, a definitive diagnosis by PCR is useful for the early initiation of antiviral treatment.

The VZV PCR protocol described is a real-time duplex assay that simultaneously detects part of the ORF 63 gene of VZV and an internal control. The internal control is a recombinant DNA construct containing the sequences of the forward and the reverse primer at either end.

M. Ratnamohan (✉)
Molecular Biology, Centre for Infectious Diseases and Microbiology, Institute for Clinical Pathology and Medical Research, Westmead Hospital, Westmead, NSW 2145, Australia
e-mail: Mala.Ratnamohan@swahs.health.nsw.gov.au

M. Schuller et al. (eds.), *PCR for Clinical Microbiology*,
DOI 10.1007/978-90-481-9039-3_55,

55.2 Accepted Specimens

The most commonly used sample for the diagnosis of VZV meningitis is CSF, which is used without centrifugation. Occasionally brain tissue following autopsy is available for testing.

55.3 Sample Extraction

DNA is extracted using the GenElute Mammalian DNA kit (Sigma-Aldrich Co.) with minor modifications to the manufacturer's instructions. In brief, 200 μl of CSF is combined with 200 μl of Lysis C solution, 20 μl of proteinase K (20 mg/ml), 10 μl of 10% polyvinyl pyrrolidine and 1 μl of the internal control (approximately 500 copies) and incubated for 10 min at 55°C. 200 μl of 100% ethanol is then added and the solution centrifuged through a prepared silica spin column. The membrane is washed twice with GenElute Wash solution and the DNA is eluted in 50 μl of Elution buffer. DNA extractions performed on automated instruments such as NucliSENS® easyMAG® (bioMérieux) and EZ1 Advanced XL (Qiagen Pty Ltd) are also suitable for this assay providing the internal control is added to the lysis step.

55.4 Primer and Probe Sequences

The primer pair VZVF (5′-TTA TGT ATT GCA CAG TCG TCG-3′) and VZVR (5′-TCG TCG CTA TCG TCT TCA C-3′) are used to amplify a 296 bp fragment of the ORF 63 gene (Genbank accession AF206304) which is detected using the HybProbe probes VZLC1 (5′-CAG TGC GCT CCT ATG CAA A-fluo-3′) and VZLC2 (5′-LC640-AAC ATT CGG CGC CTC AAT-phos-3′). The internal control primer pair betaF (5′-ACA CAA CTG TGT TCA CTA GC-3′) and betaR (5′-GGA AAA TAG ACC AAT AGG C-3′) are used to amplify a 248 bp fragment of the human beta globin gene (GenBank accession AY260740) which is detected using the HybProbe probes bglLC1 (5′-CAT GGT GCA CCT GAC TCC TG-fluo-3′) and bglLC2 (5′-LC705- GAG AAG TCT GCC GTT ACT GCC-phos-3′). The primer and probe sequences were selected and checked for their suitability using the Primer 3 software program [6].

55.5 PCR Amplification and Product Detection

The PCRs are set up in a 20 μl volume containing 1X LightCycler® FastStart DNA Master Plus HybProbe (containing FastStart TaqDNA polymerase, reaction buffer, $MgCl_2$ and dNTP mix) (Roche Applied Science Australia), 0.5 μM of each primer

(VZVF, VZVR, betaF and betaR), 0.2 μM of each probe (VZLC1, VZLC2, bg1LC1 and bg1LC2) and 5 μl of extracted DNA. Amplification is performed in 20 μl glass capillaries on a LightCycler® 2.0 instrument under the following conditions: 95°C for 10 min followed by 40 cycles of 95°C for 10 s, 55°C for 15 s and 72°C for 12 s, followed by a melt analysis whereby the amplicons are heated to 95°C for 0 s, cooled to 40°C and then heated to 80°C at a ramp rate of 0.1°C/s. Fluorescence data is acquired at the end of every annealing step (55°C) during PCR amplification and continuously during melt curve analysis. Positive amplification signals are registered as the crossing point (Cp) at which the fluorescent reading is higher than the background and begins to accumulate in the exponential phase. For analysis of the VZV and internal control amplification curves, cross talk between the two fluorescent channels (640 and 705 nm, respectively) is corrected through the use of a colour compensation file. A colour compensation file is required for all dual colour LightCycler® experiments and is created using the LightCycler® Colour Compensation set (Roche Applied Science Australia) and performing amplification as described above. Samples generating positive amplification signals will also have a melting point of 57°C, ±2°C. Generally CSF samples do not contain high copies of VZV DNA and Cp values are usually lower than that for the high positive control.

55.6 Quality Control and Validation Data

A negative control water blank is included in each batch of extractions to monitor contamination through the extraction process. A low copy number positive (at the limit of detection) and a high copy number positive are included with every run to monitor the efficiency of amplification. The inclusion of an internal control monitors the efficiency of DNA extraction and PCR amplification for each sample. A negative result for the internal control indicates the presence of inhibitors in the sample and the sample should be re-tested. The repeat testing is performed on neat and diluted DNA. Some samples that test VZV positive have negative results for the internal control due to competitive inhibition.

The real time assay described was validated against a nested gel based assay [1, 2] using vesicular swabs, conjunctival swabs and CSF specimens. The most commonly used specimen for diagnosis of chicken pox or shingles is vesicular fluid from skin lesions. CSF is most often tested as part of differential diagnosis on patients presenting with aseptic meningitis or in a small number of cases from patients with neurological complications following varicella zoster. As such the patient clinical data is not well defined to permit the calculation of the negative predictive value. The positive predictive value and specificity of the assay is 100%.

When the PCR assay was compared to VZV culture in human embryonic fibroblasts, five of the VZV isolates showed a higher sensitivity of 3 $\log_{10}$ by PCR. This is equivalent to 0.001 $TCID_{50}$ of VZV.

References

1. Cooper C, Wojtulewicz J, Ratnamohan VM et al (2000) Congenital varicella syndrome diagnosed by polymerase chain reaction-scarring of the spinal cord, not the skin. J Paediatr Child Health 36:186–188
2. Dlugosch D, Eis-Hubinger AM, Kleim JP et al (1991) Diagnosis of acute and latent varicella-zoster infections using the polymerase chain reaction. J Med Virol 35:136–141
3. Hausfater P, Fillet A, Rozenberg F et al (2004) Prevalence of viral infection markers by polymerase chain reaction amplification and interferon-alpha measurements among patients undergoing lumbar puncture in an emergency department. J Med Virol 73:137–146
4. Ihekwaba UK, Kudesia G, McKendrick W (2008) Clinical features of viral meningitis in adults: significant differences in cerebrospinal fluid findings among herpes simplex virus, varicella zoster virus and enterovirus infections. CID 47:783–789
5. Kleinschmidt-DeMasters BK, Gilden DH (2001) Varicella-zoster infections of the nervous system – clinical and pathologic correlates. Arch Pathol Lab Med 125:770–780
6. Rozen S, Skaletsky HJ (2000) Primer3 on the WWW for general users and for biologist programmers. In: Krawetz S, Misener S (eds) Bioinformatics methods and protocols: methods in molecular biology, Humana Press, Totowa, NJ, pp 365–386. http://frodo.wi.mit.edu/primer3/. Accessed 11 Nov 2009

Chapter 56
Eight Commonly Recognised Respiratory Viruses

David M. Whiley

56.1 Summary of Methods

This protocol describes three multiplex assays (Flu-TM; Para-TM; Ad-RSV-TM) and one monoplex assay (hMPV-TM) for detection of influenza viruses A and B, parainfluenza viruses 1, 2 and 3, adenovirus, respiratory syncytial virus and human metapneumovirus. The assays utilise 5′ nuclease-based real-time PCR comprising two primers and one probe for each organism.

56.2 Acceptable Specimens

Acceptable specimens are respiratory samples, including nasal swabs and nasopharyngeal aspirates (NPA).

56.3 Sample Extraction

Swabs are rotated 15 times in 400 μl of phosphate buffered saline (PBS) to dislodge cells and virus. 200 μl of PBS or NPA specimen are then extracted manually using the High Pure Viral Nucleic Acid kit (Roche Applied Science, Australia), or mechanically using the MagNAPure instrument (Roche Applied Sciences, Australia), according to the manufacturer's instructions. The final volume of extract is 50 μl.

56.4 Primers and Probe Sequences

The primers and probes used for each assay are listed in Table 56.1.

D.M. Whiley (✉)
Queensland Paediatric Infectious Diseases Laboratory, Sir Albert Sakzewski Virus Research Centre, Queensland Children's Medical Research Institute, Children's Health Service District, Brisbane, QLD 4029, Australia
e-mail: d.whiley@uq.edu.au

M. Schuller et al. (eds.), *PCR for Clinical Microbiology*,
DOI 10.1007/978-90-481-9039-3_56, © Springer Science+Business Media B.V. 2010

Table 56.1 Oligonucleotides used for the respiratory real-time PCR assays

Assay	Virus	Oligonucleotide name	Sequence (5′–3′)	References
Flu-TM	IFA	fluA-F	CTTCTAACCGAGGTCGAAACGTA	[2, 4]
		fluA-R	GGTGACAGGATTGGTCTTGTCTTTA	[2, 4]
		fluA-Probe	FAM-TCAGGCCCCCTCAAAGCCGAG-BHQ1	[2, 4]
	IFB	fluB-F	GCATCTTTTGTTTTTTATCCATTCC	[2]
		fluB-R	CACAATTGCCTACCTGCTTTCA	[2]
		fluB-Probe	YakimaYellow-TGCTAGTTCTGCTTTGCCTTCTCC ATCTTCT-BHQ1	[2]
Para-TM	PIV1	para1-F	TTTAAACCCGGTAATTTCTCATACCT	[2]
		para1-R	CCCCTTGTTCCTGCAGCTATT	[2]
		para1-Probe	FAM-TGACATCAACGACAACAGGAAA TCATGTTCTG-BHQ1	[2]
	PIV2	para2-F	AGAGTTCCAACATTCAATGAATCAGT	[2]
		para2-R	CTCAAGAGAAATGTCATTCCCATCT	[2]
		para2-Probe	YakimaYellow-CCTCTGTATTGCTCATGCATA GCACGGA-BHQ1	[2]
	PIV3	para3-F	CGGTGACACAGTGGATCAGATT	[2]
		para3-R	AGGTCATTTCTGCTAGTATT CATTGTTATT	[2]
		para3-Probe	cy5-TCAATCATGCGGTCTCAACA GAGCTTG-BHQ3	[2]

Table 56.1 (continued)

Assay	Virus	Oligonucleotide name	Sequence (5′–3′)	References
Ad-RSV-TM	ADV	Adeno-F	GCCACGGTGGGGTTTCTAAACTT	[1]
		Adeno-R	GCCCCAGTGGTCTTACATGCACAT	[1]
		Adeno-Probe	FAM-TGCACCAGACCCGGGCTCAGGTACTCCGA – bhq1	[1]
	RSV	RSV-F	AGTAGACCATGTGAATTCCCTGC	[2]
		RSV-R	GTCGATATCTTCATCACCATACTTTTCTGTTA	[2]
		RSV-Probe	FAM-TCAATACCAGCTTATAGAAC-mgb-BHQ1	[2]
hMPV-TM	HMPV	hMPV-F	CATATAAGCATGCTATATTAAAAGAGTCTC	[3]
		hMPV-R	CCTATTTCTGCAGCATATTTGTAATCAG	[3]
		hMPV-Probe	FAM-TGYAATGATGAGGGTGTCACTGCGGTTG-TAMRA[a]	[3]

[a](Y = C or T)

IFA = influenza A virus; IFB = influenza B virus; PIV = parainfluenza virus; ADV = adenovirus; RSV = respiratory syncytial virus; HMPV = human metapneumovirus; BHQ = black hole quencher; MGB = minor groove binder

56.5 PCR Amplification and Product Detection

Perform PCR reactions in 0.2 ml or 0.1 ml reaction tubes using the QIAGEN One-Step RT-PCR Kit (QIAGEN, Australia). Each reaction mix comprises 0.4 μM of each primer and 0.32 μM of each TaqMan probe (as per Table 56.1), except for the influenza A TaqMan probe (fluA-Probe) in the Flu-TM assay which is used at 0.08 μM. The total reaction volume is 25 μl, including 5 μl of nucleic acid extract. Amplification and detection are performed on RotorGene 3000 or 6000 instruments (QIAGEN, Australia) with the following conditions: initial holds at 50°C for 20 min and 95°C for 15 min followed by 45 cycles at 95°C for 15 s and 60°C for 60 s (fluorescence acquisition). The separate reactions in each multiplex assay are distinguished through the use of different fluorophores on the TaqMan probes.

56.6 Quality Control and Validation Data

Extracted nucleic acid from viral isolates or from PCR-positive samples may be used as a positive control. The positive control should be diluted to a concentration providing a cycle-threshold (Ct) value of approximately 25 cycles in the real-time PCR. PCR-grade water is subject to the extraction procedure and is used as a negative control.

These assays will detect the presence of these viruses in clinical specimens at a sensitivity of 10–100 copies/reaction, with clinical sensitivities and specificities ranging from 90 to 100% [2]. (PPV and NPV data are not available).

56.7 Assay Limitations

The particular design of any real time PCR probe should take into account the consensus sequences of the target region prior to using it for diagnostics purposes, hence the failure of some probes to detect natural variants of any microorganism or virus (A. Franco, personal communication). It has been found over the last 4 years that the clinical sensitivity for RSV using the primer/probe combination described above has diminished with time due to natural mutation. Subsequent examination of this phenomenon has highlighted a potential limitation for the use of MGB probes when mutation under the probe binding site seems likely, as is the case with the evolution of RSV over consecutive seasons [5]. For this reason we sequence the target sequence of RSV circulating in our population annually, and redesign the probe if necessary.

Recently, we have included a second RSV-specific real-time PCR assay in our diagnostic algorithm to ensure all infected patients are adequately diagnosed. This assay is described by Harnett and Chidlow in Chapter 53.

References

1. Heim A, Ebnet C, Harste G et al (2003) Rapid and quantitative detection of human adenovirus DNA by real-time PCR. J Med Virol 70:228–239
2. Lambert SB, Whiley DM, O'Neill NT et al (2008) Comparing nose-throat swabs and nasopharyngeal aspirates collected from children with symptoms for respiratory virus identification using real-time polymerase chain reaction. Pediatrics 122:e615–620
3. Maertzdorf J, Wang CK, Brown JB et al (2004) Real-time reverse transcriptase PCR assay for detection of human metapneumoviruses from all known genetic lineages. J Clin Microbiol 42:981–986
4. Whiley DM, Sloots TP (2005) A 5′-nuclease real-time reverse transcriptase-polymerase chain reaction assay for the detection of a broad range of influenza A subtypes, including H5N1. Diagn Microbiol Infect Dis 53:335–337
5. Whiley DM, Sloots TP (2006) Sequence variation can affect the performance of minor groove binder TaqMan probes in viral diagnostic assays. J Clin Virol 35:81–83

Chapter 57
Detection of CMV, HSV, VZV, EBV and Enterovirus by Multiplex PCR

Christopher McIver and Ian Carter

Cytomegalovirus (CMV) is a *viral genus* of the *Herpesviruses* group: in humans it is commonly known as HCMV. Human CMV infections are frequently associated with *salivary glands*, though they may be found throughout the body. HCMV infection can also be life threatening for patients who are *immunocompromised* (e.g. patients with *HIV*, *organ transplant* recipients, or neonates). About 58.9% of individuals aged 6 and over are infected with CMV while 90.8% of individuals aged 80 and over are positive for HCMV antibodies [1]. HCMV is also the *virus* most frequently transmitted to a developing child before birth.

Enteroviruses are small RNA viruses. Enteroviruses can be found in respiratory secretions, such as saliva, sputum or nasal secretions, and in the faeces of infected persons. Subjects may become infected by direct contact with secretions from an infected person, or by contact with contaminated objects such as drinking and eating utensils. Transmission also may occur if an infected individual coughs or sneezes directly in the face of another person. These viruses can be transmitted by contact with faeces, such as when persons changing diapers of infants and toddlers do not wash their hands thoroughly. Subjects with symptoms of illness who are infected with an enterovirus can infect other persons who may or may not become ill after they become infected. For persons who become ill with an enterovirus, most develop symptoms of a cold, an influenza-like illness with fever and muscle aches, or an illness with a rash. Less commonly, some people develop meningitis caused by an enterovirus. Rarely, enterovirus infections can cause inflammation of the heart muscle or inflamation of the brain.

Epstein-Barr virus, frequently referred to as EBV, is a member of the herpesvirus family and one of the most common human viruses. Most people become infected with EBV sometime during their lives. Many children become infected with EBV,

C. McIver (✉)
SEALS Microbiology Division, Prince of Wales Hospital, Randwick, NSW 2031, Australia
e-mail: christopher.mciver@sesiahs.health.nsw.gov.au

M. Schuller et al. (eds.), *PCR for Clinical Microbiology*,
DOI 10.1007/978-90-481-9039-3_57, © Springer Science+Business Media B.V. 2010

and these infections usually cause no symptoms or are indistinguishable from the other mild, brief illnesses of childhood. When infection with EBV occurs during adolescence or young adulthood, it causes infectious mononucleosis.

Herpes simplex virus (HSV) is one of the most common agents infecting humans of all ages. Neonatal *herpes simplex virus infections* can result in serious morbidity and mortality. Disseminated neonatal *herpes simplex virus infection* is characterized by progressive multiple organ failure and high mortality rates.

Varicella-zoster virus (Herpes zoster) causes chickenpox which is an acute, localized infection with a painful, blistering rash. Herpes zoster, or shingles, is caused by the same virus that causes chickenpox. After an episode of chickenpox, the virus becomes dormant in the body. Herpes zoster occurs as a result of the virus re-emerging after many years.

57.1 Principle of Procedure

Detection of these viruses by polymerase chain reaction (PCR) has been used to diagnose such diseases as chicken pox, keratoconjuctivitis, acute respiratory disease, gastroenteritis, disseminated neonatal HSV infection as well as many skin infections. The universal nature of PCR technology makes it possible to diagnose infections caused by these viruses.

This assay is based on the amplification of a specific nucleic acid sequence using PCR. The amplicon if present is visualized by agarose gel electrophoresis under ultraviolet light. A biotin labeled probe is used to confirm and differentiate the PCR product. The table below contains primer and probe sequences.

57.2 Specimen Collection

Copan Viral swab – break swab into a 2.0 ml microtube containing 750 μl of saline; CSF, body fluids, urines – place 1.0 ml of specimen into a 2.0 ml microtube; Biopsies – sample is ground in saline to a final volume of 500 μl; EDTA plasma – placed in a 2.0 ml microtube.

All prepared specimens for PCR are stored at –20°C prior to extraction.

57.3 Sample Extraction

Extract using MagNa Pure LC automated extractor (refer to MagNa Pure Method). Use MagNa Pure LC Total Nucleic Acid Isolation Kit. Place 200 μl of each specimen into MagNa Pure specimen tray.

57.4 Primer and Probe Sequences

Pathogen	Description	Sequence	Product size (bp)
CMV	Outer sense	5′-AAG GTT CGA GTG GAC ATG GT -3′	396
	Outer antisense	5′-CAG CCA TTG GTG GTC TTA GG-3′	
(Jiwa, 1989)	Inner sense	5′-GAG CCT TTC GAG GAG ATG AA -3′	229
	Inner antisense	5′-GGC TGA GTT CTT GGT AAA GA-3′	
	Probe	Biotin-5′-AGG CCC GTG CTA AAA AGG AT-3′	
HSV	Outer sense	5′-ATC CGA ACG CAG CCC CGC TG -3′	382
	Outer antisense	5′-TCC GGS GGC AGC AGG GTG CT -3′ [a]	
(Read, 1997)	Inner sense	5′-GCG CCG TCA GCG AGG ATA AC -3′	280
	Inner antisense	5′-AGC TGT ATA SGG CGA CGG TG -3′	
	Probe HSV-1	Biotin-5′-AAG GGC TCC TGT AAG TAC ACC C-3′	
	Probe HSV-2	Biotin-5′-CGC GCC TCC TGC AAG TAC GCT C -3′	
VZV	Outer sense	5′-ACG GGT CTT GCC GGA GCT GGT-3′	272
	Outer antisense	5′-AAT GCC GTG ACC ACC AAG TAT AAT-3′	
(Read, 1997)	Inner sense	5′-ACC TTA AAA CTC ACT ACC AGT-3′	208
	Inner antisense	5′-CTA ATC CAA GGC GGG TGC AT-3′	
	Probe	Biotin-5′-GAG AAC GGT TTG GGT TTT CA-3′	
Enterovirus	Inner sense	5′-CAA GCA CTT CTG TTT CCC CGG-3′	479
	Inner antisense	5′-CAC YGG ATG GCC AAT CCA A-3′ [a]	
	Outer sense	5′-TCC TCC GGC CCC TGA ATG CG-3′	156
	Outer antisense	5′-ATT GTC ACC ATA AGC AGC CA-3′	
	Probe	Biotin-5′-AAA CAC GGA CAC CCA AAG TA	
EBV	Outer sense	5′-ATC AGA AAT TTG CAC TTT CTT TGC -3′	482
	Outer antisense	5′-CAG CTC CAC GCA AAG TCA GAT TG-3′	
	Inner sense	5′-TTG ACA TGA GCA TGG AAG AC - 3′	363
	Inner antisense	5′-CTC GTG GTC GTG TTC CCT CAC - 3′	
	Probe	Biotin-5′-AAC TAC CCG CAA TGA AAT GG - 3′	

[a] Universal code: S = C or G, Y = C or T

57.5 PCR Amplification and Product Detection

57.5.1 First Round PCR Reaction Mix

Reagents	Final concentration	Volume/tube (μl) 1x
Nuclease free water	–	10
Amperase		1
5x Qiagen one-step buffer	1x	10
Qiagen one-step dNTP	0.4 mM	2
Combined outer primers	0.1 pmol/μl	5
Qiagen one-step enzyme	0.5 U	2
Total		30

57.5.2 First Round PCR Amplification

Combine 30 μl of above Master Mix with 20 μl of extracted DNA and label as Round 1. The 50 μl sample is ready to amplify. Load samples onto Thermal Cyclers. Cycle using Bio-Rad MyCycler Thermal Cycler or Applied Biosystems Gene Amp PCR System 2400 under the following amplification conditions: 50°C × 30 min, 95°C × 15 min, 35 cycles of 94°C × 45 s, 57°C × 45 s, 72°C × 1 min followed by 72°C × 7 min and 4°C × ∞.

57.5.3 Second Round PCR Reaction Mix

Reagents	Final concentration	Volume/tube (μl)
Nuclease free water	–	17.8
Combined Inner primers	0.1 pmol/μl	5
Amplitaq Gold 2x mix (Applied Biosystems)	0.5 U	25
DIG-11-dUTP (1 nmol/μl)	0.4 nm	0.2
Total		48

57.5.4 Second Round PCR Amplification

Aliquot 48 μl of above Master Mix into 2.0 ml sterile PCR tubes and label as round 2. Add 2.0 μl of first round amplification product into the tubes and load samples onto thermal cyclers. Amplify under the following conditions: 95°C × 5 min, 33 cycles of 94°C × 20 s, 57°C × 20 s, 72°C × 20 s followed by 72°C × 10 min.

57.6 Agarose Gel Electrophoresis

Prepare a 2% agarose gel using the following: 200 ml 1x TBE, 4.0 g Agarose (Promega Cat No V312A). Store solidified agarose in glass 500 ml Schott bottle at room temperature. Melt solidified agarose – microwave at full power for 2 min. Mix and microwave a further 1 min. Add 10 μl of SYBR Safe DNA gel stain (10,000 × concentration in DMSO) to the melted 200 ml agarose. Pour 100 ml of melted agarose into the gel-casting chamber containing specimen combs. Allow to set for 20 min. Place prepared gel into electrophoresis tank containing running buffer. Add 10 μl of PCR product to 2.0 μl loading dye. Mix well and load onto gel. Use 10.0 μl of a 1:10 dilution (300 mg) of 50 bp Promega as the marker. Run gel at 110 V for 35 min. View gel under ultraviolet light and photograph.

57.7 EIA (DIG) Detection of PCR Products

Allow PCR ELISA (Dig Detection) Kit (Roche Diagnostics) to stand at room temperature for 1 h before use. Add equal volumes of denaturation solution (protocol A solution in kit) to each tube containing amplicon, aspirate several times, and incubate at room temperature for 10 min. Add selected probe to the following volumes of Hybridization solution.

Strip	Hybridisation sol (μl)	Probe (μl)
1	1000	5
2	2000	10

Add 85 μl of Probe/Hybridisation solution to the respective well of the streptavidin microtitre plate. Add 15 μl of denatured amplicon to appropriate well(s). Cover with plate sealer and incubate at 40°C for 40 min shaking at 400 rpm. Wash five times with wash buffer, with 10 s soak between each cycle. Dilute Anti-DIG-POD conjugates 1:10 in conjugate dilution buffer and add 100 μl to each well. Incubate at 40°C for 30 min shaking at 400 rpm. Wash five times as above.

Add 100 μl of TMB (Sigma) HRP substrate to each well and incubate at 37°C for 10 min with cover on plate. Add 100 μl stop solution (1 N H_2SO_4) and read at 450 nm on a spectrophotometer. Check correlation with Gel results. Gel Doc positive bands should correlate with ELISA OD >0.80. If there is any discrepancy, the specimen must be re-tested.

Reference

1. McIver CJ et al (2005) Development of multiplex PCRs for detection of common viral pathogens and agents of congenital infections. J Clin Microbiol 43:5102–5110

Part V
PCR Protocols for Fungal and Parasitic Pathogens

Chapter 58
Aspergillus Species

Catriona Halliday

58.1 Summary of Methods

Invasive aspergillosis (IA) is a major cause of infection related morbidity and mortality in immunocompromised patients unless treatment is initiated early in the course of infection and the pathogen is accurately identified. Conventional laboratory diagnosis, which depends on positive histology and/or culture from normally sterile sites, lacks sufficient sensitivity and specificity for early diagnosis. Efforts to improve diagnosis of IA in recent years have concentrated on the detection of circulating surrogate markers such as galactomannan or fungal DNA by PCR-based assays. The majority of published PCR assays target parts of the multi-copy ribosomal DNA gene cluster (rDNA) comprising the 18S, 28S and 5.8S genes and the intervening internal transcribed spacer regions (ITS1 and ITS2). The *Aspergillus* PCR assay described below is a nested qualitative real-time PCR assay, which targets a specific region of the multi-copy 18S rRNA gene that is highly conserved in all *Aspergillus* spp.

58.2 Acceptable Specimens

Specimens accepted for testing by *Aspergillus* PCR include BAL, EDTA whole blood, serum, fresh and paraffin embedded tissue, vitreous fluid, ascitic fluid and CSF.

58.3 Unacceptable Specimens

Sputum and plasma specimens are unacceptable for testing by *Aspergillus* PCR.

C. Halliday (✉)
Centre for Infectious Diseases and Microbiology, Westmead Hospital, Westmead, NSW 2145, Australia
e-mail: Catriona.Halliday@swahs.health.nsw.gov.au

M. Schuller et al. (eds.), *PCR for Clinical Microbiology*,
DOI 10.1007/978-90-481-9039-3_58, © Springer Science+Business Media B.V. 2010

58.4 Sample Extraction

The High Pure PCR Template Preparation kit (Roche Applied Science, Australia) is used for DNA extraction from BAL, CSF, fluids and tissue specimens. All fluid specimens are initially concentrated to ~200 μl by centrifugation at 13,200 rpm for 10 min. All specimens are then incubated in 200 μl of Tissue Lysis buffer and 40 μl of proteinase K at 55°C for 60 min-overnight. DNA is then extracted according to the manufacturer's instructions and eluted in a final volume of 200 μl.

The GenElute Mammalian DNA kit (Sigma-Aldrich Co.) is used for DNA extraction from whole blood and serum specimens. Whole blood samples (500 μl) are lysed with 3 volumes of erythrocyte lysis buffer (0.155 M NH_4Cl, 0.01 M NH_4HCO_3 and 0.1 mM EDTA [pH 7.4]) [3] for 10 min at –20°C and centrifuged at 6,600 rpm for 10 min. Serum samples (600 μl) are centrifuged at 13,000 rpm for 10 min. Supernatants from both blood and serum samples are discarded and the pellets re-suspended in 200 μl of sorbitol buffer (1 M sorbitol, 100 mM EDTA and 0.1% 2-mercaptoethanol) [4] and 200 U of lyticase (Sigma-Aldrich, Australia). Following incubation at 37°C for 60 min, spheroplasts are precipitated by centrifugation at 7,600 rpm for 5 min, re-suspended in 180 μl of Lysis T solution T and 20 μl of proteinase K (20 mg/ml) and incubated at 55°C for 60 min. DNA is then extracted according to the manufacturer's instructions with a final elution volume of 120 μl.

58.5 Primers and Probe Sequences

Primer pairs AFU 7S (5′-CGG CCC TTA AAT A-G CCC G) and AFU 7AS (5′-GAC CGG GTT TGA CCA ACT TT) are used in the first step of the assay to form a 405 bp product. Primer pairs AFU 5S (5′-AGG G-C CAG CGA GTA CAT CAC CTT G) and AFU 5AS (5′-GGG RGT CGT TGC CAA C-Y CYC CTG A) are used in the second step of the assay to form a 236 bp product [3]. The *Aspergillus*-specific probe (5′ 6FAM-TTG TTA AAC CCT GT-C GTG CTG GGG ATA GAG-TAMRA 3′) [2] is included in the second round PCR reaction mix.

58.6 PCR Amplification and Product Detection

The first round PCR is performed in a 25 μl volume consisting of 1X PCR buffer (10 mM Tris-HCl [pH 8.3], 50 mM KCl, 1.5 mM $MgCl_2$, 0.001% gelatin) (Applied Biosystems, Australia), 1 mM MgAc, 200 mM dNTPs (Roche Applied Science, Australia), 1.2 μM AFU 7S and AFU 7AS, 1.25 U *Taq* DNA polymerase (Applied Biosystems, Australia) and 10 μl of DNA. Amplification is performed on a standard block-based thermocycler under the following amplification conditions: 94°C for 2 min, followed by 30 amplification cycles of 94°C for 20 s, 65°C for 30 s, and 72°C for 30 s, and a final extension at 72°C for 5 min.

The second-round PCR amplification is performed in glass capillaries (Roche Applied Science, Australia) in a 20 μl final volume using the LightCycler (LC) system (Roche Applied Science, Australia). The reaction mix consists of 1X LC FastStart reaction mix (containing FastStart *Taq* DNA polymerase, reaction buffer, dNTPs, and 1 mM $MgCl_2$) (Roche Applied Science, Australia), 3.5 mM $MgCl_2$, 0.5 μM AFU 5S and AFU 5AS, 0.1 μM *Aspergillus*-specific probe, and 2 μl of the first-round PCR product. Cycling parameters on the LC machine are 95°C for 10 min, followed by 35 cycles of 95°C for 0 s, 65°C for 5 s, and 72°C for 15 s, and a final cooling at 40°C for 30 s. Fluorescence data are collected at the end of each 65°C annealing step with a single fluorescence acquisition for each capillary. Positive samples are detected by an exponential increase in fluorescence; negative samples have no change in fluorescence.

58.7 Quality Control and Validation Data

With the exception of DNA from blood/serum which is only tested undiluted, DNA extracts are tested both neat and diluted 1:100 to ensure both strong and weak positive specimens are detected. An inhibition control composed of an equal mixture of specimen DNA and positive control DNA is included for each sample to exclude the presence of inhibitory substances. A negative control of molecular biology grade water is included during each DNA extraction and PCR set-up to act as contamination controls. A serial dilution of *Aspergillus fumigatus* (ATCC 204305) positive control DNA is included with every run to ensure the limit of detection of the assay remains constant over time.

The sensitivity of the assay is equivalent to 1–10 CFU/ml of *Aspergillus* conidia per ml of blood. The usefulness of the *Aspergillus* PCR assay as a screening tool for high-risk haematology patients was determined from 998 blood samples from 65 patients during 95 episodes of febrile neutropenia [1]. Using the criterion of ≥ 2 positive PCR results to indicate a "PCR positive" episode, the assay had sensitivity, specificity, PPV, NPV of 100%, 75.4%, 46.4% and 100%, respectively.

58.8 Assay Limitations

The *Aspergillus* PCR assay is used as a screening test for blood samples and a diagnostic test for all other specimen types. It is recommended that twice weekly screening of blood for *Aspergillus* DNA be performed in patients at high risk for IA. DNA from some *Penicillium* spp. can also be amplified by the assay. *Penicillium* spp. is phylogenetically very closely related to *Aspergillus* spp. but is not a human pathogen and should not be detectable in sterile site specimens. Positive PCR results from non-sterile sites should always be interpreted in the appropriate clinical context.

References

1. Halliday C, Hoile R, Sorrell T et al (2006) Role of prospective screening of blood for invasive aspergillosis by polymerase chain reaction in febrile neutropenic recipients of haematopoietic stem cell transplants and patients with acute leukaemia. Br J Haematol 132:478–486
2. Halliday C, Wu Q, James G et al (2005) Development of a nested qualitative real-time PCR assay to detect *Aspergillus* DNA in clinical specimens. J Clin Microbiol 43:5366–5368
3. Skladny H, Buccheidt D, Baust C et al (1999) Specific detection of *Aspergillus* species in blood and bronchoalveolar lavage samples of immunocompromised patients by two-step PCR. J Clin Micrbiol 37:3865–3871
4. Williamson ECM, Leeming JP, Palmer HM et al (2000) Diagnosis of invasive aspergillosis in bone marrow transplant recipients by polymerase chain reaction. Br J Haematol 108:132–139

Chapter 59
Cryptosporidium

Damien Stark

59.1 Summary of Methods

Laboratory diagnosis of cryptosporidiosis traditionally relies on special staining techniques, such as modified acid-fast, Kinyoun's and Giemsa stains as oocysts are difficult to detect using basic light microsopy [4]. Other alternative diagnostic techniques have also been employed. The use of ELISA has been described for the detection of *Cryptosporidium* antigen in stools [7]. Various PCR assays, including conventional PCR, nested and RT-PCR have also been developed for the detection of *Cryptosporidium* species in stool specimens [2, 5, 6, 15]. Whilst being more expensive and time consuming, PCR has shown superior sensitivity for the detection of *Cryptosporidium* species compared to conventional staining, microscopy and ELISA [9, 12].

59.2 Organism

Cryptosporidium species have a worldwide distribution, and the ability to infect a large range of vertebrate hosts [13]. *Cryptosporidium parvum* and *Cryptosporidium hominis* are the species most commonly associated with human cryptosporidiosis [8] though infections with other species such as *Cryptosporidium felis* and *Cryptosporidium meleagridis* [14] have been reported [8, 11]. Infection is acquired via the faecal-oral route and *C. parvum* has been recognized as the cause of large waterborne and food-borne outbreaks of gastroenteritis.

Patients tend to present with a self limiting diarrhoea which may last for several weeks to months, even in immunocompetent individuals, with the highest burden of disease occurring in children under 5 years of age [1]. In immunosuppressed patients the disease is often more severe and is usually associated with chronic diarrhoea, wasting, and can be life threatening [1].

D. Stark (✉)
Microbiology Department, St Vincents Hospital, Darlinghurst, NSW, Australia
e-mail: dstark@stvincents.com.au

M. Schuller et al. (eds.), *PCR for Clinical Microbiology*,
DOI 10.1007/978-90-481-9039-3_59, © Springer Science+Business Media B.V. 2010

59.3 Acceptable Specimens

Stool samples – fresh or frozen.

59.4 Sample Extraction

Qiagen stool kit.

59.5 Primer and Probe Sequences

The nested PCR developed by Coupe et al. [3] uses primer pairs encompassing the polymorphous region of the 18S rRNA gene located between nucleotides 179 and 271.

Initial amplification
SCL1 CTGGTTGATCCTGCCAGTAG
CPB-DIAGR TAAGGTGCTGAAGGAGTAAGG
Second amplification
SCL2 CAGTTATAGTTTACTTGATAATC
SCR2 CAATACCCTACCGTCTAAAG

59.6 Nested PCR

59.6.1 PCR Amplification and Product Detection

Reactions are carried out in a volume of 50 μl, using 10x GeneAmp PCR Buffer (Applied Biosystems) 0.4 μM of each primer and 5 μl of template DNA. The second round PCR utilizes 5 μl of the initial amplification product as the template.

The first round amplification conditions are as follows: 5 min at 94°C, initial denaturation for 30 s at 94°C, and 39 cycles of amplification (annealing for 45 s at 60°C, extension for 90 s at 72°C, and denaturation for 30 s at 94°C). The final extension step lasts 10 min. The second-round PCR amplifies a 214-bp fragment. The optimized conditions for second-round PCR are the same as those for the first round except that the annealing lasted 45 s at 58°C and extension 60 s at 72°C.

The PCR product was analysed by electrophoresis on 1.0% Agarose Gels.

59.6.2 Quality Control and Validation Data

The specificities of each of these primer pairs were evaluated by amplifying human, *G. lamblia*, *Encephalitozoon intestinalis*, *Toxoplasma gondii*, *Trichophyton rubrum*, and *E. coli* DNA. No cross-amplification was detected with these DNAs or with DNA extracted from stools containing either no parasites or parasites other than *Cryptosporidium*.

PCR sensitivity was tested on serial 10-fold dilutions of purified DNA. The detection limit was 5×10^{-5} ng of pure *Cryptosporidium* DNA per reaction mix, which corresponds approximately to one oocyst. Sensitivity was then tested with fresh oocyst dilutions in water and human stools. Under both conditions the nested PCR procedure was able to detect the equivalent of one oocyst. According to the dilution and stool sample size used for PCR, 40 oocysts/ml of stools could be detected.

59.7 Real Time PCR Using FRET Probes

59.7.1 Primer and Probe Sequences

Limor et al. [10] developed a real-time PCR targeting the most polymorphic region of the SSU rRNA gene (~820 bp) which was amplified from samples by PCR carried out in a LightCycler (Roche Molecular Biochemicals, Indianapolis, Ind.) using hybridization probes:

Forward primer GGAAGGGTTGTATTTATTAGATAAAG
Reverse primer AAGGAGTAAGGAACAACCTCCA
Probe 1 CCGTCTAAAGCTGATAGGTCAGAAACTTGAATG-flourescein
Probe 2 LCred705-GTCACATTAATTGTGATCCGTAAAG

59.7.2 PCR Amplification and Product Detection

The PCR mixture contains 2 μl of Perkin-Elmer (Norwalk, Conn.) 10x buffer, 4 mM $MgCl_2$, 100 μM (each) deoxynucleoside triphosphate, 400 nM forward and reverse primers, 200 nM (each) hybridization probes, 0.5 μl of nonacetylated bovine serum albumin (10 mg/ml), 1 U of *Taq* polymerase, and 1 μl of DNA template in a total of 20 μl. Alternatively the manufacturer-suggested Master Hybridization Probe Kit (Roche) may be used.

Each PCR mixture is subjected to 55 cycles of denaturation at 94°C for 2 s, annealing at 50°C for 10 s, and extension at 72°C for 15 s, with an initial denaturation at 95°C for 3 min. The PCR was performed on the Lightcycler instrument.

Detection of the fluorescent signal is made after each cycle's annealing phase.

59.7.3 Quality Control and Validation Data

The real-time PCR described can detect and differentiate all five common *Cryptosporidium* parasites that are pathogenic for humans. The sensitivity was similar to that of conventional PCR, but the specificity was increased because of the use of two hybridization probes during the detection phase and melting curve analysis during the differentiation phase. Real-time PCR has a more rapid turnaround time and can be quantitative, both of which are important in investigations of waterborne outbreaks of cryptosporidiosis.

References

1. Abubakar I, Aliyu SH, Arumugam C et al (2008) Prevention and treatment of cryptosporidiosis in immunocompromised patients (Review). Cochrane Database Syst Rev 2007, Issue 1. Art. No.: CD004932. DOI: 10.1002/14651858.CD004932.pub2
2. Balatbat AB, Jordan GW, Tang YJ et al (1996) Detection of *Cryptosporidium parvum* DNA in human feces by nested PCR. J Clin Microbiol 34:1769–1772
3. Coupe S, Sarfati C, Hamane S et al (2005) Detection of cryptosporidium and identification to the species level by nested PCR and restriction fragment length polymorphism. J Clin Microbiol 43:1017–1023
4. Fotedar R, Stark D, Beebe N et al (2007) PCR detection of *Entamoeba histolytica*, *Entamoeba dispar*, and *Entamoeba moshkovskii* in stool samples from Sydney, Australia. J Clin Microbiol 45:1035–1037
5. Garcia LS (2001) Diagnostic medical parasitology. ASM Press, Washington, DC.
6. Gile M, Warhurst DC, Webster KA et al (2002) A multiplex allele specific polymerase chain reaction (MAS-PCR) on the dihydrofolate reductase gene for the detection of Cryptosporidium parvum genotypes 1 and 2. Parasitology 125:35–44
7. Higgins JA, Fayer R, Trout JM et al (2001) Real-time PCR for the detection of *Cryptosporidium parvum*. J Microbiol Methods 47:323–337
8. Jayalakshmi J, Appalaraju B, Mahadevan K (2008) Evaluation of an enzyme-linked immunoassay for the detection of *Cryptosporidium* antigen in fecal specimens of HIV/AIDS patients. Indian J Pathol Microbiol 51:137–138
9. Joachim A (2004) Human cryptosporidiosis: an update with special emphasis on the situation in Europe. J Vet Med B Infect Dis Vet Public Health 51:251–259
10. Kaushik K, Khurana S, Wanchu A et al (2008) Evaluation of staining techniques, antigen detection and nested PCR for the diagnosis of cryptosporidiosis in HIV seropositive and seronegative patients. Acta Trop 107:1–7
11. Limor JR, Lal AA, Xiao L (2002) Detection and differentiation of *Cryptosporidium* parasites that are pathogenic for humans by real-time PCR. J Clin Microbiol 40:2335–2338
12. Morgan U, Weber R, Xiao L et al (2000) Molecular characterization of *Cryptosporidium* isolates obtained from human immunodeficiency virus-infected individuals living in Switzerland, Kenya, and the United States. J Clin Microbiol 38:1180–1183
13. Morgan UM, Pallant L, Dwyer BW et al (1998) Comparison of PCR and microscopy for detection of *Cryptosporidium parvum* in human fecal specimens: clinical trial. J Clin Microbiol 36:995–998
14. O'Donoghue PJ (1985) Cryptosporidium infections in man, animals, birds and fish. Aust Vet J 62:253–258
15. Pedraza-Diaz S, Amar CFL, McLauchlin J et al (2001) *Cryptosporidium meleagridis* from humans: molecular analysis and description of affected patients. J Infect 42:243–250
16. Zhu G, Marchewka MJ, Ennis JG et al (1998) Direct isolation of DNA from patient stools for polymerase chain reaction detection of *Cryptosporidium parvum*. J Infect Dis 177:1443–1446

Chapter 60
Cyclospora

Damien Stark

60.1 Organism

Cyclospora cayetanensis is a coccidian protozoan restricted to developing regions of the world with humans the only known host for this parasite [3]. Infection is via the faecal oral route, though oocysts are not infective immediately after excretion in the faeces, and require sporulation in the environment to become infective.

Generally patients present with a rapid onset self limiting diarrhoea [5]. With progressive immune suppression (CD4$^+$ T cells < 200 cells/μl in HIV infected individuals) prolonged carriage occurs, resulting in frequent severe relapses [1, 2, 4]. These recurrences can lead to severe malnutrition and significant morbidity and mortality in immunosuppressed patients.

60.2 Summary of Methods

Diagnosis of *Cyclospora* oocysts may be problematic as most laboratories fail to recognise them in direct faecal smears. Special stains such as modified acid-fast auramine, or modified iron-haematoxylin are usually required for definitive diagnosis. Other methods used are autofluorescence under UV epifluorescence [3]. PCR assays have been developed and been shown to be useful in detecting the parasite.

60.3 Acceptable Specimens

Stool samples – fresh or frozen

D. Stark (✉)
Microbiology Department, St Vincents Hospital, Darlinghurst, NSW, Australia
e-mail: dstark@stvincents.com.au

M. Schuller et al. (eds.), *PCR for Clinical Microbiology*,
DOI 10.1007/978-90-481-9039-3_60, © Springer Science+Business Media B.V. 2010

60.4 Sample Extraction

Extraction method – Qiagen stool kit

60.5 Primer and Probe Sequences

Verweij et al. [6] developed a real-time PCR assay targeting the SSU rRNA gene.

Cyclo250F TAGTAACCGAACGGATCGCATT
Cyclo350R AATGCCACGTAGGCCAATA
HEX-CCGGCGATAGATCATTCAAGTTTCTGACC-DABCYL

60.6 PCR Amplification and Product Detection

25 μl PCR buffer (10x real-time reaction buffer: Eurogenetic, Belgium), 3.5 mM $MgCl_2$ each nucleotide 200 μM, 50 pmol of each primer, 25 pmol double-labelled probe, 1U of Taq polymerase and 2 μl sample DNA. Amplification consists of 10 min at 95°C followed 40 cycles of 15 s at 95°C, 60 s at 60°C. Fluorescence is measured at the end of each extension step.

60.7 Quality Control and Validation Data

Sensitivity of the assay determined was by serially diluting 2000 *C. cayetanensis* oocysts in faeces. Detection limit of the assay was estimated at 0.5 oocysts. No cross-reaction was found with other protozoan parasites including *Entamoeba histolytica*, *Entamoeba dispar*, *Giardia intestinalis*, *Enterocytozoon bieneusi*, *Encephalitozoon intestinalis* or *Cryptosporidium parvum* DNA.

References

1. Añé MS, Fernández FAN, Avila FP et al (2000) Emergence of a new pathogen: Cyclospora cayetanensis in patients infected with human immunodeficiency virus. Rev Cubana Med Trop 52:66–69
2. Fleming CA, Caron D, Gunn JE et al (1998) A foodborne outbreak of Cyclospora cayetanesis at a wedding: clinical features and risk factors for illness. Arch Intern Med 158:1121–1125
3. Garcia LS (2001) Diagnostic medical parasitology. ASM Press. Washington, D.C.
4. Hodge CW, Shilm DR, Ghimire M et al (1995) Placebo-controlled trial of cotrimoxazole for Cyclospora infections among travellers and foreign residents in Nepal. Lancet 345:691–693
5. Lewthwaite P, Gill GV, Hart CA et al (2005) Gastrointestinal parasites in the immunocompromised. Curr Opin Infect Dis 18:427–435
6. Verweij JJ, Laeijendecker D, Brienen EA et al (2003) Detection of *Cyclospora cayetanensis* in travellers returning from the tropics and subtropics using microscopy and real-time PCR. Int J Med Microbiol 293:199–202

Chapter 61
Dientamoeba fragilis

Damien Stark

61.1 Summary of Methods

Traditional diagnosis of *Dientamoeba fragilis* is based on prompt fixation and permanent staining as trophozoites degenerate within hours of being passed with demonstration of the characteristic nuclear structure achieved by permanently stained preparations only. Such techniques are time consuming and require experienced personnel to interpret the stained smears. Newer molecular techniques, conventional and real-time PCR, targeting the 18S rDNA have been developed for the diagnosis of *D. fragilis* and offer greater sensitivity and specificity than microscopy [2, 3].

61.2 Organism

Dientamoeba fragilis is a trichomonad parasite with a worldwide distribution. Humans are probably the definitive host of this parasite even though *D. fragilis* trophozoites have been reported in non-human primates. *Dientamoeba fragilis* is now considered a pathogenic protozoan and has been shown to cause acute gastrointestinal disease with chronic infections also documented. The most frequent clinical symptoms associated with *D. fragilis* are diarrhoea and abdominal pain [4].

61.3 Acceptable Specimens

Stool samples – fresh or frozen

D. Stark (✉)
Microbiology Department, St Vincents Hospital, Darlinghurst, NSW, Australia
e-mail: dstark@stvincents.com.au

M. Schuller et al. (eds.), *PCR for Clinical Microbiology*,
DOI 10.1007/978-90-481-9039-3_61, © Springer Science+Business Media B.V. 2010

61.4 Sample Extraction

Qiagen stool kit.

61.5 Conventional PCR

61.5.1 Primer Sequences

PCR developed by Stark et al. [2] targeting the SSU rDNA

DF400 TATCGGAGGTGGTAATGACC
DF1250 CATCTTCCTCCTGCTTAGACG

61.5.2 PCR Amplification and Product Detection

PCR amplifications (25 μl) are performed using pureTaq Ready-To-Go^{TM} (Amersham Pharmacia Biotech) PCR beads (each containing ~1.5 units Taq DNA polymerase, 10 mM Tris–HCl at pH 9, 50 mM KCl, 1.5 mM $MgCl_2$, 200 mM of each dNTP and stabilisers, including BSA), 1.0 μl of genomic DNA extract and 0.5 μM of each PCR primer.

PCR amplification is performed under the following conditions: 3 min denaturation hold at 94°C; 30 cycles of 1 min at 94°C, 1.5 min at 57°C, 2 min at 72°C.

The PCR product is analysed by electrophoresis on 1.0% Agarose Gels.

61.6 Real-Time PCR Using TaqMan Probes

61.6.1 Primer and Probe Sequences

PCR developed by Stark et al. [3] targeting the SSU rDNA.

DF3 GTTGAATACGTCCCTGCCCTTT
DF4 TGATCCAATGATTTCACCGAGTCA
Probe FAM-CACACCGCCCGTCGCTCCTACCG-TAMRA

61.6.2 PCR Amplification and Product Detection

RT-PCR is performed using a LightCycler (Roche) in a 20 μl reaction volume in a glass capillary tube containing 2 μl of FastStart reaction mix hybridization probes (a component of the Faststart DNA master hybridization probe kit:Roche

Diagnostics), 3 mM $MgCl_2$, 0.25 μM forward and reverse primer, 0.25 μM dual labeled fluorescent probe, and 2 μl of DNA extract.

PCR amplification is as follows: 10 min at 95°C followed by 35 cycles of 95°C, 10 s at 58°C.

The fluorescent signal is measured at the end of each extension plateau.

61.6.3 Quality Control and Validation Data

Peek et al. [1] was the first to report the development of a conventional PCR by amplification of the small-subunit rRNA gene to detect *D. fragilis*; however the sensitivity and specificity of the assay was not determined in this study. In this study the detection limit of PCR was the equivalent of approximately 0.1 *D. fragilis* trophozoites per sample. A more recent conventional PCR and real-time PCR assay based on the small-subunit ribosomal RNA gene of *D. fragilis* for the specific detection of *D. fragilis* DNA in fresh unpreserved stool samples was developed [2]. To determine the sensitivity of both conventional and real-time PCR, the entire SSUrRNA gene was cloned and a known number of copies were then amplified. This showed that the detection limit was 100 plasmid copies or an equivalent of approximately one *D. fragilis* trophozoite for conventional PCR. The detection limit for the real-time PCR was determined at one plasmid copy (a crossing point of 27.87) of the SSUrRNA gene which is equivalent to approximately 0.01 *D. fragilis* trophozoite. This shows that the real-time PCR was 100 times more sensitive than the conventional PCR. When the conventional and RT-PCR assays were compared when run on clinical specimens the conventional PCR was shown to have a sensitivity of 88.9% and a specificity of 100%, while the real-time PCR was shown to be 100% sensitive and specific [3].

References

1. Peek R, Reedeker FR, van Gool T (2004) Direct amplification and genotyping of *Dientamoeba fragilis* from human stool specimens. J Clin Microbiol 42:631–635
2. Stark D, Beebe N, Marriott D et al (2005) Detection of *Dientamoeba fragilis* in fresh stool specimens using PCR. Int J Parasitol 35:57–62
3. Stark D, Beebe N, Marriott D et al (2006) Evaluation of three diagnostic methods, including real-time PCR, for detection of *Dientamoeba fragilis* in stool specimens. J Clin Microbiol 44:232–235
4. Stark DJ, Beebe N, Marriott D et al (2006) Dientamoebiasis: clinical importance and recent advances. Trends Parasitol 22:92–96

Chapter 62
Entamoeba histolytica

Damien Stark

62.1 Summary of Methods

Entamoeba histolytica is morphologically identical to the non-pathogenic species *Entamoeba dispar*, and *Entamoeba moshkovskii* though genetic differences have confirmed the separation of these three as independent species [4, 13]. Due to this conserved morphology, stained smears of stool specimens are insufficient for differentiation of these species. While staining of fixed faecal smears can determine the presence of the *Entamoeba histolytica/dispar/moshkovskii* complex within a stool specimen, other techniques such as PCR or ELISA must be employed for differentiation. A number of PCR assays are available for detection and/or differentiation of *Entamoeba* species [3, 6, 11].

Immunological assays are also useful and often employed. Commercial antigen capture and antibody detecting ELISA kits are also available [7–9]. While some authors suggest that the commercial antigen capture ELISA kits are preferable to the use of PCR due to the rapidity and simplicity of these kits [5], other authors suggest that PCR is more useful for detection of *E. histolytica* in stools rather than the antigen capture ELISA due to the higher sensitivities observed in PCR and the reduced chance of cross reactivity with other *Entamoeba* species.

62.2 Organism

Entamoeba histolytica is a pathogenic amoeboid protozoan parasite for which humans are the primary reservoir [14]. It is an invasive pathogen and the causative agent of amoebiasis, with approximately 50 million cases annually acquired in the developing world [17]. Clinical presentation can range from asymptomatic carriage

D. Stark (✉)
Microbiology Department, St Vincents Hospital, Darlinghurst, NSW, Australia
e-mail: dstark@stvincents.com.au

M. Schuller et al. (eds.), *PCR for Clinical Microbiology*,
DOI 10.1007/978-90-481-9039-3_62, © Springer Science+Business Media B.V. 2010

to gastrointestinal and invasive disease. Following an incubation period that can vary greatly, individuals infected with *E. histolytica* present with symptoms of abdominal pain, tenderness and diarrhoea. Invasive or extraintestinal disease is uncommon, with the liver being the most common site involved (>50%) [15]. The lungs are the second most common site of invasive amoebiasis [12]. Invasive amoebiasis involving the heart [12], brain [1] and the genitourinary tract [2] have also been reported. Generally, when clinical symptoms develop they are limited to the gastrointestinal tract.

62.3 Acceptable Specimens

Stool samples – fresh or frozen.

62.4 Sample Extraction

Qiagen stool kit.

62.5 Conventional PCR

62.5.1 Primer Sequences

This PCR assay is based on that developed by Troll et al., for specific detection of pathogenic *E. histolytica* derived from the small subunit (18S) rRNA gene for amplification of an 880 bp fragment from *E. histolytica* only [16].

Eh5 GTAACTTACTTAACCGGTAAAACATG
Eh3 TCTCTTCGTAACAAAGATCTAGACTC

62.5.2 PCR Amplification and Product Detection

The PCR reaction mix contains 1x PCR buffer, 3 mM $MgCl_2$, 200 μM dNTPs, 18 pmol of each primer (Eh5 and Eh3), 1.0 μl purified DNA, and distilled water (up to 50 μl). Include a tube containing the PCR mixture but not template DNA as negative control.

PCR amplification is performed in a thermocycler under the following cycling program: 1 cycle of 94°C for 1 min; 40 cycles of 94°C for 1 min, 60°C for 1 min, and 72°C for 2 min; 1 cycle of 72°C for 10 min.

PCR products are separated on 1% agarose with 0.5 μg/ml ethidium bromide, visualized and photographed.

62.6 Conventional PCR Adapted for SYBR Green Real-Time PCR

62.6.1 Primer Sequences

Adapted from Qvarnstrom et al. [10] targeting the small subunit (18S) rRNA gene for amplification of an 877 bp amplicon.

PSP5 GGCCAATTCATTCAATGAATTGAG
PSP3 CTCAGATCTAGAAACAATGCTTCTC

62.6.2 PCR Amplification and Product Detection

PCR mixture is prepared using HotStar Taq polymerase/QuantiTect SYBR Green PCR Master Mix (QIAGEN, Valencia, Calif.) in accordance with the manufacturer's instructions utilizing 0.1 μM of each primer and 2.0 μl of purified sample DNA.

PCR amplification is as follows: 95°C for 15 min, followed by 50 cycles of 95°C for 15 s, 60°C for 1 min, 72°C for 1.5 min, 80°C for 30 s + melting curve on a real-time platform.

Fluorescence is measured at the end of the 80°C incubation plateau and continuously during the melting curve analysis.

62.7 Real Time PCR Using TaqMan Probes

62.7.1 Primer Sequences

Developed by Verweij et al. [18] targeting the small subunit (18S) rRNA gene for amplification of an 231 bp amplicon.

Ehd-239F ATTGTCGTGGCATCCTAACTCA
Ehd-88R GCGGACGGCTCATTATAACA
histolytica-96T FAM-UCAUUGAAUGAAUUGGCCAUUU-BHQ1

62.7.2 PCR Amplification and Product Detection

PCR mixture is prepared using iTaq DNA polymerase/IQ Supermix (BioRad, Hercules, CA) in accordance with the manufacturer's instructions utilizing 0.5 μM of each primer/0.1 μM of probe and 1.0 μl of purified sample DNA.

PCR amplification is as follows: 95°C for 3 min, followed by 40 cycles of 95°C for 15 s, 60°C for 30 s, 72°C for 30 s.

The fluorescent signal is measured at the end of each extension plateau.

62.7.3 Quality Control and Validation Data

A number of distinct real-time PCR (RT-PCR) protocols have been published for the identification and detection of *E. histolytica*. RT-PCR offers several advantages over conventional PCR including elimination of post PCR analysis reducing the risk of amplicon contamination, and faster cycling times resulting in reduced assay turnaround times. RT-PCR also allows for the quantitation of organisms. These benefits do come at a cost with RT-PCR significantly more expensive than conventional PCR. Like conventional PCR assays the majority of the RT-PCR assays have targeted the rDNA. An evaluation study conducted by Qvarnstrom et al. [10] compared three *E. histolytica* real-time PCR techniques (LightCycler assay targeting the 18S rDNA, a Taqman assay targeting episomal repeats and a Taqman assay targeting the 18S rDNA) and a SYBR green RT-PCR method. The limits of detection and efficiency of each real-time PCR assay were determined and major differences in detection limits and assay performance were observed among the evaluated tests. The probe-based assays (i.e., the LightCycler and the two TaqMan assays) were very sensitive, with detection limits of less than 10 cells per ml of spiked stool. The SYBR Green assay was the least sensitive of the real-time PCR assays, but it still had approximately 10-fold-higher sensitivity than conventional PCR. Two of the assays could not reliably distinguish *E. histolytica* from *E. dispar*: the LightCycler assay and the TaqMan assay targeting episomal repeats. The LightCycler assay was found to be prone to false positives and clearly illustrated a lack of specificity of the primers. The TaqMan assay targeting episomal repeats also produced false results.

This study identified the TaqMan assay targeting the 18S rRNA gene as a superior real-time PCR assay for specific and quantitative diagnosis of amebiasis. The SYBR Green assay also performed well and offered a good alternative to the TaqMan assay.

References

1. Becker GL Jr, Knep S, Lance KP et al (1980) Amebic abscess of the brain. Neurosurgery 6:192–194
2. Calore EE, Calore NM, Cavaliere MJ (2002). Salpingitis due to *Entamoeba histolytica*. Braz J Infect Dis 6:97–99
3. Cruz-Reyes JA, Spice VM, Rehman T et al (1992) Ribosomal DNA sequences in the differentiation of pathogenic and non-pathogenic isolates of *Entamoeba histolytica*. Parasitology 104 (Pt 2):239–246
4. Diamond LS, Clark CG (1993) A redescription of *Entamoeba histolytica* Schaudinn, 1903 (Emended Walker, 1911) separating it from *Entamoeba dispar* Brumpt, 1925. J Eukaryot Microbiol 40:340–344
5. Evangelopoulos A, Legakis N, Vakalis N (2001) Microscopy, PCR and ELISA applied to the epidemiology of amoebiasis in Greece. Parasitol Int 50:185–189
6. Hamzah Z, Petmitr S, Mungthin M et al (2006) Differential detection of *Entamoeba histolytica, Entamoeba dispar*, and *Entamoeba moshkovskii* by a single-round PCR assay. J Clin Microbiol 44:3196–3200
7. Haque R, Mollah NU, Ali IK et al (2000) Diagnosis of amebic liver abscess and intestinal infection with the TechLab *Entamoeba histolytica* II antigen detection and antibody tests. J Clin Microbiol 38:3235–3239

8. Haque R, Neville LM, Hahn P et al (1995) Rapid diagnosis of Entamoeba infection by using Entamoeba and *Entamoeba histolytica* stool antigen detection kits. J Clin Microbiol 33:2558–2561
9. Knappik M, Borner U, Jelinek T (2005) Sensitivity and specificity of a new commercial enzyme-linked immunoassay kit for detecting *Entamoeba histolytica* IgG antibodies in serum samples. Eur J Clin Microbiol Infect Dis 24:701–703
10. Qvarnstrom Y, James C, Xayavong M et al (2005) Comparison of real-time PCR protocols for differential laboratory diagnosis of amebiasis. J Clin Microbiol 43:5491–497
11. Sanchez-Guillen Mdel C, Perez-Fuentes R, Salgado-Rosas H et al (2002) Differentiation of *Entamoeba histolytica/Entamoeba dispar* by PCR and their correlation with humoral and cellular immunity in individuals with clinical variants of amoebiasis. Am J Trop Med Hyg 66:731–737
12. Shamsuzzaman SM, Hashiguchi Y (2002) Thoracic amebiasis. Clin Chest Med 23:479–492
13. Som I, Azam A, Bhattacharya A et al (2000) Inter- and intra-strain variation in the 5.8S ribosomal RNA and internal transcribed spacer sequences of *Entamoeba histolytica* and comparison with *Entamoeba dispar*, *Entamoeba moshkovskii* and *Entamoeba invadens*. Int J Parasitol 30:723–728
14. Stauffer W, Abd-Alla M, Ravdin JI (2006) Prevalence and incidence of *Entamoeba histolytica* infection in South Africa and Egypt. Arch Med Res 37:266–269
15. Stauffer W, Ravdin JI (2003) *Entamoeba histolytica*: an update. Curr Opin Infect Dis 16:479–485
16. Troll H, Marti H, Weiss N (1997) Simple differential detection of *Entamoeba histolytica* and *Entamoeba dispar* in fresh stool specimens by sodium acetate-acetic acid-formalin concentration and PCR. J Clin Microbiol 35:1701–1705
17. van Hal SJ, Stark DJ, Fotedar R et al (2007) Amoebiasis: current status in Australia. Med J Aust 186:412–416
18. Verweij JJ, Vermeer J, Brien EA et al (2003) *Entamoeba histolytica* infections in captive primates. Parasitol Res 90:100–103

Chapter 63
Giardia

Damien Stark

63.1 Summary of Methods

Various diagnostic modalities are available for the detection of *Giardia* in clinical samples. *Giardia* infection can be diagnosed microscopically by identification of cysts and trophozoites in stained or unstained fecal smears. *Giardia* cysts and trophozoites have a unique morphology, different to most other protozoa, and as such can be identified by trained lab staff in a simple wet preparation of the fecal specimen. Enzyme immunoassays, immunochromatographic and direct fluorescence assays for detection of *G. intestinalis* in stool specimens have been available in the form of commercial kits for several years [7]. These kits are commonly used in diagnostic laboratories. When compared to microscopy, the copro-antigen assays are less time-consuming and easier to perform, but are generally less sensitive than conventional microscopic methods [4, 7]. A number of PCR assays are also available for the detection of *Giardia* in stool specimens and show excellent sensitivity and specificity [2, 3, 6].

63.2 Organism

Giardia intestinalis is a common and ubiquitous flagellated protozoan parasite, with a worldwide distribution and is considered one of the main non viral causes of diarrhoea in industrialized countries. *Giardia* species have a wide host range and are capable of infecting mammals and other animals including reptiles and birds [1]. Humans become infected by ingestion of cysts, which develop into trophozoites after excystation. Infections occur in both developed and developing regions of the world [5]. In developed regions *Giardia* infections are the second most common protozoan parasite detected in stool samples after *Blastocystis hominis*.

D. Stark (✉)
Microbiology Department, St Vincents Hospital, Darlinghurst, NSW, Australia
e-mail: dstark@stvincents.com.au

M. Schuller et al. (eds.), *PCR for Clinical Microbiology*,
DOI 10.1007/978-90-481-9039-3_63, © Springer Science+Business Media B.V. 2010

Clinical presentation ranges from asymptomatic carriage to acute and chronic gastrointestinal infections. After infection the majority of patients clear *Giardia* without any untoward effects, some will shed cysts asymptomatically while the remainder develop an acute and/or chronic infection [1]. The acute infection lasts for days to weeks and is accompanied by nausea and the sudden onset of explosive, watery, foul smelling diarrhoea. The acute phase is often followed by a sub-acute or chronic phase. Chronic symptoms may last for months and be continuous, intermittent, sporadic or recurrent with episodes of diarrhoea or loose stools.

63.3 Acceptable Specimens

Stool samples – fresh or frozen.

63.4 Sample Extraction

Qiagen stool kit.

63.5 Nested PCR

63.5.1 Primer Sequences

The nested PCR developed by Ghosh et al. [2], amplifies the Intergenic spacer (IGS) region of the multicopy rRNA gene.

First round PCR
AS1 CGACCGGGAGACACGCCC
AS2 AGGACTGCATATCACGGC
Second round PCR
SG3 AGAGCAGCCGATCCCCCG
SG4 AATTGGAGGCTGACTGTG

63.5.2 PCR Amplification and Product Detection

The PCR reaction mix consists of 10 mM Tris-HCl (pH 9), 50 mM KCl, 1.5 M $MgCl_2$, 0.01% (w/v) gelatin, 250 μM deoxyribonucleoside triphosphate mix, 3 U Taq polymerase and 200 ng of each primer. PCR carried out in 25 μl reaction volume containing 5 μl of DNA extract for the first PCR or 1 μl of the first round PCR product for the nested PCR. Both PCR conditions are identical.

PCR amplification is as follows: 30 cycles with denaturation at 95°C for 60 s, annealing at 55°C for 60 s and extension at 72°C for 60 s.

The PCR product is visualized on a 1.5% (w/v) agarose gel.

63.5.3 Quality Control and Validation Data

This nested PCR has a detection limit of 10 parasites which is approximately 200 times more sensitive than conventional PCR for the detection of *Giardia* [2].

63.6 Real Time PCR

63.6.1 Primer Sequences

This RT-PCR assay developed by Verwiej et al. [6] targets the SSU rDNA.

Forward primer GACGGCTCAGGACAACGGTT
Reverse primer TTGCCAGCGGTGTCCG
Probe FAM-CCCGCGGCGGTCCCTGCTAG-TAMRA

63.6.2 PCR Amplification and Product Detection

The PCR reaction mix is performed in a 50 μl volume consisting of PCR buffer (2x QIAgen, Hotstar), 5 mM $MgCl_2$, 50 pmol of each specific primer, 25 pmol double labeled probe, and 5 μl of DNA isolated from stool samples.

PCR amplification is carried out on a BioRad-I-cycler system as follows: 15 min at 95°C followed by 50 cycles of 15 s at 95°C, 30 s at 60°C, and 30 s at 72°C.

63.6.3 Quality Control and Validation Data

The sensitivity of this assay was found to be 98.1%, and a specificity of 100% when compared to other diagnostic methods [6].

References

1. Ali SA, Hill DR (2003) *Giardia intestinalis*. Curr Opin Infect Dis 16:453–460
2. Ghosh S, Debnath A, Sil A et al (2000) PCR detection of *Giardia lamblia* in stool: targeting intergenic spacer region of multicopy rRNA gene. Mol Cell Probes 14:181–189
3. Guy RA, Xiao C, Horgen PA (2004) Real-time PCR assay for detection and genotype differentiation of *Giardia lamblia* in stool specimens. J Clin Microbiol 42:3317–3320

4. Oster N, Gehrig-Feistel H, Jung H et al (2006) Evaluation of the immunochromatographic CORIS Giardia-Strip test for rapid diagnosis of *Giardia lamblia*. Eur J Clin Microbiol Infect Dis 25:112–115
5. Savioli L, Smith H, Thompson A (2006) *Giardia* and *Cryptosporidium* join the 'Neglected Diseases Initiative'. Trends Parasitol 22:203–208
6. Verweij JJ, Schinkel J, Laeijendecker D et al (2003) Real-time PCR for the detection of *Giardia lamblia*. Mol Cell Probes 17:223–225
7. Weitzel T, Dittrich S, Mohl I et al (2006) Evaluation of seven commercial antigen detection tests for *Giardia* and *Cryptosporidium* in stool samples. Clin Microbiol Infect 12:656–659

Chapter 64
Malaria (*P. falciparum*, *P. vivax*, *P. malariae*, *P. ovale*)

Seweryn Bialasiewicz, David M. Whiley and Theo P. Sloots

64.1 Summary of Methods

In Australia, all four human *Plasmodium* species are routinely detected, with cases typically arising in migrants, travellers, soldiers, and refugees [2]. Traditionally, light microscopy has been utilized in *Plasmodium* species detection; however, the last decade has seen the introduction of nucleic acid amplification-based diagnostic assays [5], including rapid real-time PCR methods [3, 4].

The real-time PCR method described here for the detection of four *Plasmodium* species utilises three separate PCR reactions involving primers and TaqMan probes targeting the 18S gene. These are (1) primers and probe for the detection of *P. falciparum*, (2) primers and probe for *P. vivax* and (3) consensus primers for *P. malaria* and *P. ovale* with a separate TaqMan probe for each of those targets.

64.2 Acceptable Specimens

Whole blood samples are the only specimens suitable for the detection of *Plasmodium* parasites using this assay.

64.3 Sample Extraction

Nucleic acids were extracted from each blood specimen using the High Pure viral nucleic acid kit (Roche Diagnostics, Australia) or the QIAmp DNA Blood Mini kit (Qiagen, Australia) by following the manufacturer's protocol.

S. Bialasiewicz (✉)
Queensland Paediatric Infectious Diseases Laboratory, Sir Albert Sakzewski Virus Research Centre, Queensland Children's Medical Research Institute, Children's Health Service District, Brisbane, QLD 4029, Australia
e-mail: seweryn@uq.edu.au

M. Schuller et al. (eds.), *PCR for Clinical Microbiology*,
DOI 10.1007/978-90-481-9039-3_64, © Springer Science+Business Media B.V. 2010

64.4 Primers and Probe Sequences

Name	Sequence (5′–3′)
P. falciparum [3]	
Pfal-F	CTTTTGAGAGGTTTTGTTACTTTGAGTAA
Pfal-R	TATTCCATGCTGTAGTATTCAAACACA
Pfal-probe	Fam-TGTTCATAACAGACGGGTAGTCATGATTGAGTTCA-BHQ1
P. vivax [3]	
Pviv-F	ACGCTTCTAGCTTAATCCACATAACT
Pviv-R	ATTTACTCAAAGTAACAAGGACTTCCAAGC
Pviv-probe	Yakima yellow-TTCGTATCGACTTTGTGCGCATTTTGC – BHQ1
P. malariae and P. ovale [4]	
Rplas-F	GTTAAGGGAGTGAAGACGA
Rplas-R	TCAGA AACCCAAAGACTTTGATTTC TCATAA
Rplas-Malprobe:	FAM-CTATCTAAAA GAAACACTCAT-MGB-BHQ
Rplas-Ovaprobe:	VIC-CGAAAGGAATTTTCTTATT–MGB-BHQ

64.5 PCR Amplification and Product Detection

PCR was performed using 25-μl reaction mixtures containing 5 μl of nucleic acid extract, 10 pmol of each primer, 4 pmol of each probe, and 12.5 μl of QIAGEN QuantiTect Probe PCR master mix (Qiagen, Australia). PCR cycling was performed on a RotorGene 3000 cycler (Qiagen, Australia), with an initial activation at 95°C for 15 min and 45 cycles of 95°C for 15 s and 60°C for 60 s.

NOTE: *Plasmodium falciparum* and *malariae* are detected by the FAM channel whereas *Plasmodium vivax* and *ovale* are detected by the JOE channel.

64.6 Quality Control and Validation Data

Positive controls for the assays were prepared from parasites in whole blood provided by the malaria reference laboratory. These were typed for individual species by light microscopy. The limit of detection of the *P. falciparum* PCR was determined by testing serial dilutions of parasites in blood. The detection limit was determined to be 1.4 parasites per microliter of blood. Similar limits of detection were determined for the three other malarial parasites.

To validate the real-time PCR method, 279 blood samples, from patients with suspected malaria were tested by both PCR and light microscopy. Overall, the results of both assays were in agreement for 270 of the 279 specimens, with discrepant results in 9 (3.2%) of the 279 samples tested. For six of these samples, the microscopy results were negative and PCR was positive, and in a further three specimens, *Plasmodium* parasites were detected by both methods, but the identification results were discordant. The discordant results in these samples were resolved by sequencing of the PCR product, which showed the PCR results to be correct. Negative results were obtained for 54 (19.3%) blood specimens by both assays.

64.7 Assay Limitations

During the validation of these assays, it was noted that in 10 (3.6%) specimens two *Plasmodium* species were present simultaneously. The accurate detection of two targets in a single sample using consensus primers may be compromised due to competitive inhibition [1]. This could be of potential concern in the real-time PCR described above for *P. malariae* and *P. ovale* involving patients in whom both these parasites may be present.

References

1. Bialasiewicz S, Whiley DM, Nissen MD et al (2007) The impact of competitive inhibition and sequence variation upon the sensitivity of malaria PCR. J Clin Microbiol 45:1621–1623
2. Charles DM, Wendy JH, Davis A et al (2005) Notifications of imported malaria in Western Australia, 1990–2001: incidence, associated factors and chemoprophylaxis. Med J Aust 182:164–167
3. Perandin F, Manca N, Calderaro A et al (2004) Development of a real-time PCR assay for detection of *Plasmodium falciparum*, *Plasmodium vivax*, and *Plasmodium ovale* for routine clinical diagnosis. J Clin Microbiol 42:1214–1219
4. Rougemont M, Van Saanen M, Sahli R et al (2004) Detection of four Plasmodium species in blood from humans by 18S rRNA gene subunit-based and species-specific real-time PCR assays. J Clin Microbiol 42:5636–5643
5. Whiley DM, LeCornec GM, Baddeley A et al (2004) Detection and differentiation of Plasmodium species by polymerase chain reaction and colorimetric detection in blood samples of patients with suspected malaria. Diag Microbiol Infect Dis 49:25–29

Chapter 65
Microsporidia

Damien Stark

65.1 Summary of Methods

Diagnosis of enteric Microsporidia relies on microscopy, either by staining of fixed smears using traditional fluorescent stains. PCR assays have been developed for the diagnosis of *E. bieneusi* and *E. intestinalis* infection [3, 4, 6].

65.2 Organism

The term microsporidia are a group of obligate intracellular parasites belonging to the phylum Microsporidia. Over one thousand, two hundred species of microsporidia belonging to 143 genera have been described infecting a wide range of vertebrate and invertebrate hosts. Of these, *E. bieneusi* and *E. intestinalis* are the two most common causes of human enteric disease.

Enterocytozoon bieneusi is the most common microsporidian in humans [8] and the second most prevalent cause of diarrhoea in immunocompromised patients after *Cryptosporidium* [1]. However, in countries with access to HAART, the prevalence of microsporidia infections has declined [7]. In contrast, in developing countries with limited access to HAART the incidence of microsporidiosis remains high [2].

65.3 Acceptable Specimens

Stool samples – fresh or frozen.

D. Stark (✉)
Microbiology Department, St Vincents Hospital, Darlinghurst, NSW, Australia
e-mail: dstark@stvincents.com.au

M. Schuller et al. (eds.), *PCR for Clinical Microbiology*,
DOI 10.1007/978-90-481-9039-3_65, © Springer Science+Business Media B.V. 2010

65.4 Sample Extraction

Qiagen Stool Kit

65.5 Conventional PCR

65.5.1 Primer Sequences

This conventional PCR assay targets the conserved region of SSU rRNA gene of *E. cuniculi, E. hellem, E. intestinalis, E. bieneusi*. Differentiation of species is based on amplicon size with *E. bieneusi* producing a 250 bp product, while *E. intestinalis* produces a 270 bp product [5, 10].

Forward primer CACCAGGTTGATTCTGCCTGAC
Reverse primer CCTCTCCGGAACCAAACCCTG

65.5.2 PCR Amplification and Product Detection

PCR is performed using, 0.5 μM primers, 2.5 μl of DNA in a final volume of 50 μl with Gen Amp kit (Perkin Elmer) according to the manufacturer's instructions.

PCR amplification is as follows: initial denaturation 94°C for 10 min; followed by 35 cycles of 94°C for 30 s, 60°C for 30 s, and 72°C for 30 s, and a final hold at 72°C for 10 min for primer extension.

PCR products are detected by agarose gel electrophoresis and ethidium bromide staining.

65.5.3 Quality Control and Validation Data

A blinded, externally controlled multicenter evaluation of light microscopy and PCR for detection of microsporidia in stool specimens found that PCR had greater sensitivity and specificity when compared to microscopy [9]. The average overall sensitivities were 89% for the PCR laboratories and 80% for the light microscopy laboratories on patient samples. Specificities were 98 and 95%, respectively. PCR detection limits as low as 10^2 spores/g faeces were found when compared to a detection limit of between 10^4 and 10^6 spores per g of stool for light microscopy [9].

65.6 Real Time Multiplex PCR

65.6.1 Primer and Probe Sequences

Verweij et al. [11] developed a multiplex real-time PCR assay targeting the SSU rDNA to detect *E. bieneusi* and *Encephalitozoon intestinalis.*

E. bieneusi

TGTGTAGGCGTGAGAGTGTATCTG
CATCCAACCATCACGTACCAATC
FAM-CACTGCACCCACATCCCTCACCTT-eclipse

Encephalitozoon intestinalis

CACCAGGTTGATTCTGCCTGAC
CTAGTTAGGCCATTACCCTAACTACCA
Yamika Yellow-CTATCACTGAGCCGTCC-eclipse

65.6.2 PCR Amplification and Product Detection

Amplification is performed in a volume of 25 μl with PCR buffer (Hotstar Taq master mix, QIAGEN), 5 mmol/L $MgCl_2$, 2 pmol of each *E. bieneusi* specific primer, 6.25 pmol of each *Encephalitozoon intestinalis* specific primer, 2.5 pmol of each double labeled probe, and 5 μl of the DNA sample.

Amplification consisted of 15 min at 95°C, followed by 50 cycles of 15 s at 95°C, 30 s at 60°C and 30 s at 72°C. Amplification and detection is performed with the I-cycler real time detection system

Fluorescence of FAM and Yakima Yellow is measured at their respective wavelengths during the annealing step of each cycle.

65.6.3 Quality Control and Validation Data

This real time assay provides excellent sensitivity and specificity and has estimated detection limit of 1 microsporidial spore [11].

References

1. Brasil P, de Lima DB, de Paiva DD et al (2000) Clinical and diagnostic aspects of intestinal microsporidiosis in HIV-infected patients with chronic diarrhea in Rio de Janeiro, Brazil. Rev Inst Med Trop Sao Paulo 42:299–304

2. Breton J, Bart-Delabesse E, Biligui S et al (2007) New highly divergent rRNA sequence among biodiverse genotypes of *Enterocytozoon bieneusi* strains isolated from humans in Gabon and Cameroon. J Clin Microbiol 45:2580–2589
3. Da Silva AJ, Slemenda SB, Visvesvara GS et al (1997) Detection of Septata intestinalis (Microsporidia) Cali et al. 1993 using polymerase chain reaction primers targeting the small submit subunit ribosomal RNA coding region. Mol Diagn 2:47–52
4. Dowd SE, Gerba CP, Enriquez FJ et al (1998) PCR amplification and species determination of microsporidia in formalin-fixed feces after immunomagnetic separation. Appl Environ Microbiol 64:333–336
5. Fedorko DP, Nelson NA, Cartwright CP (1995) Identification of microsporidia in stool specimens by using PCR and restriction endonucleases. J Clin Microbiol 33:1739–1741
6. Franzen C, Muller A, Hartmann P et al (1998) Polymerase chain reaction for diagnosis and species differentiation of microsporidia. Folia Parasitol (Praha) 45:140–148
7. Gazzard B (2005) AIDS and the gastrointestinal tract. Medicine 33:24–26
8. Hart CA, Beeching NJ, Duerden BI et al (2000) Infections in AIDS. J Med Microbiol 49: 947–967
9. Rinder H, Janitschke K, Aspock H et al (1998) Blinded, externally controlled multicenter evaluation of light microscopy and PCR for detection of microsporidia in stool specimens. The Diagnostic Multicenter Study Group on Microsporidia. J Clin Microbiol 36:1814–1818
10. Stark D, van Hal S, Barratt J et al (2009). Limited genetic diversity among genotypes of *Enterocytozoon bieneusi* strains isolated from HIV-infected patients from Sydney, Australia. J Med Microbiol 58:355–357
11. Verweij JJ, Ten Hove R, Brienen EA et al (2007) Multiplex detection of *Enterocytozoon bieneusi* and *Encephalitozoon* spp. in fecal samples using real-time PCR. Diagn Microbiol Infect Dis 57:163–167

Chapter 66
Pneumocystis jirovecii

Rebecca Fonte and Ian D. Kay

66.1 Background

Pneumocystis jirovecii, formerly known as *Pneumocystis carinii* f. sp. *hominis*, is a fungal pathogen that causes *Pneumocystis* pneumonia (PCP). This is a serious opportunistic infection in immunocompromised patients, traditionally those infected with human immunodeficiency virus (HIV) [2]. However, with the increasing number of organ transplants, PCP is also a significant risk for transplant recipients [4]. *Pneumocystis* may also be present in individuals showing no clinical signs or symptoms of PCP. The detection of *Pneumocystis* in these individuals is referred to as colonisation, carriage or sub-clinical infection [2].

Traditional methods of detecting *Pneumocystis* depended on the visualisation of *Pneumocystis* organisms. Respiratory specimens were prepared using either cytochemical or immunofluorescent stains and examined microscopically. The diagnosis of PCP utilising these methods was suitable when the organism burden was relatively high, as in individuals with acquired immunodeficiency syndrome (AIDS). Diagnosing PCP in non-HIV immunocompromised individuals, where the organism burden is much lower, is more difficult when using traditional staining techniques due to sensitivity limitations of the assay [2].

PCR techniques have been developed for the detection of *P. jirovecii*. These methods have reportedly improved sensitivity and comparable specificity in relation to cytochemical and immunofluorescent stains.

66.2 Acceptable Specimens

Suitable specimens for testing include sputum, induced sputum, bronchial washings, bronchial lavages and lung biopsies.

I.D. Kay (✉)
Department of Microbiology and Infectious Diseases, PathWest Laboratory Medicine, Royal Perth Hospital, Perth, WA, 6847, Australia
e-mail: Ian.Kay@health.wa.gov.au

M. Schuller et al. (eds.), *PCR for Clinical Microbiology*,
DOI 10.1007/978-90-481-9039-3_66, © Springer Science+Business Media B.V. 2010

66.3 Sample Extraction

2.5 μl of Dithiothreitol (DTT) solution (1 M) is added to 500 μl of specimen in an Eppendorf tube. The sample is vortexed and incubated at 37°C for 10 min. Following incubation, the specimen is vortexed again and then centrifuged at 13,000 rpm for 10 min at room temperature. The supernatant is removed and the pellet resuspended in 200 μl of sterile distilled water and vortexed again. 200 μl of this suspension is removed and DNA is extracted using the Roche Diagnostics High Pure PCR Template Preparation Kit according to the manufacturer's instructions, except that an elution volume of 50 μl is used for the PCP PCR assay [3].

Tissue specimens, such as lung biopsies, are processed using the Roche Diagnostics High Pure PCR Template Preparation Kit according to the manufacturer's instructions, except that an elution volume of 50 μl was used for the PCP PCR assay.

66.4 Primers and Probe Sequences

The primer sequence targets the 5S rDNA conserved region of *P. jirovecii*.

The PCP PCR assay uses oligonucleotide primers previously described [1]. The primer sequences are: Forward primer (5′ AgT TAC ggC CAT ACC TCA gA 3′) and reverse primer (5′ AAA gCT ACA gCA CgT CgT AT 3′).

The PCP PCR assay uses Fluorescein labelled (5′ gAA gTC AAg CTC TgA Agg gCg T Fluorescein 3′) and LightCycler Red 640 (5′ LC640 gTC AgT ACT ATA gTg ggT gAC CAT ATg g PH 3′) hybridisation probes.

66.5 PCR Amplification and Product Detection

The prepared master mix is loaded into Roche LightCycler glass capillary tubes containing 10 μl of DNA extract, 2 μl of 10x LightCycler FastStart Hybridisation Probe mix (Roche Applied Science, Australia), 1.2 μl 25 mM $MgCl_2$ solution, 0.4 μl of *P. jirovecii* oligonucleotide primer 1 (50 μM), 0.4 μl of *P. jirovecii* oligonucleotide primer 2 (50 μM), 0.2 μl of *P. jirovecii* specific Fluorescein labelled probe (20 μM), 0.2 μl of *P. jirovecii* specific LightCycler Red 640 labelled probe (20 μM) and 5.6 μl water (LightCycler Hybridisation Probes Kit – Roche Applied Science, Australia).

PCR amplification on the Roche LightCycler is as follows an initial denaturation at 95°C for 10 min to activate the FastStart *Taq* DNA polymerase, followed by a 48 cycle program consisting of heating at 20°C/s to 95°C with a 15 s hold, cooling at 10°C/s to 55°C with a 10 s hold (fluorescence acquisition) and heating at 20°C/s to 72°C with a 10 s hold. To improve the specificity of the assay, the PCR is immediately followed by a melting curve analysis to determine the melting point (Tm) of the PCR product. This cycle consists of heating at 20°C/s to 95°C with a 0 s hold, cooling at 20°C/s to 55°C with a 10 s hold, heating at 0.1°C/s to 95°C with a 0 s hold (continual fluorescence acquisition) and finally cooling at 20°C/s to 40°C with a 60 s hold.

The LightCycler assay uses FRET Hybridisation for the detection of the specific targets.

66.6 Quality Control and Validation Data

The positive control used is a *P. jirovecii* PCR product cloned into a pGEM52f(+) plasmid via blunt end ligation. A reagent control is also used, consisting of sterile distilled water, which should be negative for *P. jirovecii*. The analytical sensitivity of this assay has not been determined.

66.7 Assay Limitations

The limitations of this assay are that it can only be used as a qualitative assay, and due to the nature of the samples, it is difficult to determine the analytical sensitivity.

References

1. Kitada K, Oka S, Kimura S et al (1991) Detection of *Pneumocystis carinii* sequences by polymerase chain reaction: animal models and clinical applications to noninvasive specimens. J Clin Microbiol 29:1985–1900
2. Morris A, Wei K, Afshar K et al (2008) Epidemiology and clinical significance of *Pneumocystis* colonization. J Infect Dis 197:10–17
3. Palladino S, Kay I, Fonte R et al (2001) Use of real-time PCR and the LightCycler system for the rapid detection of *Pneumocystis carinii* in respiratory specimens. Diagn Microbiol Infect Dis 39.233–236
4. Su YS, Lu JJ, Perng CL et al (2008) *Pneumocystis jirovecii* pneumonia in patients with and without human immunodeficiency virus infection. J Microbiol Immunol Infect 41:478–482

Chapter 67
Trichomonas vaginalis

Tuckweng Kok

67.1 Summary of Methods

Two *T. vaginalis* real-time PCR assays, targeting the β-tubulin and 18S rRNA genes, are described. Targeting two different genetic regions of *T. vaginalis* provides increased test specificity [2].

67.2 Acceptable Specimens

Urine and genital swabs

67.3 Sample Extraction

Extraction of 200 μl of processed sample can be performed by automated MagNA Pure LC (Roche Applied Science, Australia) or by manual column DNA kit according to the manufacturer's instructions.

67.4 Primer and Probe Sequences

Primer and probe sequences used in the two real-time PCR assays are as follows:

Targeting the β-tubulin gene [1, 2]: primers; BTUB3f: 5′- TCC AAA GGT TTC CGA TAC AGT-3′ and BTUB_bkmt: 5′- GTT GTG CCG GAC ATA ATC ATG-3′ with hybridisation probes; BTUB FL: 5′-CCG TAC ACT CAA GCT CAC AAC ACC AAC-fluorescein-3′ and BTUB LC: 5′- LCred 640-CGG CGA TCT TAA CCA CCT TGT TTC C-phosphate-3′.

T. Kok (✉)
Microbiology-IMVS, SA Pathology, Adelaide, SA 5000, Australia
e-mail: Tuckweng.Kok@adelaide.edu.au

M. Schuller et al. (eds.), *PCR for Clinical Microbiology*,
DOI 10.1007/978-90-481-9039-3_67,

Targeting the 18S rRNA gene [2]: primers; TV16Sf-2: 5′- TGA ATC AAC ACG GGG AAA C-3′ and TV16Sr-2: 5′- ACC CTC TAA GGC TCG CAG T-3′ with hybridisation probes; TV16_FL: 5′- CAG GTG GAA GAG GGT AGC AAT AAC A-fluorescein-3′ and TV16_LC: 5′- LC red 640-TCC GTG ATG CCC TTT AGA TGC TCT G-phosphate-3′.

67.5 PCR Amplification and Product Detection

Perform PCR for *T. vaginalis* β-tubulin and 18S rRNA genes in separate LightCycler capillaries.

T. vaginalis β-tubulin assay**:** Each capillary contains 1 x LC FastStart DNA Master hybridisation probe buffer (Roche Applied Science, Australia), 4 mM $MgCl_2$ (Roche Applied Science, Australia), 0.5 μM of each primer (BTUB3f and BTUB_bkmt), 0.2 μM of FRET probe BTUB FL, 0.4 μM of FRET probe BTUB LC, and 2.0 μl of template DNA, in a final reaction volume of 20 μl. Thermocycling is conducted in a LightCycler instrument (Roche Applied Science, Australia) with an initial 10 min at 95°C followed by 50 cycles of 95°C for 10 s, 55°C for 10 s (fluorescence acquisition) and 72°C for 10 s. This is followed by melting-curve analysis at 95°C for 0 s, 40°C for 60 s then increasing to 80°C with a transition rate of 0.1°C/s (continual fluorescence acquisition). A final 40°C/30 s cooling step is then performed.

T. vaginalis 18S rRNA assay**:** Each capillary contains 1x LC FastStart DNA Master hybridisation probe buffer (Roche Applied Science, Australia), 4 mM $MgCl_2$ (Roche Applied Science, Australia), 0.5 μM of each primer (TV16Sf-2 and TV16Sr-2) and 0.2 μM of both FRET probes (TV16_FL and TV16_LC), and 2.0 μl of template DNA, in a final reaction volume of 20 μl. Thermocycling is performed as per the above β-tubulin assay except that 45 cycles of PCR amplification (instead of 50) is used and the PCR extension phase is performed for 13 s (instead of 10 s) [2].

67.6 Quality Control and Validation Data

Extracted DNA from a clinical specimen that is positive in both wet-mount and PCR is used as a *T. vaginalis* positive control. Extracted DNA from a genital swab confirmed negative for *T. vaginalis* DNA or PCR-grade water is used as a negative control.

The *T. vaginalis* 18S rRNA and β-tubulin PCR assays will detect the presence of approximately 10 and 14 copies/reaction of *T. vaginalis* DNA, respectively. Test sensitivity, specificity, PPV and NPV of these assays are >95% [2].

For result interpretation the following criteria apply:

T. vaginalis β-tubulin assay: A specimen is considered positive if a sample provides a characteristic amplification curve crossing above background fluorescence and double melting peaks at 59°C and 68°C are observed upon melting curve analysis.

T. vaginalis 18S rRNA assay: As per above except that a single melting peak at 65°C is observed upon melting curve analysis.

67.7 Assay Limitations

Tritrichomonas foetus may provide a positive amplification curve with a single melting peak at 54.5°C in the β-tubulin assay. This non-specific reaction is distinguished from the specific β-tubulin test with its characteristic double melting peaks at higher temperatures [2]. The sensitivity of the real-time PCR may be lower when using urine compared to genital swab specimens.

References

1. Hardick J, Yang S, Lin S et al (2003) Use of the Roche LightCycler instrument in a real-time PCR for *Trichomonas vaginalis* in urine samples from females and males. J Clin Microbiol 41:5619–5622
2. Simpson P, Higgins G, Qiao M et al (2007) Real-time PCR for detection of *Trichomonas vaginalis* beta-tubulin and 18S rRNA genes in female genital specimens. J Med Microbiol 56:772–777

Chapter 68
Toxoplasma gondii

Neisha Jeoffreys

68.1 Summary of Methods

Traditionally, *Toxoplasma gondii* encephalitis was diagnosed using brain radioimaging techniques, culture and serology. Mouse inoculation is the 'gold standard' but is rarely performed these days and culture in cell lines is relatively slow and lacks sensitivity [3]. Serology is often used but it may be difficult to differentiate between past and current infection [6] and immunocompromised hosts may have a variable serological response. The detection of ring enhancing lesions in the brain by radioimaging is often used as an indicator of toxoplasmosis, however these signs may also be seen in lymphoma [4] and there are atypical presentations which make diagnosis difficult [5]. PCR has become the method of choice for laboratory diagnosis of toxoplasmosis, with the multicopy B1 gene the most commonly used target. Other targets include the AF146527 gene and 18S rDNA [1, 2]. This protocol describes a real-time PCR targeting the B1 gene. Real-time PCR has the advantage that results can be obtained rapidly and contamination is minimized by the closed tube format.

68.2 Acceptable Specimens

CSF ($\geq$200 μl) and brain tissue (fresh, formalin fixed or paraffin embedded (PE)) are the most appropriate specimens for identification of *T. gondii* infection in the CNS. Vitreous fluid or aqueous humor ($\geq$200 μl) is acceptable for the diagnosis of ocular toxoplasmosis. Whole blood specimens may be tested, but with caution as the transient nature of *T. gondii* in the circulatory system can lead to false negative results.

N. Jeoffreys (✉)
Molecular Biology, Centre for Infectious Diseases and Microbiology, Institute for Clinical Pathology and Medical Research, Westmead Hospital, Westmead, NSW 2145, Australia
e-mail: Neisha.Jeoffreys@swahs.health.nsw.gov.au

M. Schuller et al. (eds.), *PCR for Clinical Microbiology*,
DOI 10.1007/978-90-481-9039-3_68,

68.3 Unacceptable Specimens

CSF supernatant is unacceptable for testing as the *T. gondii* parasites will be retained in the discarded pellet. Plasma and serum samples are also unsuitable.

68.4 Sample Extraction

DNA is extracted from CSF, vitreous fluid and fresh tissue samples using the GeneElute Genomic DNA extraction kit (Sigma-Aldrich Co.) according to the manufacturer's instructions. In brief, ground tissue or vitreous samples are combined with 180 μl of Lysis T solution and 20 μl proteinase K (20 mg/ml) and incubated at 55°C overnight. 200 μl of Lysis C solution is then added and the mixture incubated for a further 10 min at 70°C before it is transferred to a prepared silica spin column. The column is centrifuged and the eluate discarded. The membrane is washed twice with GeneElute Wash solution and the DNA eluted with 200 μl of Elution solution and stored at –20°C prior to use.

CSF and blood samples are processed in a similar manner, but do not require incubation with Lysis T. Instead they are combined with 200 μl Lysis C solution and 20 μl proteinase K, incubated at 55°C for 10 min then transferred to the spin column, as above.

For PE tissue specimens, 10–20 slices (10 μm thickness) are combined with 5 ml of histolene (Sigma-Aldrich Co.) to dissolve the paraffin. The tissue is then collected by centrifugation at 4,800 rpm and washed twice with 100% ethanol, followed by a third wash with 70% ethanol. The ethanol is then removed and the tissue pellet air dried before the addition of Lysis T as described for fresh tissues.

68.5 Primer and Probe Sequences

The primers TM1 (5′-TTC GAC AGA AAG GGA GCA AGA-3′) and TM2 (5′-TCC GTT GCG ATA GGA CAG AAC-3′) are used to amplify a 141 bp region of the B1 gene which is detected using the hydrolysis probe (5′ 6-FAM-AGG GCT GAC TCG AAC CAG ATG TGC T-BHQ1-3′).

68.6 PCR Amplification and Product Detection

PCR is performed in a 20 μl volume containing 1X LightCycler® FastStart DNA Master Plus Hybprobes reagent (Roche Applied Science Australia), 250 nM each primer (TM1 & TM2), 100 nM hydrolysis probe, 2.5 mM $MgCl_2$and 10 μl of DNA. Amplification is performed on a LightCycler® 2.0 instrument (Roche Applied

Science Australia) using the following cycling conditions: 95°C for 10 min, followed by 50 cycles of 95°C for 10 s, 55°C for 10 s, and 72°C for 5 s, with a final cooling step of 40°C for 30 s. Fluorescence is measured after the annealing step (55°C) of each amplification cycle. Positive samples are detected by an exponential increase in fluorescence during the first 40 cycles and negative samples have no change in fluorescence.

68.7 Quality Control and Validation Data

All samples are tested both neat and diluted (1:100) to ensure both strong and weak positive specimens are detected. Each sample is also tested in parallel with an inhibition control and two negative water blanks which act as contamination controls during DNA extraction and PCR set-up. The inhibition control contains both dilute sample and positive control material (1:1). A negative result in this tube indicates inhibition in the sample.

A serial dilution of the positive control to include weak and strong positives is also included with every run to ensure the limit of detection of the assay remains constant over time. The assay has a limit of detection of 1–10 parasites and a sensitivity and specificity of 100% in 102 clinical samples.

68.8 Assay Limitations

The sensitivity of the assay is most limited by the quality and type of specimens received. PE and formalin fixed tissues have a significant impact on assay sensitivity and false negative results are common with these specimen types. The transport conditions can also adversely affect PCR sensitivity, with a 10–100 fold reduction in sensitivity seen in samples which have been frozen or stored at 4°C for 48 h before DNA extraction [3]. It is recommended that negative PCR results from poor quality specimens be confirmed by serological testing to exclude false negative results.

References

1. Filisetti D, Gorcii M, Pernot-Marino E et al (2003) Diagnosis of congenital toxoplasmosis: comparison of targets for detection of *Toxoplasma gondii* by PCR. J Clin Microbiol 41:4826–4828
2. Homan WL, Vercammen M, De Braekeleer J et al (2000) Identification of a 200- to 300-fold repetitive 529 bp DNA fragment in *Toxoplasma gondii*, and its use for diagnostic and quantitative PCR. Int J Parasitol 30:69–75
3. James GS, Sintchenko VG, Dickeson DJ et al (1996) Comparison of cell culture, mouse inoculation, and PCR for detection of *Toxoplasma gondii*: effects of storage conditions on sensitivity. J Clin Microbiol 34:1572–1575

4. Masamed R, Meleis A, Lee EW et al (2009) Cerebral toxoplasmosis: case review and description of a new imaging sign. Clin Radiol 64:560–563
5. Ramsey RG, Gean AD (1997) Neuroimaging of AIDS. I. Central nervous system toxoplasmosis. Neuroimag Clin N Am 7:171–186
6. Remington JS, Thulliez P, Montoya JG (2004) Recent developments for diagnosis of toxoplasmosis. J Clin Microbiol 42:941–945

Chapter 69
Multiplex PCR for Protozoan Parasites

Damien Stark

69.1 Summary of Methods

Verweij et al. [3] developed a multiplex real-time PCR method for the detection of *G. lamblia*, *C. parvum* and *E. histolytica.* PCR primers target the small-subunit (SSU) rRNA gene sequences for *E. histolytica* and *E. dispar* such that DNAs for *E. histolytica* and *E. dispar* should be amplified. The *E. histolytica* minor groove binding (MGB) TaqMan probe was designed with the same sequence such that amplified *E. histolytica* DNA should be detected specifically. The *E. histolytica*- and *E. dispar*-specific primers amplified a 172-bp fragment inside the SSU rRNA gene. The MGB TaqMan probe was used to detect *E. histolytica*-specific amplification. *G. lamblia*-specific PCR primers and a detection probe were based on SSU RNA gene sequence for *G. lamblia* such that a 62-bp fragment within the SSU RNA gene should be amplified and detected specifically for *G. lamblia* [2]. The *G. lamblia*-specific primers and probe set consisted of forward primer Giardia-80F, reverse primer Giardia-127R, and the *G. lamblia*-specific double-labeled probe Giardia-105T . The *C. parvum*-specific primers and detection probe amplify a 138-bp fragment inside the *C. parvum*-specific 452-bp fragment. Specific DNA amplification was detected with the *C. parvum*-specific double-labeled probe Crypto (Biolegio).

An internal control using a PhHV-1 specific primer and probe [1] set consisted of forward primer PhHV-267 s, reverse primer PhHV-337as, and the specific double-labeled probe PhHV-305tq.

D. Stark (✉)
Microbiology Department, St Vincents Hospital, Darlinghurst, NSW, Australia
e-mail: dstark@stvincents.com.au

M. Schuller et al. (eds.), *PCR for Clinical Microbiology*,
DOI 10.1007/978-90-481-9039-3_69,

69.2 Primer and Probe Sequences:

E. histolytica (5′–3′)

Ehd-239F ATTGTCGTGGCATCCTAACTCA
Ehd-88R GCGGACGGCTCATTATAACA
histolytica-96T VIC-TCATTGAATGAATTGGCCATTT-nonfluorescent quencher

G. lamblia (5′–3′)

Giardia-80F GACGGCTCAGGACAACGGTT
Giardia-127R TTGCCAGCGGTGTCCG
Giardia-105T FAM-CCCGCGGCGGTCCCTGCTAG-BHQ1

C. parvum (5′–3′)

CrF CGCTTCTCTAGCCTTTCATGA
CrR CTTCACGTGTGTTTGCCAAT
Crypto Texas Red-CCAATCACAGAATCATCAGAATCGACTGGTATC-BHQ2

PhHV-1 (5′–3′)

PhHV-267s GGGCGAATCACAGATTGAATC
PhHV-337as GCGGTTCCAAACGTACCAA
PhHV-305tq Cy5-TTTTTATGTGTCCGCCACCATCTGGATC-BHQ2

69.3 PCR Amplification and Product Detection

Amplification reactions are performed in a volume of 50 μl with PCR buffer (HotstarTaq master mix; Qiagen), 5 mM $MgCl_2$, 6.25 pmol of each *E. histolytica*-specific primer, 6.25 pmol of each *G. lamblia*-specific primer, 25 pmol of each *C. parvum*-specific primer, 15 pmol of each PhHV-1-specific primer, 1.75 pmol of *E. histolytica*-specific MGB-TaqMan probe, 2.5 pmol of *G. lamblia*-specific double-labeled probe, 8.75 pmol of *C. parvum*-specific double-labeled probe, 2.5 pmol of PhHV-1-specific double-labeled probe, and 5 μl of the DNA sample.

PCR amplification is as follows: 15 min at 95°C followed by 40 cycles of 15 s at 95°C, 30 s at 60°C, and 30 s at 72°C.

Amplification, detection, and data analysis is carried out on an iCycler real-time detection system (Bio-Rad). Fluorescence is measured during the annealing step of each cycle.

69.4 Quality Control and Validation Data

The multiplex real-time assay for the detection of *E. histolytica*, *G. lamblia*, and *C. parvum* achieved a sensitivity and specificity of 100% when tested against a range of well defined DNA and stool samples as controls.

References

1. Haque R, Neville LM, Hahn P et al (1995) Rapid diagnosis of Entamoeba infection by using Entamoeba and *Entamoeba histolytica* stool antigen detection kits. J Clin Microbiol 33:2558–2561
2. Knappik M, Borner U, Jelinek T (2005) Sensitivity and specificity of a new commercial enzyme-linked immunoassay kit for detecting *Entamoeba histolytica* IgG antibodies in serum samples. Eur J Clin Microbiol Infect Dis 24:701–703
3. Verweij JJ, Blange RA, Templeton K et al (2004) Simultaneous detection of *Entamoeba histolytica*, *Giardia lamblia*, and *Cryptosporidium parvum* in fecal samples by using multiplex real-time PCR. J Clin Microbiol 42:1220–1223

Chapter 70
Universal Detection and Identification of Fungi by PCR and DNA Sequencing

Todd M. Pryce

70.1 Summary of Methods

This assay was developed to rapidly (within 24 h) detect and identify fungi from positive blood cultures and from clinical samples where initial laboratory evidence of fungal infection is suspected or the clinical pre-test probability of invasive fungal disease is high. Two PCR-DNA sequence based methods (PCRS) are described. The first method (PCRS-B) is used for the identification of fungi from culture plates and positive blood cultures. The second method (PCRS-D) is used for the detection and identification of fungi from microscopy-positive clinical samples such as fresh tissue or fluid. Requests for fungal PCR directly from clinical specimens are screened by a Clinical Microbiologist to determine whether it is worth testing based on laboratory and clinical evidence of invasive fungal disease.

Following sample extraction, DNA is amplified using a broad-range fungal PCR that targets the internal transcribed spacer (ITS) region of the rRNA. Amplified PCR products are purified and sequenced using automated methods and the derived sequence identified following comparison with reference sequences [3, 4].

70.2 Acceptable Specimens

Microscopy-positive clinical samples such as synovial fluid, peritoneal fluid, vitreous fluid, tissue biopsies and paraffin embedded sections are acceptable although fresh tissue or fluid in a sterile container are preferred. Additionally, the assay has been validated for use from a range of enrichment media used for the isolation of fungi in blood culture systems. These include BacT/ALERT FA media (BioMérieux, Marcy l'Etoile, France), BACTEC Plus Aerobic/F media (Becton

T.M. Pryce (✉)
Department of Microbiology and Infectious Diseases, PathWest Laboratory Medicine WA, Royal Perth Hospital, Wellington Street Campus, Box X2213 GPO, Perth, WA, 6847, Australia
e-mail: Todd.Pryce@health.wa.gov.au

M. Schuller et al. (eds.), *PCR for Clinical Microbiology*,
DOI 10.1007/978-90-481-9039-3_70, © Springer Science+Business Media B.V. 2010

Dickinson, Microbiology Systems, Sparks, Md.), BACTEC Lytic/10 Anaerobic/F media (Becton Dickinson), BACTEC Peds Plus/F media (Becton Dickinson) and BBL MGIT media (Becton Dickinson) [4].

70.3 Unacceptable Specimens

Superficial swabs, or samples considered 'non-sterile' are not suitable for testing.

70.4 Sample Extraction

Nucleic acids are isolated from blood culture enrichment media using a universal DNA extraction procedure described previously [4].

The High Pure PCR Template Preparation Kit (Roche Diagnostics, Mannheim, Germany) is used to extract DNA from clinical samples following the manufacturer's instructions. Nucleic acids are eluted in 50 μl of Elution buffer and stored at –70°C until use.

70.5 Primer Sequences

The primers V9D (5′-TTAAGTCCCTGCCCTTTGTA-3′) and LS266 (5′-GCATTCCCAAACAACTCGACTC-3′) [2] are used amplify an 800-1300 bp fragment that encompasses a portion of the 18S and 28S rRNA genes and the entire intervening ITS1, 5.8S and ITS2 rRNA gene regions. These primers bind to conserved regions, with corresponding positions to *Saccharomcyes cerevisiae* 18S (1609–1627) and 28S (287–266) rRNA genes.

70.6 PCR Amplification and Product Detection

Each PCR reaction is set-up in a 50 μl volume containing 1X FastStart Taq PCR Reaction Buffer (500 mM Tris-HCl, 100 mM KCl, 50 mM $(NH_4)_2SO_4$, pH 8.3) (Roche Diagnostics, Australia), 1.5 mM $MgCl_2$, 0.6 μM of each primer (V9D and LS266), 200 μM dNTPs, 2.25 U of FastStart Taq DNA Polymerase (Roche Diagnostics, Australia), and 5 μl of DNA template. PCR amplification is performed in a MyCycler (Bio-Rad Laboratories, Australia) under the following conditions: 95°C for 9 min, followed by either 33 (PCRS-B) or 35 (PCRS-D) cycles of 95°C for 30 s, 62°C for 60 s and 72°C for 2 min, and a final extension of 72°C for 5 min. PCR-amplified products are detected by gel electrophoresis using a 2% (w/v) agarose gel stained with ethidium bromide.

70.7 DNA Sequencing

For both PCRS-B and PCRS-D, DNA sequencing is performed using primers V9D, LS266 and the internal primers ITS1 (5′-TCCGTAG-GTGAACCTGCGG-3′) and ITS4 (5′-TCCTCCGCTTATTGATATGC-3′) [2]. Sequencing primer concentrations are adjusted to 1 μM and all PCR products are sequenced using standard automated methods.

70.8 Sequence Assembly and Editing

Sequence electropherograms are visualised and edited using SeqScape Software Version 2.0 (Applied Biosystems) following DNA sequencing analysis and interpretation guidelines [5]. Sequence similarity searching is performed using the NCBI BLAST [1]. Test sequences are assigned a species identification using previously published criteria based on sequence length, similarity, type strain information, and current nomenclature [3, 4].

70.9 Quality Control and Validation Data

All DNA samples are tested both neat and diluted (1:10 with Elution buffer). A plasmid DNA construct derived from *Candida albicans* (ATCC 14053) amplified DNA is used as the positive control and Elution buffer is used as the negative control for both assays.

The analytical sensitivity of the PCRS-B and PCRS-D assays are 10 pg of *C. albicans* DNA (2×10^3copies/μl) and 1 pg of *C. albicans* DNA (2×10^2copies/μl), respectively.

The PCRS-B has successfully identified a wide range of fungal pathogens from 140 blood cultures, including *C. albicans* (n=14), *Candida parapsilosis* (n=8), *C. glabrata* (n = 7), *Candida krusei* (n = 2), *Scedosporium prolificans* (n = 4), and one each of *Candida orthopsilosis*, *Candida dubliniensis*, *Candida kefyr*, *Candida tropicalis*, *Candida guilliermondii*, *S. cerevisiae*, *Aspergillus fumigatus*, *Histoplasma capsulatum*, and *Malassezia pachydermatis*. All molecular identifications were in agreement with the morphological identification. Additionally, all culture negative blood cultures were negative by PCR and no PCR inhibition was detected (n=95).

The PCRS-D assay has been validated on a range of clinical specimens (including tissue, vitreous fluid and peritoneal dialysis fluid) where initial laboratory evidence of fungal infection is suspected, or the clinical pre-test probability of invasive fungal disease is high. A variety of fungal pathogens have been detected and identified including *C. albicans*, *C. parapsilosis*, *H. capsulatum*, *Cryptococcus neoformans* complex, *S. prolificans*, *Pseudoallescheria boydii*, *Aspergillus terreus*, *A. flavus*, *Rhizopus microsporus*, *Cunninghamella bertholletiae*, *Absidia corymbifera*, *Alternaria infectoria*, *Aureobasidium pullulans* and *Malassezia globosa*.

70.10 Assay Limitations

It is possible that DNA from some fungi may not be amplified using these broad range PCR assays due to polymorphisms in primer binding domains. To date we have found the assays cannot amplify *C. neoformans* DNA. Consequently when *C. neoformans* is suspected, a second PCR is performed (PCRS-C) and the primers V9D and LS266 are substituted with ITS1 and ITS4. A negative PCRS result cannot exclude the presence of fungi in a clinical sample. Other factors that may influence results include sampling, PCR inhibitors, contamination and non-specific interactions with human DNA. Non-specific interactions with human DNA can sometimes occur when the concentration of human DNA is high (tissues or fluids with a high inflammatory cell content i.e. pus), particularly in microscopy-negative samples where there may be no specific fungal target. In such cases, primers may bind non-specifically to human DNA and produce a high molecular weight PCR product. In these situations the presence of fungal DNA cannot be excluded until sequencing is performed. Finally, a direct PCRS result should always be interpreted in conjunction with other laboratory evidence or clinical evidence of disease.

References

1. Altschul SF, Madden TL, Schaffer AA et al (1997) Gapped BLAST and PSI-BLAST: a new generation of protein database search programs. Nucleic Acids Res 25:3389–3402
2. de Hoog GS, Guarro J, Gene J et al (2000) Atlas of Clinical Fungi, 2nd ed. Centraalbureau voor Schimmelcultures/Universitat Rovira i Virgili
3. Pryce TM, Palladino S, Kay ID et al (2003) Rapid identification of fungi by sequencing the ITS1 and ITS2 regions using an automated capillary electrophoresis system. Med Mycol 41:369–381
4. Pryce TM, Palladino S, Price DM et al (2006) Rapid identification of fungal pathogens in BacT/ALERT, BACTEC, and BBL MGIT media using the polymerase chain reaction and DNA sequencing of the internal transcribed spacer regions. Diagn Microbiol Infect Dis 54: 289–297
5. Taylor G (2003) CMGS best practice guidelines for DNA sequencing analysis and interpretation. Clinical Molecular Genetics Society [Online] http://www.cmgs.org/BPGs/Pdfs%20current%20bpgs/Sequencingv2.pdf. Accessed April 2008

Part VI
Susceptibility Screening

Chapter 71
An Introduction to Antimicrobial Susceptibility Screening

Graeme R. Nimmo

The development of rapid molecular diagnostics has coincided with a period of increasing challenges in hospital infection control and with the emergence of antimicrobial resistance as a major issue in community infections. The "golden age" of antibiotics from the 1940s to 1970s promised the cure of all bacterial infections with a drug for every bug. Predictions were even made of the demise of Clinical Microbiology as a diagnostic discipline [1]. They were somewhat overstated. The reality we now face is very different with the emergence over the last quarter century of an impressive variety of antimicrobial resistance mechanisms in an ever increasing number of bacterial pathogens. This situation has arisen due to selective pressure provided by the widespread use (and, to varying degrees, misuse) of antibiotics in both the hospital and the community and due to the ability of bacteria to transfer genetic material coding for resistance horizontally by multiple mechanisms. Fortunately, the rapid evolution of bacterial genomics and of molecular diagnostic methods has provided us with the tools for rapid response to this emerging problem.

Cross infection within hospitals due to healthcare-adapted strains of pathogenic bacteria resistant to multiple antimicrobial agents (so-called multi-resistant organisms or MROs) has progressively increased during the antibiotic era to create a major burden that threatens the safe, efficient and cost-effective provision of services. The increasing sophistication and cost of hospital facilities and of in-patient medical services over recent decades have put pressure on healthcare budgets, particularly in the public sector. This has resulted in policies that have tended to limit the number of available hospital beds, especially in expensive tertiary facilities, and this trend in turn has increased bed occupancy to very high levels. The desired outcome of these changes has been to maximise the utility of these expensive facilities by increasing patient throughput and decreasing lengths of stay. However, increased bed occupancy has resulted in difficulties in bed management, with consequent frequent movement of patients between beds and wards. High occupancy and high

G.R. Nimmo (✉)
Department of Microbiology, Pathology Queensland Central Laboratory, Herston, QLD 4029, Australia
e-mail: Graeme_nimmo@health.qld.gov.au

M. Schuller et al. (eds.), *PCR for Clinical Microbiology*,
DOI 10.1007/978-90-481-9039-3_71,

throughput has increased workloads for clinical staff. Both of these outcomes militate against effective and routine application of good infection control principles [2]. Therefore, an unintended outcome has been an increase in cross-infection due to MROs.

Paradoxically the consequence of this has been significantly prolonged lengths of stay for infected patients. Health administrators are now recognising this creates a chronic "bed-block" that detracts from healthcare outcomes and increases costs. Effective infection control measures are central in alleviating this situation. As a result there is now an imperative to identify rapidly those patients who are colonised or infected with MROs so that measures can be implemented to reduce the risk of spread.

Methicillin-resistant *Staphylococcus aureus* (MRSA) has presented the major challenge due to its ability to colonise patients, survive on fomites and spread readily on the hands of clinical staff and its pathogenicity which is generally greater than that of other MROs. Other agents such as vancomycin-resistant enterococci (VRE), *Clostridium difficile* and multi-resistant *Acinetobacter baumannii* (MRAB) have also come to prominence. Laboratories have been called on to provide sensitive and specific screening for MROs to assist in infection control interventions and to guide treatment when necessary. Culture based screening methods have relied on isolation of suspect colonies on selective media incorporating antibiotics and chemical indicators followed by subculture for phenotypic identification and susceptibility testing. This process involves multiple over-night incubations and so is inherently slow. This slowness may be further compounded if an initial broth enrichment step is undertaken to enhance sensitivity. Proprietary chromogenic screening agar formulations have reduced the turn around times for screening for some MROs but confirmation of identity and susceptibility is still required. Specificity of phenotypic methods is another major issue for infection control screening. Due to the substantial costs and disruption involved in dealing with an MRO infected or colonised patient, infection control practitioners are reluctant to act unless they can have absolute faith in the laboratory result. Therefore, tests with very high specificity and positive predictive values are highly desirable. Phenotypic methods may be difficult to interpret and result in significant numbers of false positive calls. For example, some strains of MROs may have minimum inhibitory concentrations of key antibiotics close to susceptibility breakpoints making visual interpretation of results problematic. The ability of molecular methods to provide rapid, unequivocal and readily interpretable results make them inherently suitable for application to infection control screening.

Some laboratories have responded to the challenge of rapid identification of MROs by the development and implementation of in-house molecular assays. Successful assay development requires the application of a considerable degree expertise and the availability of suitable equipment and laboratory space. Australian regulatory requirements for the development and use of in-house assays are not trivial but favour production of assays of a quality suitable for their purpose [3]. Many in-house assays such as those detailed in this book by Merlino and colleagues for the detection of species specific markers, resistance genes and virulence genes were developed prior to the introduction of commercial real-time assays. While commercial assays have made PCR assays available to a wider range of clinical bacteriology laboratories, high equipment and consumable costs, limited assay menus and relatively long lead times for commercial assay development all favour the continued use of in-house assays and the maintenance of responsive assay development capabilities in larger laboratories.

The epidemiology of MROs is evolving rapidly and varies markedly geographically. Therefore, the indications for molecular screening, identification and typing will not be uniform among health services and laboratories. Budgetary and resource issues will also influence choice of methods. The methods detailed in this section are probably best suited to larger laboratories and the latter tend to be associated with the tertiary institutions that have borne the brunt of emerging MROs. More in-house methods will be required to deal with newly emerging problems. Multi-resistant *Enterobacteriaceae* for example are not dealt with in these chapters but are on the rise and will undoubtedly be the subject of method development. New technologies with the ability to increase capacity and to reduce turnaround are already being introduced and will be taken up by laboratories with method development capabilities. The place of molecular methods in smaller laboratories is perhaps more problematic. Without dedicated space and human resources, most smaller laboratories refer specimens for molecular confirmatory and typing tests. However, the recent advent of commercial tests with low technical requirements and "scalable" equipment opens the way for wider dissemination of molecular testing with the potential for substantial reductions in turnaround times. Nonetheless, while commercially provided kit and "black box" tests have the potential for broad application in clinical microbiology, in-house molecular tests will continue in the forefront of responses to emerging challenges for the foreseeable future.

References

1. Isenberg HD (2003) Clinical microbiology: past, present, and future. J Clin Microbiol 41(3):917–918
2. Clements A, Halton K, Graves N, Pettitt A, Morton A, Looke D et al (2008) Overcrowding and understaffing in modern health-care systems: key determinants in meticillin-resistant *Staphylococcus aureus* transmission. Lancet Infect Dis 8(7):427–434
3. (2007) National Pathology Accreditation Advisory Council. Requirements for the development and use of in-house in vitro diagnostic devices (IVDS). Commonwealth of Australia, Canberra

Chapter 72
PCR Assays in Detecting Methicillin Resistance in Staphylococci: Coagulase Negative Staphylococci (CNS), *S. aureus*, and MRSA with the PVL Gene

John Merlino, Ian D. Kay, Geoff Coombs, and Silvano Palladino

72.1 Introduction

Methicillin resistance, in particular hetero-methicillin resistance, in *S aureus* and the coagulase negative staphylococci (CNS) can be difficult to detect by phenotypic methods. Subsequently detection of the mecA gene for determining methicillin resistance in staphylococci is generally considered to be the "Gold Standard". In Australian diagnostic microbiology laboratories, molecular assays used for the detection of the mecA gene have generally been designed in-house. These assays have either been single primer PCR assays targeting the mecA gene or a duplex primer PCR assay with primers targeting the mecA gene and a gene specific for *S. aureus* such as the nuc (nuclease) or fem gene. In some laboratories a Panton Valentine-leukocidin (PVL) gene primer has been incorporated with the duplex PCR assay allowing the detection of PVL virulence genes in *S aureus*.

More recently, methods for the extraction and detection of staphylococcal DNA directly from clinical specimens and enrichment broths have been described in the literature. These extraction methods will be discussed elsewhere.

In Australia, quality assurance programs, such as the Royal College of Pathology of Australia, are available to ensure laboratories performing mecA and/or nuc/fem gene molecular assays provide results that are both accurate and standardised.

This chapter addresses the following sections and provides methods and oligonucleotides which have been commonly described:

Single primer assay specific targeting the *mecA* gene for methicillin resistance in both *S. aureus* and CNS

Duplex primer assays for MRSA detection - *S. aureus* identification for the *nuc* or *femA* gene and detection of the *mecA* for methicillin resistance

J. Merlino (✉)
Department of Microbiology and Infectious Diseases, Concord Hospital, Hospital Rd, Concord, NSW 2139, Australia; Department of Microbiology and Infectious Diseases, PathWest Laboratory Medicine, Nedlands, WA 6909, Australia
e-mail: john.merlino@sswahs.nsw.gov.au

M. Schuller et al. (eds.), *PCR for Clinical Microbiology*,
DOI 10.1007/978-90-481-9039-3_72,

Triplex Real-time Probe assay for MRSA and the PVL (Panton-Valentine Leukocidin) virulence gene
Commercial Assays

72.2 Single Primer Assay Specific Targeting the *mecA* Gene for Methicillin Resistance in Both *S. aureus* and CNS

72.2.1 Gel-Based Single Primer Assay for the mecA *Gene*

The primers amplify a segment of the mecA gene. F: (5′ -AAA ATC GAT GGT AAA GGT TGG C -3′) and R: (5′ - AGT TCT GCA CTA CCG GAT TTG C -3′) to produce a 533 bp product that is analysed by gel electrophoresis [4, 8].

72.2.1.1 PCR Amplification and Product Detection

Amplify the target region of DNA for 30 cycles in a programmed thermal cycler as follows: denaturation at 94°C for 60 s, annealing at 50°C for 30 s, and extension at 72°C for 90 s, with a final extension at 72°C for 5 min. Amplified product is detected electrophoretically on a 1.5% or 2% agar gel after staining. PCR product for the mecA gene should be 533 bp under UV light.

72.2.2 Real-Time Single Primer Probe Based Assay Targeting the mecA Gene

The primers amplify a segment of the mecA gene. F: (5′ - CAT TGA TCG CAA CGT TCA ATT T -3′) and R: (5′ - TGG TCT TTC TGC ATT CCT GGA - 3′) that is detected using a hydrolysis probe (5′- JOE -TGG AAG TTA GAT TGG GAT CAT AGC GTC AT-TAMRA - 3′) [1, 6].

72.2.2.1 PCR Amplification and Product Detection

On a Rotor- Gene (QIAGEN) hold 95°C for 15 min; amplify the target region of DNA for 35 cycles: Step 1. 95°C hold for 15 s Step 2. 60°C hold for 30 s. Select JOE channel and the alleleic discrimination key as instructed by the manufacturer. Set Threshold at 0.05 and read curves between 15 and 35 cycles.

72.2.3 Real-Time Single Primer SYBR Green Based with a Melting Curve Analysis for the mecA Gene

The primers amplify a segment of the mecA gene. F: (5′ - GCA ATC GCT AAA GAA CTA AG -3′) and R: (5′ - GGG ACC AAC ATA ACC TAA TA - 3′) that is confirmed by Sybr Green melt curve (Tm) analysis [5].

72.2.3.1 PCR Amplification and Product Detection

Following an initial denaturation at 95°C for 10 min to activate the FastStart Taq DNA polymerase, the 32-cycle amplification program consists of heating at 20°C/s to 95°C with a 0-s hold, cooling at 20°C/s to 55°C with a 5-s hold, and heating at 20°C/s to 72°C with an 8-s hold. Then the one-cycle melting curve program consists of heating at 20°C/s to 95°C with a 0-s hold, cooling at 20°C/s to 58°C with a 60-s hold, and heating at 0.1°C/s to 95°C with a 0-s hold. Finally, the protocol ends with one cycle of cooling at 20°C/s to 35°C with a 30-s hold. The fluorescence channel is set at F1 (530 nm). Samples with a T_m of 77.50 to 79.00°C are mecA positive.

72.3 Duplex Primer Assays for MRSA Detection – *S. aureus* Identification for the *nuc* or *femA* Gene and Detection of the *mecA* for Methicillin Resistance

72.3.1 Duplex Primer Gel-Based Assay for nuc *and* mecA *Gene Using a Traditional Thermocycler*

Two sets of primers are used to amplify specific segments of the respective target genes. For the mecA gene F: (5′ - AAA ATC GAT GGT AAA GGT TGG C -3′) and R: (5′ - AGT TCT GCA CTA CCG GAT TTG C - 3′) primers produce a 533 bp product that is analysed by gel electrophoresis. The nuc gene F: (5′ - GCG ATT GAT GGT GAT ACG GTT -3′) and R: (5′ - AGC CAA GCC TTG ACG AAC TAA AGC - 3′) primers produce a 367 bp product [3, 4, 8].

72.3.1.1 PCR Amplification and Product Detection

Amplify the target for 30 cycles consisting of 60 s at 94°C for denaturation, 30 s at 50°C for annealing and 90 s at 72°C for primer extension. The PCR product is then analysed by agarose gel electrophoresis 1.5% or 2% agarose prepared in TAE (1 mM EDTA/40 mM Tris acetate, pH 8) buffer. Gels are stained and photographed under UV light (Fig. 72.1).

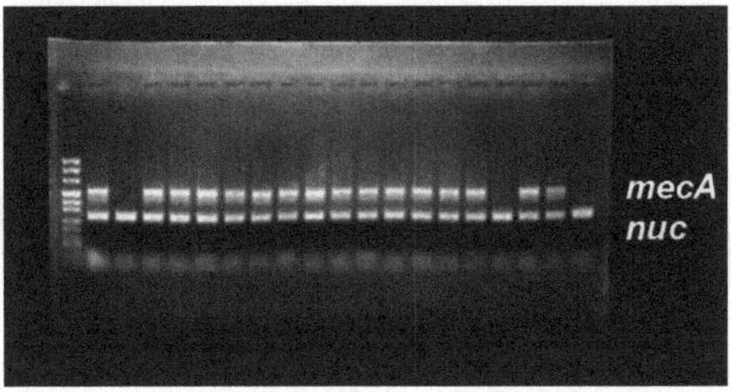

Fig. 72.1 Agar gel showing Multiplex PCR analysis for the *mecA* and *nuc* genes [8]

72.3.2 Duplex Real-Time DNA Probe Based Assay for mecA and femA Gene

Two sets of primer/probe combinations amplify and detect respective gene targets [1, 6].

Gene(s) target	Primer/probe sequences (5′ →3′)
mecA gene	F: CAT TGA TCG CAA CGT TCA ATT T
	R: TGG TCT TTC TGC ATT CCT GGA
	Probe: JOE -TGG AAG TTA GAT TGG GAT CAT AGC GTC AT -TAMRA
femA gene	F: GCG ATT GAT GGT GAT ACG GTT
	R: AGC CAA GCC TTG ACG AAC TAA AGC
	Probe: FAM -TCA TTT CAC GCA AAC TGT TGG CCA CTA TG-TAMRA

72.3.2.1 PCR Amplification and Product Detection

Rotor- Gene (QIAGEN): Hold 95°C for 15 min; amplify the target region of DNA for 35 cycles: Step 1. 95°C hold for 15 s. Step 2. 60°C hold for 30 s. Select JOE channel and follow the alleleic discrimination key as instructed by the manufacturer. Set Threshold at 0.05 and read curves between 15 and 35 cycles, see Fig. 72.2.

On the Rotor-Gene the allelic discrimination is based on channel probe fluorescence detection to differentiate MSSA, MRSA, MRSE, and MSSE as shown below.

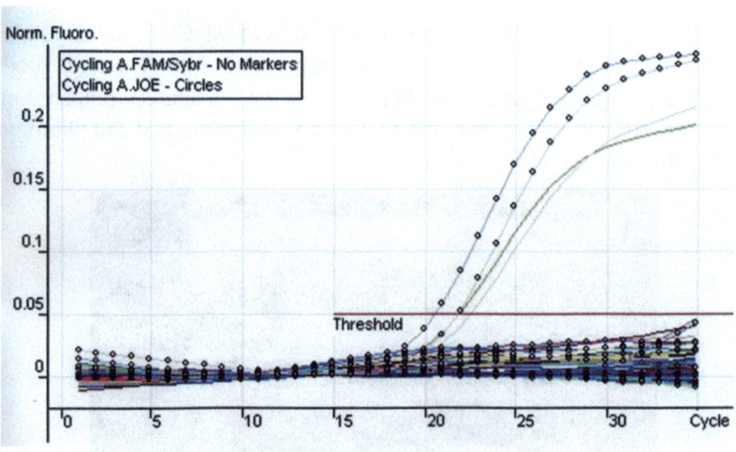

Fig. 72.2 Amplification plots showing increases in fluorescence from positive samples on a Rotor-gene. Methicillin resistant *mecA* positive samples with circles and *femA* with no markers or circles. Anything below the 0.05 threshold is negative for both *mecA* and *femA*

Organisms	FAM Channel femA or nuc gene	JOE Channel mecA gene
MSSA	Pos	Neg
MRSA	Pos	Pos
MRSE	Neg	Pos
MSSE	Neg	Neg

72.3.3 *Real-Time Duplex SYBR Green Assay with a Melting Curve Analysis for* mecA *and* nuc *Gene*

PCR product is detected and confirmed by SYBR Green fluorescence and DNA melt curve (Tm) analysis [5].

Gene(s) target	Primer Sequences (5′ →3′)
mecA gene	F: GCA ATC GCT AAA GAA CTA AG
	R: GGG ACC AAC ATA ACC TAA TA
nuc gene	F: GCG ATT GAT GGT GAT ACG GTT
	R: AGC CAA GCC TTG ACG AAC TAA AGC-3

72.3.3.1 PCR Amplification and Product Detection

The cycling program is the same as that used for detection of mecA using the Light Cycler above except for 35 cycles of amplification. One peak is specific for the mecA gene with a Tm of 77.50 to 79.00°C, and one is specific for the nuc gene with a Tm of 79.90–80.60°C. MSSA strains have only a nuc peak, MRSE strains have only a mecA peak, and MSSE strains have no peak. See Fig. 72.3.

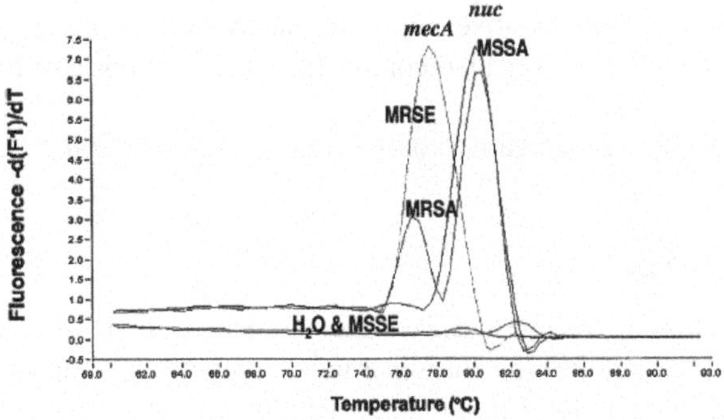

Fig. 72.3 Melting curves (Tm) for MRSA, MSSA, MRSE, and MSSE for *mecA* and *nuc* gene in the duplex real-time PCR assay with Sybr green as described by Fang and Hedin [5] on a Light Cycler (Roche Molecular Biochemicals, Mannheim, Germany)

72.4 Triplex Real-Time Probe Assay for MRSA and the PVL (Panton-Valentine Leukocidin) Virulence Genes

In this PCR assay, three gene targets are amplified in the same reaction and are detected by a hydrolysis probe specific for the respective target gene [7].

Gene(s) target	Primer/Probe Sequences (5′ →3′)
mecA gene	F: GGCAATATTACCGCACCTCA R: GTCTGCCACTTTCTCCTTGT Probe:FAM-AGATCTTATGCAAACTTAATTGGCAAATCC- TAMRA
nuc gene	F: CAAAGCATCAAAAAGGTGTAGAGA R: TTCAATTTTCTTTGCATTTTCTACCA Probe:VIC-TTCAATTTTCTTTGCATTTTCTACCA- TAMRA
PVL genes	F: ACACACTATGGCAATAGTTATTT R: AAAGCAATGCAATTGATGTA Probe:TET-ATTTGTAAACAGAAATTACACAGTTAAATATGA-TAMRA

72.4.1 PCR Amplification and Product Detection

Thermal cycling on a ABI 7500 PCR system (Applied Biosystems) is performed under the following conditions – 2 min at 50°C and 10 min at 95°C, followed by 40 cycles of 95°C for 15 s and 60°C for 1 min. PVL is detected as a triplex assay or as single-target TaqMan PCR assay by using a dual-labeled probe with TET (6-carboxytetrachlorofluorescein) as the reporter and TAMRA (6-carboxytetramethylrhodamine) as the quencher.

72.4.2 Quality Control Strains

MRSA ATCC 43300: Positive control for mecA and nuc or fem genes.
MSSA ATCC 25923: Negative control for mecA but positive for nuc or fem genes.
MSSE ATCC 12228: Negative control for mecA, nuc or fem genes.

72.4.2.1 Master Mix

In such assays commercial master mixes are complete ready to go heat-activated 2x reaction-mix which requires the consumer to add only water, template and primers, then pre-heated to 95°C for 7 min to successfully carry out polymerase amplification. An activation step (7–10 min) eliminates the presence of non-specifics such as primer-dimers and mis-primed products, by ensuring the enzyme is inactive at low temperatures. Most master mixes have been optimised for a wide variety of templates however some include additional $MgCl_2$ solution if fine adjustment is required. Ensure that the correct master mix is used for the PCR instrument used by contacting the manufacturer or technical support. Such master mixes dramatically reduce the time needed to set up reactions, thereby reducing the risk of contamination. Greater reproducibility is ensured, by reducing the number of pipetting errors.

SYBR Green PCR master mix kits: These SYBR green kits can provide accurate real-time quantification of DNA and cDNA targets. The fluorescent dye SYBR Green in the master mix enables the analysis of many different targets without having to synthesize target-specific labelled probes. High sensitivity and specificity in PCR are achieved by the use of a hot-start enzyme HotStarTaq DNA Polymerase together with a specialized buffer. The buffer may also contain ROX dye, which allows fluorescence normalization on certain cyclers. Such kits have been optimised for most real-time cyclers, including instruments Applied Biosystems, Bio-rad/MJ Research, SmartCyclers - Cepheid, Rotor-Gene – QIAGEN, Eppendorf, Light Cyclers – Roche, and Stratagene.

72.5 Commercial Assays

In addition to the duplex and triplex PCR assays described above, qualitative in vitro commercial assays for the rapid direct detection of MRSA nasal colonisation have recently been developed. In Australia the BD GeneOhm^{TM} MRSA assay (Becton Dickinson Diagnostics, San Diego CA) assay and the Cepheid Xpert MRSA^{TM} assay (Cepheid, Sunnyvale, CA, USA) are available. Both assays use the SmartCycler^{TM} real-time PCR system (Cepheid, Sunnyvale, CA, USA) and rather than amplifying the mecA gene amplify the junction region between the SCCmec element and the orfX thereby allowing MRSA to be detected in specimens that may also contain methicillin susceptible S. aureus and methicillin resistant CNS. These assays can be completed with 2 h from sample collection and both have been reported to have excellent sensitivity and specificity [2, 9, 10].

Finally, in the near future, it may be likely that in clinical microbiology laboratories the chip arrays, which are currently used mainly in a research setting, will be developed for the rapid mass screening of bacterial isolates for both identity and antimicrobial resistance genes. Currently, it remains doubtful as to whether it would be economically viable to have a very small chip that could be used purely for the detection of MRSA in screening swabs.

References

1. Adams N (2005) Shortcut method for extraction of *Staphylococcus aureus* DNA from blood cultures and conventional cultures for use in real-time PCR assays. J Clin Microbiol 43:2932–2933
2. Boyce JM, Havill NL (2008) Comparison of BD GeneOhm methicillin-resistant *Staphylococcus aureus* (MRSA) PCR versus the CHROMagar MRSA assay for screening patients for the presence of MRSA strains. J Clin Microbiol 46:350–351

3. Brakstad OG, Aasbakk K et al (1992) Detection of *Staphylococcus aureus* by polymerase chain reaction amplification of the nuc gene. J Clin Microbiol 30:1654–1660
4. Coombs G, Kay I et al (1999) The role of multiplex mecA/nuc PCR for routine detection of methicillin resistance in Staphylococci. Aust Soc Antimicrob Newslett 1:3–4
5. Fong H, Hedin G (2003) Rapid screening and identification of methicillin resistant *Staphylococcus aureus* from clinical samples by selective-broth and real-time PCR assay. J Clin Microbiol 41:2894–2899
6. Francois P, Pittet D et al (2003) Rapid detection of methicillin-resistant *Staphylococcus aureus* directly from sterile or non-sterile clinical samples by a new molecular assay. J Clin Microbiol 41:254–260
7. MacDonald RR, Antonshyn NA et al (2005) Development of a triplex real-time PCR assay for detection of Panton-valentine leukocidin toxin genes in clinical isolates methicillin–resistant *Staphylococcus aureus*. J Clin Microbiol 43:6147–6149
8. Merlino J, Watson J et al (2002) Detection and expression of methicillin/oxacillin resistance in multi-drug and non-multidrug-resistant *Staphylococcus aureus* in Central Sydney, Australia. J Antimicrob Chemother 49:793–801
9. Paule SM, Hacek DM et al (2008) MRSA detection: comparison of two molecular methods (BD GeneOhm® PCR assay and Easy-Plex) with two selective MRSA agars (MRSA-ID and Oxoid MRSA) for nasal swabs. Eur J Clin Microbiol Infect Dis 45:2993–2998
10. Wolk DM, Picton E et al (2009) Multicenter evaluation of the cepheid xpert methicillin-resistant *Staphylococcus aureus* (MRSA) test as a rapid screening method for detection of MRSA in nares. J Clin Microbiol 47:758–764

Chapter 73
Detection of VRE: vanA and vanB Genes by PCR

John Merlino, Ian D. Kay, Geoff Coombs, and Silvano Palladino

73.1 Introduction

In clinical laboratories the phenotypic detection of low-level vancomycin resistance in enterococci is problematic. Rapid detection of vancomycin resistant enterococci (VRE), in particular those that carry the *vanA* or *vanB* genes, may minimize transmission within a hospital and therefore reduce the risk of colonization and infection.

Screening patients for VRE colonization using traditional culture and susceptibility methods can take up to five days. However, the results from *vanA* and *vanB* PCR assays performed on broth enrichment culture or following isolation from solid culture media (eg. specific chromogenic plate media for *E. faecalis* and *E. faecium*) may be available within one day. Therefore using PCR to screen patients should allow early implementation of contact isolation precautions, which in turn may improve VRE infection control efforts within a healthcare facility.

The following methods describe simple real-time polymerase chain reaction (PCR) assays used for the detection of the *vanA* and/or *vanB* genes from enterococcal isolates or directly from a selective enrichment broth.

This chapter describes two methods:

Real-time analysis of *vanA* and *vanB* genes in eneterococcal isolates by Sybr green and melting curve analysis

The *vanA* and *vanB* VRE gene Light Cycler assay for use with rectal swabs or stool samples incubated in Enterococcosel broth

J. Merlino (✉)
Department of Microbiology and Infectious Diseases, Concord Hospital, Hospital Rd, Concord, NSW 2139, Australia; Department of Microbiology and Infectious Diseases, PathWest Laboratory Medicine, Nedlands, WA 6909, Australia
e-mail: john.merlino@sswahs.nsw.gov.au

M. Schuller et al. (eds.), *PCR for Clinical Microbiology*,

DOI 10.1007/978-90-481-9039-3_73,

73.2 Real-Time Analysis of *vanA* and *vanB* Genes in Enterococcal Isolates by SYBR Green and Melting Curve Analysis

The following method describes a simple real-time polymerase chain reaction (PCR) assay for the detection of the *vanA* or *vanB* genes from enterococcal isolates using SYBR green (Qiagen, Australia) and a melting-curve analysis.

The performance of this real-time PCR method was compared in two institutions, Concord Hospital and Nepean Hospital to that of a reference laboratory, Royal North Shore Hospital, Sydney, using a gel-based PCR system previously described with 100% sensitivity and specificity [1, 3].

73.2.1 Primer Sequences

Real-time primer sequences for *vanA* and *vanB* gene target [1, 2] are shown:

Gene(s) target	Primer sequences (5′→3′)
VanAB	F: GTAGGCTGCGATATTCAAAGC
VanA	R: CGATTCAATTGCGTAGTCCAA
VanB	R: GCCGACAATCAAATCATCCTC

Working stock primer concentrations: 10 μM.

73.2.2 Reaction Mixture

Total volume 12 μl, containing 5 μl DNA, 5 μl QuantiTectSYBR green (Qiagen, Australia) and 1 μl of VanA F+R primer or vanB F+ R primer in separate tubes.

73.2.3 Amplification Profile

Hold 95°C for 15 min; amplify the target region of DNA for 40 cycles: Step 1. 95°C hold for 10 s Step 2. 55°C hold for 15 s and Step. 3 72°C hold 20 s.

73.2.4 Melt Curve Settings

Melt (78°C–85°C) hold 45 s on step 1 and 5 s next step.

73.2.5 Results

On the Rotor-Gene positive *vanA* VRE isolates have a melting peak (Tm) around 81 to 81.5 and *vanB* VRE isolates around 83 to 83.5 as shown on Fig. 73.1.

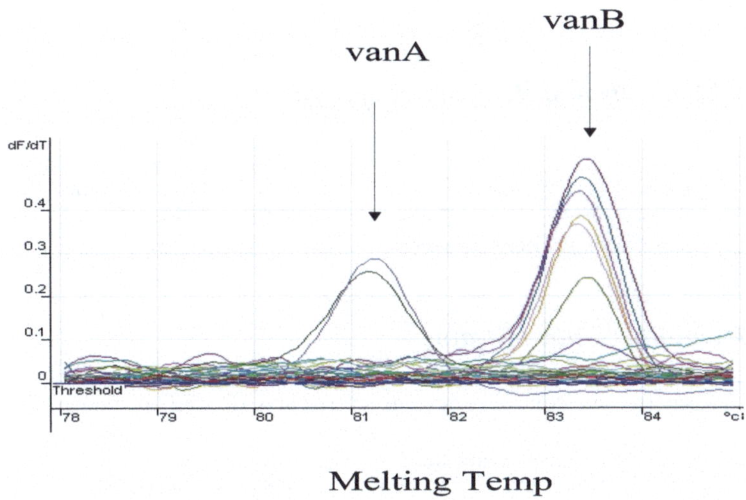

Fig. 73.1 Melting curve (Tm) analysis peaks showing *vanA* and *vanB* positive isolates on a Rotor-gene

73.2.6 Quality Control

Negative control – *E. faecalis* ATCC 29212
vanA positive control – *E. faecium* ATCC 27637
vanB positive control – *E. faecalis* ATCC 51299

73.3 The *vanA* and *vanB* VRE Gene Light Cycler Assay for Use with Rectal Swabs or Stool Samples Incubated in Enterococcosel Broth

73.3.1 Introduction

The Light Cycler provides a suitable platform to perform this multiplex assay which is capable of rapidly and accurately detecting the *vanA* and *vanB* genes, and can be easily incorporated into the molecular diagnostic workflow of a clinical microbiology laboratory [4].

73.3.2 Specimens

The *vanA* and *vanB* VRE gene Light Cycler assay is for use with rectal swabs or stool samples incubated in Enterococcosel broth plus vancomycin 8 mg/L (EVB). Enterococci colonies growing on laboratory media that are less than 72 h old can also be tested.

73.3.3 Materials and Reagents Required

Item	per isolate	stored
Van gene PCR reaction mix	15 μl	prepare fresh
MagNA Pure LC Total Nucleic Acid Isolation Kit		RT
10% Chelex 100 Resin Solution	50 μl	4°C

73.3.4 Quality Control Strains

For each assay run two control organisms (stored at 4°C), and one reagent control.

(i)	*E. faecium* (wild strain)	
	Expected results:	*vanA gene* – DETECTED
		vanB gene – Not Detected
(ii)	*E. faecalis* (ATCC 51299)	
	Expected results:	*vanA gene* – Not Detected
		vanB gene – DETECTED
(iii)	Reagent Control: EVB broth	
	Expected results:	*vanA gene* – Not Detected
		vanB gene – Not Detected

73.3.5 Extraction Procedures

73.3.5.1 VRE Broths

(i) Aliquot 200 μl of well mixed EVB broth controls and EVB broth samples to each well in the sample cartridge.

(ii) Set up the MagNA Pure LC. Refer to the MagNA Pure LC Total Nucleic Acid Isolation Kit Insert for reagent set-up using the Total NA Variable elution volume protocol. A 200 μl sample volume with a 50 μl elution volume is used.

73.3.5.2 Control Strains

(i) Using a 1 μl disposable loop collect a bacterial colony from each control and suspend in separate EVB broths. Include an un-inoculated EVB broth as a reagent control.

(ii) Incubate broths at 37°C for 24 h.

73.3.5.3 Test Organism

(i) In the reagent preparation biohazard cabinet dispense 10% Chelex-100 Resin Solution into 50 μl aliquots into 1.5 ml eppendorf tubes.
(ii) Using a 1 μl disposable loop collect a bacterial colony and suspend in the Chelex solution.
(iii) Incubate tubes at 95°C for 10 min.
(iv) Centrifuge tubes at 13,000 rpm in a biofuge for 5 min.
(v) 5.0 μl of supernatant is added to the PCR mix.

73.3.5.4 Amplification Procedures

For information on the LightCycler, refer to Roche Molecular Biochemicals LightCycler Operators Manual (Version 3.5).

(i) Add 5 μl of DNA extract to PCR mix.
(ii) Centrifuge carousel containing the loaded capillaries.
(iii) Use the LightCycler VRE programme

Activation	95°C	10 min	x1 cycles
Amplification	95°C	15 s	
	54°C	15 s	x40 cycles
	72°C	32 s	
Melt	95°C	0 s	x1 cycles
	50°C	10 s	
	95°C	0 s	
Cool	40°C	60 s	

73.3.5.5 Interpretation and Reporting van gene PCR Results: Interpretation of Light Cycler Data

To be considered positive for *vanA* or *vanB* gene detection a control/specimen must fulfil the following criteria:

(i) demonstrate an exponential increase in fluorescence.
(ii) generate a characteristic melting temperature (*Tm*)

Control	**Target**	***Tm* (°C)**
E. faecium	*vanA*	63–65
E. faecalis	*vanB*	62–64
Specimen	**Target**	**Tm (°C)**
Broth/isolate	*vanA*	63–65
Broth/isolate	*vanB*	58–60

For further information on Quantification Data Analysis, refer to Roche Molecular Biochemicals LightCycler Operators Manual (Version 3.5).

73.3.6 Reagent Preparation for van gene PCR

73.3.6.1 Van Gene PCR Master Mix

Van Gene PCR Master Mix is prepared fresh and aliquotted into capillaries prior to loading specimen extracts.

Oligonucleotide Sequence

Van A F : 5′ ggg AAAA CgA CAA TTg C 3′
Van A R : 5′ gTA CAA TgC ggC CgT TA 3′
Van B F : 5′ ATg ggA AgC CgA Tag TC 3′
Van B R : 5′ gAT TTC gTT CCT CgA CC 3′

Hybridisation Probe Sequence

Van A-FL : 5′ ACg CAg TTA TAA CCg TTC CCg X 3′
Van A-705 : 5′ LC Red705-AgA CC TTT CAg CAg Agg AgCg p 3′
Van B-FL : 5′ CAC ggT CAg gTT CgT CCT X 3′
Van B-640 : 5′ LC Red640-Tgg CgT AAC CAA AgT AAA CAg TAC g p 3′
X = fluoroscein p = phosphorylation

Reagents	μl per capillary
Fast Start 10 X Reaction Mix Hybridisation probes	2.0
$MgCl_2$25 mM	1.6
Van A F Primer * (20 μM)	1.0
Van A R Primer * (20 μM)	1.0
Van B F Primer * (20 μM)	1.0
Van B R Primer * (20 μM)	1.0
Van A-FL Probe (20 μM)	0.2
Van A-705 Probe (20 μM)	0.2
Van B-FL Probe (20 μM)	0.2
Van B-604 Probe (20 μM)	0.2
Water (supplied in Light Cycler Reagent Kit)	6.6
Dispensed Volume	15
Extracted template	5
Total Reaction Volume	20

10% Chelex Solution. Concentration required : 10% ∴ 2 g in 20 ml
Add 2 g of Chelex-100 Resin to 20 ml of Tris Buffer (0.1 M, pH 7.5)
Dispense into 1.0 ml aliquots in Eppendorf tubes
Label Chelex and preparation date
Store at 4°C

References

1. Ballard SA, Grabsch EA et al (2005) Comparison of three PCR primer sets for identification of *vanB* gene carriage in feces and correlation with carriage of vancomycin-resistant enterococci: interference by *vanB*-containing anaerobic bacilli. Antimicrob Agents Chemother 49:77–81
2. Bell J, Paton JC, Turnidge J (1998) Emergence of vancomycin-resistant enterococci in Australia: phenotypic and genotypic characteristics of isolates. J Clin Microbiol 36:2187–2190
3. Dukta-Malen S, Evers S, Courvalin P (1995) Detection of glycopepetide resistance genotypes and identification to the species level of clinical relevant enterococci by PCR. J Clin Microbiol 33:24–27
4. Palladino et al (2003) Rapid detection of vanA and vanB genes directly from clinical specimens and enrichment broths by real-time multiplex PCR assay. J Clin Microbiol 41:2483–2486

Chapter 74
Metallo β Lactamases Gene bla_{imp}, bla_{spm} and bla_{vim} Detection by Multiplex Real-Time TaqMan Assay on the Smartcycler

Ian Carter

74.1 Introduction

Gram-negative bacteria have a range of resistance mechanisms that they can use to evade the actions of carbapenems and other beta-lactams. The common form of resistance is either through lack of drug penetration (i.e., outer membrane protein (OMP) mutations and efflux pumps), hyperproduction of an AmpC-type beta-lactamase, and/or carbapenem-hydrolyzing beta-lactamases. Based on molecular studies, two types of carbapenem-hydrolyzing enzymes have been described: serine enzymes possessing a serine moiety at the active site, and metallo-β-lactamases (MBLs), requiring divalent cations, usually zinc, as metal cofactors for enzyme activity. MBLs, like all β-lactamases, can be divided into those that are normally chromosomally mediated and those that are encoded by transferable genes. The early studies on chromosomally mediated MBLs mainly centered around *Bacillus cereus* (BC II), and *Stenotrophomonas maltophilia* (L1). However, primarily due to genomic sequencing, increasingly more chromosomally mediated genes are being discovered but are often found across a range of bacterial genera.

Over the last decade there have been several articles summarizing the levels of MBLs in bacteria. However, in the past 3–4 years many new transferable types of MBLs have been studied and appear to have rapidly spread [3, 4]. In some countries, *P. aeruginosa* possessing MBLs constitute nearly 20% of all nosocomial isolates, whereas in other countries the number is still comparatively small. In recent years MBL genes have spread from *P. aeruginosa* to *Pseudomonas* spp., *Acinetobacter* spp., and Enterobacteriaceae, and a clinical scenario appears to be developing that could simulate the global spread of extended-spectrum beta-lactamases. Moreover, given that MBLs will hydrolyse virtually all classes of beta-lactams and that we

I. Carter (✉)
SEALS Microbiology Division, Prince of Wales Hospital, Randwick, NSW 2031, Australia
e-mail: ian.carter@sesiahs.health.nsw.gov.au

M. Schuller et al. (eds.), *PCR for Clinical Microbiology*,
DOI 10.1007/978-90-481-9039-3_74,

are several years away from the implementation of a 'therapeutic inhibitor', their continued spread could be a clinical catastrophe.

74.2 Summary of Methods

This method focuses on the ability to multiplex in one reaction tube for detection of all three types of MBL genes; bla_{imp4}, bla_{spm} *and* bla_{vim}.

The assay utilises 5′ nuclease-based real-time PCR comprising two primers and one probe for each MBL gene target, each labelled with a different fluorophore. bla_{imp4} has been chosen as this appears to be the predominate MBL present in the Australian bacterial population [2].

74.3 Acceptable Specimens

Pure bacterial cultures are the preferred samples for this PCR.

74.4 Sample Extraction

Plugged tips must be used for all pipetting steps. Change powder-free gloves regularly. DNA extraction is performed in the Class II Biological Safety Cabinet. Only trained staff with demonstrated competence may perform the test.

Extract DNA following the Instagene matrix (BioRad Laboratories) procedure for bacterial isolates. Briefly, to 200 μl of Instagene matrix (mix well prior to addition) is added a small loop of colony of the organism of interest. Incubate at 56°C for 15–30 min with shaking in the Eppendorf thermomixer. Place the tube in a 99°C heat block with shaking (second thermomixer) for 8 min. Spin at 13,000 rpm for 2–3 min (this pellets the Chelex resin and cellular material). Use the supernatant in the PCR reaction.

For each assay run, one positive bla_{imp}, bla_{spm} *and* bla_{vim} control and one reagent negative control and no DNA control (5 μl of sterile distilled autoclaved water) would be typical. After extraction each nucleic acid suspension is held at 4°C for up to 4 h or frozen at –20°C or lower until required. They may be thawed for use as required but repeat the last centrifugation step if reusing the Instagene DNA preparation.

74.5 MBL Multiplex PCR Master Mix

Master Mix is prepared fresh and as a bulk mix. Volumes will vary depending on the numbers of samples and controls to be tested. The master mix is added to capillaries just prior to adding DNA extract.

74.6 Reactants for 1x Mix

Fast Start DNA Master mix *PLUS* Hybridisation probes reaction mix (Roche) (10x) (10 μl 1a into 1b. Store at 4°C for 1 week)		5.0 μl
Bla_{imp4}-F	*(20 μM)*	0.625 μl
Bla_{imp4}-R	*(20 μM)*	0.625 μl
Bla_{imp4} probe (FAM) *(20 μM)*		0.25 μl
Bla_{spm} -F	*(20 μM)*	0.625 μl
Bla_{spm} -R	*(20 μM)*	0.625 μl
Bla_{spm} probe (TxR) *(20 μM)*		0.25 μl
Bla_{vim}- F	*(20 μM)*	0.625 μl
Bla_{vim} -R	*(20 μM)*	0.625 μl
Bla_{vim} probe (TET) *(20 μM)*		0.25 μl
Uracil-n-glycosylase (UNG) (LC Roche Cat. No. 03539806) (optional at 1 μl /reaction)		
Water nuclease free		*10.5 μl*
Total per reaction tube		20.0 μl
Add DNA extract		5.0 μl
Total per tube (μl)		**25.0** μl

The primers and probe are held at -30°C in small volume aliquots (5 μl probe and 10 μl each primer) to exclude repeated freeze/thaw cycles and potential probe hydrolysis. This bulk mix is made up in a sterile or nuclease free 1.5 ml eppendorf tube using a separate plugged tip for each solution. Briefly hand mix the master mix and spin for 5 s to assemble all components into the bottom of the tube. Aliquot the master mix to the capillaries and add specimen or control sample extract. Keep cool.

74.7 Primers and Probe Sequences

Bla_{imp4} – F	5′ GGCAGTATTTCCTCTCATTT 3′
Bla_{imp4}– R	5′ GCAGCTCATTAGTTAATTCAG 3′
Bla_{spm} –F	5′ GGGTACGCAAACGCTTATGG 3′
Bla_{spm} – R	5′ CCGTGCCGTCCAAATGAAAG 3′
Bla_{vim} – F	5′ CGCGGAGATTGAGAAGCAAA 3′
Bla_{vim} – R	5′ AGCCGCCCGAAGGACATC 3′

These are purchased from Proligo Australia (Lismore) at a concentration of 100 μM. Aliquot small volumes of 20 μM solutions (1:5 dilution in nuclease free water) and store labelled –20°C.

Bla_{imp4}5′ (6- FAM) CATAGTGACAGCACGGGCGGAAT (BHQ1) 3′
Bla_{spm} 5′ (Texas Red) TTGGGTGGCTAAGACTATGAAGCCGA (BHQ2) 3′
Bla_{vim} 5′ (TET) TTGGACTTCCTGTAACGCGTGCA (BHQ1) 3′

These are purchased from Biosearch Technologies, Novato CA. USA. and are diluted to a concentration of 100 μM. Aliquot small volumes of 20 μM solutions as a working solution (1:5 dilution in nuclease free water) and store labelled at –20°C.

74.8 Addition to Reaction Tubes

Add 20 μl of this bulk master mix to the lid section of the SmartCycler reaction tubes. Load 5 μl of specimen/control sample extract to the lid section and immediately close the lid. Spin the reaction tubes for 10 s at a speed of 3000 rpm in the dedicated microfuge to ensure all solution is within the reaction tube. Let stand at room temperature for 10 min to allow UNG (if incorporated) to destroy any previously amplified product (potential source of contamination).

74.9 PCR Amplification and Product Detection

This is performed on a Cepheid SmartCycler II Instrument, using SmartCycler software and dye set FTTC25 (FAM, TET, TXR, Cy5).

The reaction tubes are initially spun and then placed into the appropriate 'site' in the SmartCycler. You can use any of the sites as long as you set up the software to match. Ensure tubes are in the correct order:

The programme cycling profile is Activation 1x at 95°C × 10 min, Amplification is 30 cycles at 95°C × 10 s, 60°C × 30 s single acquisition, cool at 40°C × 30 s. A low cycle number of cycles is fine when using a culture isolate. Total cycling time is around 40 min.

74.10 Interpretation and Reporting of PCR Results

Print each of the FAM, TET, TxR dyes and analyse and report results. If any of the controls do not meet the expected results the run is considered invalid and samples must be retested and/or phenotypic analyses performed.

74.11 Quality Control and Validation Data

Extracted nucleic acid from previously PCR-positive bacterial isolates may be used as a positive control. Any positive control should be diluted to a concentration providing a cycle-threshold (Ct) value of approximately 25 cycles in the real-time PCR.

200 μl of Instagene matrix is subject to the extraction procedure and is used as a negative control.

This assay has been validated against the CDS method of antibiotic susceptibility testing [5]. It was presented at the annual ASM Conference in Melbourne 2008 as a poster.

To date all isolates tested have matched both phenotypic and multiplex PCR results 100% and bacterial isolates tested containing an MBL include *Enterobacter aerogenes*, *Enterobacter cloacae*, *Klebsiella pneumoniae*, *Klebsiella oxytoca*, *Citrobacter freundii*, *Serratia marcescens*, *Pseudomonas aeruginosa* and *Morganella* sp. isolate. The majority have contained an *imp* gene but only one has contained a *vim* gene. Allelic variants for the MBL types $bla_{\text{SPM-1}}$, $bla_{\text{GIM-1}}$, and $bla_{\text{SIM-1}}$ have not been detected yet. A SYBR green based multiplex PCR utilising melt-curve analysis has been published [1] and may prove useful for identification of MBL genes contained within resistant organisms.

References

1. Mendes R, Kiyota K, Monteiro J et al (2007) Rapid detection and identification of metallo-beta-lactamase-encoding genes by multiplex real-time PCR assay and melt curve analysis. J Clin Microbiol 45(2):544–547
2. Peleg A, Franklin C, Bell J et al (2004) Emergence of IMP-4 metallo-beta-lactamase in a clinical isolate from Australia. J Antimicrob Chemother 54:699–700
3. Walsh T, Toleman M, Laurent Poirel et al (2005) Metallo-beta-Lactamases: the quiet before the storm? Clin Micro Rev 18(2):306–325
4. Walsh T (2005) The emergence and implications of metallo-*b*-lactamases in Gram-negative bacteria. Clin Microbiol Infect 11 (Suppl 6): 2–9
5. http://web.med.unsw.edu.au/cdstest/ Section 5.5.8 (accessed 10 November 2009)

Chapter 75
PCR-Sequencing for Detection of Human Cytomegalovirus Mutations Conferring Antiviral Resistance

Gillian M. Scott

75.1 Introduction

Antiviral resistant strains will emerge in at least 10% of patients receiving long-term antiviral therapy for the prevention or treatment of cytomegalovirus (CMV) infection and disease. Prophylaxis therapy with the antiviral agent valganciclovir (the oral pro-drug of ganciclovir) is becoming a common strategy for the prevention of CMV infection in immunocompromised transplant recipients, and accordingly, requests for CMV antiviral resistance testing are also increasing. The antiviral agents cidofovir and foscarnet are also used against ongoing CMV infection and disease when ganciclovir or valganciclovir therapy has failed, but resistance to these compounds can also develop. The most accurate, sensitive and comprehensive approach for identification of antiviral resistant CMV strains is PCR-sequencing, where segments of the CMV UL97 and UL54 genes are amplified by nested PCR and the PCR products sequenced [1, 2]. Mutations that confer antiviral resistance to CMV strains occur in specific regions of UL97 and UL54, and therefore only specific segments of these genes (one segment for UL97 and two for UL54) need to be amplified and sequenced, increasing the overall sensitivity and specificity of the assay. In combination, sequencing of these three PCR products allows the identification of all mutations known to confer resistance to ganciclovir and valganciclovir, as well as foscarnet and cidofovir, including DNA polymerase mutations that can confer cross-resistance to two or more of these compounds. This information is highly valuable to clinicians trying to determine a course of action where CMV-positive patients fail to respond to the antiviral agent they have been administered.

G.M. Scott (✉)
Virology Division, Department of Microbiology, SEALS, Prince of Wales Hospital and School of Biotechnology and Biomolecular Sciences, Randwick, NSW 2031, Australia; University of New South Wales, Sydney, NSW, Australia
e-mail: gillian.scott@sesiahs.health.nsw.gov.au

M. Schuller et al. (eds.), *PCR for Clinical Microbiology*,
DOI 10.1007/978-90-481-9039-3_75, © Springer Science+Business Media B.V. 2010

75.2 Specimens

Most specimen types can be tested for CMV antiviral resistant strains. EDTA whole blood, plasma or a specimen from the site of active disease (for example, vitreous fluid for CMV retinitis or gastric biopsy for CMV gastritis) are preferable. CMV antiviral resistant strains can sometimes co-exist and circulate with antiviral sensitive strains, and resistant genotypes will only be detected if they are greater than 10% of the total CMV content.

Different CMV strains can also be present in separate compartments of an infected person. For example, it is common to have a resistant CMV strain as the predominant genotype at the site of disease, but a sensitive genotype present in the kidneys and being excreted, and urine is not a good specimen for antiviral resistance testing for this reason.

75.3 DNA Extraction

DNA is extracted using standard commercial kits such as the Qiagen DNA mini kit for a small volume of specimens, with DNA eluted in a final volume of 50 μl. For a high volume of specimens DNA is extracted using automated workstations and kits such as the MagNAPure DNA extraction system (Roche Diagnostics).

75.4 PCR Conditions

Separate reactions are carried out for UL97, UL54exo-C and UL54ii–v inner and outer PCR.

To avoid contamination of nested PCR reactions, separate rooms of the laboratory are used for reagent preparation (PCR1), extraction of DNA and addition of sample DNA to the PCR reaction (PCR2), addition of positive controls and thermal cycling (PCR3), addition of DNA from outer PCR to inner PCR reactions (PCR4) and analysis of PCR products (PCR5). Separate pipettes and tips are used at each

Table 75.1 Forward and reverse primers for UL97, UL54exo-C and UL54ii–v PCR

	Forward primer (5′–3′)	Reverse primer (5′–3′)
UL97		
Outer PCR	GTGCTCACGGTCTGGATGTCGGG	CCTTTCCCCTCAGCAACCGTC
Inner PCR	TCATCACGACCAGTGGAAGCT	GCGACACGAGGACATCTTGG
UL54exo-C		
Outer PCR	AAGGGCGGCGACATCTGAAAC	GCGTCGACTTGTGATATCGAG
Inner PCR	GCGGCGGTAGAGATGATAGCG	GCGATGTCTCCGACCTGGTG
UL54ii–v		
Outer PCR	CCGCGTGGCACGCCGTATTTC	GTTTCAGATGTCGCCGCCCTT
Inner PCR	CACGGCTGCTTCCGAGACCTC	GCAATCTGCGCCGTCGAGTCA

Table 75.2 Reagent mixes for outer and inner PCR

	Volume per reaction (μl)	Final concentration
Deionised water	13.3	–
10X buffer[a] (ABgene)	5	1X
25 mM MgC12[a] (ABgene)	5	5 mM
1 mM dNTP mix (Promega)	12.5	0.25 mM
100% DMSO	5	10%
10 μM forward primer	2	0.4 μM
10 μM reverse primer	2	0.4 μM
RedHot® Taq DNA polymerase, 5 U/μl (ABgene)	0.2	0.02 U/μl
Total	45	

[a]Supplied with RedHot® Taq DNA polymerase

station, and new gowns and gloves worn at each step primer sequences are listed in Table 75.1. Purified plasmid constructs containing UL97 or UL54 gene segments are used as positive controls at 102 copies per reaction, which is the minimum detection limit of the PCR reactions. Negative controls consist of water (no DNA).

For Outer PCR, 5 μl of purified specimen DNA, plasmid construct at 20 copies/μl (positive control tube) or deionised water (negative control tube) is added to the reaction mix (Table 75.2), and reactions subject to thermal cycling conditions consisting of an initial denaturation of 95°C for 3 min followed by 30 cycles of (94°C for 30 s, 55°C for 30 s, 72°C for 2 min).

For Inner PCR, 5 μl of outer PCR products are added to the reaction mix, and reactions subject to thermal cycling conditions consisting of an initial denaturation of 95°C for 3 min, followed by 30 cycles of (94°C for 30 s, 55°C for 30 s, 72°C for 1 min), and a final extension of 72°C for 7 min.

75.5 Analysis of PCR Products

A small amount of the final inner PCR reactions (no more than 5 μl) are mixed with SYBR green loading dye and run on a 1.2% agarose gel. Expected sizes of the PCR products are 677 bp for UL97, 1045 bp for UL54exo-C and 1080 bp for UL54ii–v. PCR products are precipitated by mixing the remaining PCR reaction with an equal

Table 75.3 2X PEG PCR mix

	Volume	Final concentration
Polyethylene glycol (PEG) 8000	6 g	26.7%
3 M sodium acetate, pH 5.2	6.15 ml	0.6 M
1 M magnesium chloride	200 μl	6.5 mM
Deionised water	to 30 ml	

Table 75.4 Common mutations conferring antiviral resistance to CMV strains

Codon position	DNA mutation	Amino acid change	Antiviral resistance[a]
UL97			
460	ATG to GTG	Methionine to Valine	Ganciclovir
592	TGC to GGC	Cysteine to Glycine	Ganciclovir
594	GCG to GTG	Alanine to Valine	Ganciclovir
595	TTG to TCG	Leucine to Serine	Ganciclovir
UL54exo-c			
501	CTC to ATC	Leucine to Isoleucine	Ganciclovir, Cidofovir
513	AAG to AAC	Lysine to Asparagine	Ganciclovir, Cidofovir
522	CCG to TCG	Proline to Serine	Ganciclovir, Cidofovir
UL54ii–v			
A809V	GCG to GTG	Alanine to Valine	Ganciclovir, Foscarnet
V812L	GTA to CTA	Valine to Leucine	Ganciclovir, Foscarnet, Cidofovir
A987G	GCT to GGT	Alanine to Glycine	Ganciclovir, Cidofovir

[a]Mutations conferring resistance to ganciclovir also confer resistance to valganciclovir

volume of 2X PEG PCR mix (Table 75.3), vortexing for one minute and centrifugation at maximum speed in a microfuge for 15 min. Precipitate is washed twice with 95% ethanol, air dried and resuspended in 20 μl of water.

Purified PCR products are sequenced using protocols recommended by the sequencing facility being utilized. Only the forward inner PCR primer in a single reaction is required for sequencing the 677 bp UL97 PCR product, unless confirmation is necessary (Table 75.1). The larger UL54exo-C and UL54ii–v PCR products will require separate sequencing reactions containing forward and reverse inner PCR primers. The UL97, UL54exo-C and UL54ii–v DNA sequences from each specimen are aligned against UL97 and UL54 sequences from AD169 (accession number X17403) to identify mutations that confer resistance to CMV strains. A single nucleotide or amino acid change in UL97 or UL54 does not always signify a resistant mutation and may be the result of natural variation. It is therefore important to be aware of the mutations known to confer antiviral resistance [3–5]. Some of the most common antiviral resistant mutations identified are listed in Table 75.4. This is only a selection of the large number of UL97 and UL54 mutations confirmed to confer antiviral resistance by marker transfer to antiviral sensitive strains. The potential for identification of new previously unrecognised mutations continues, particularly with increased use of the current antiviral agents for treatment and prophylaxis of CMV.

References

1. Iwasenko J, Scott G, Zeigler J et al (2007) Emergence and persistence of multiple antiviral-resistant CMV strains in a highly immunocompromised child. J Clin Virol 40:152–155
2. Scott GM, Isaacs MA, Zeng F et al (2004) Cytomegalovirus antiviral resistance associated with treatment induced UL97 (protein kinase) and UL54 (DNA polymerase) mutations. J Med Virol 74(1):85–93

3. Iwasenko JM, Scott GM, Rawlinson WD et al (2009) Successful valganciclovir treatment of post-transplant cytomegalovirus infection in the presence of UL97 mutation N597D. J Med Virol 81(3):510–807
4. Scott GM, Weinberg A, Rawlinson WD et al (2007) Multi-drug resistance conferred by novel DNA polymerase mutations in human cytomegalovirus isolates. Antimicrob Agents Chemother 51(1):89–94
5. Chou S, Lurain NS, Thompson KD et al (2003) Viral DNA polymerase mutations associated with drug resistance in human cytomegalovirus. J Infect Dis 188(1): 32–39

ERRATUM

Erratum to: *Mycobacterium tuberculosis* Complex

Maria Globan and Janet Fyfe

Mycobacterium Reference Laboratory
Victorian Infectious Diseases Reference Laboratory (VIDRL)
10 Wreckyn Street
North Melbourne, VIC 3051
Australia
maria.globan@mh.org.au, janet.fyfe@mh.org.au

M. Schuller et al. (eds.), *PCR for Clinical Microbiology*,
DOI 10.1007/978-90-481-9039-3_18, © Springer Science+Business Media B.V. 2010

DOI 10.1007/978-90-481-9039-3_76

In Chapter 18, Figure 3 should be replaced by the version below:

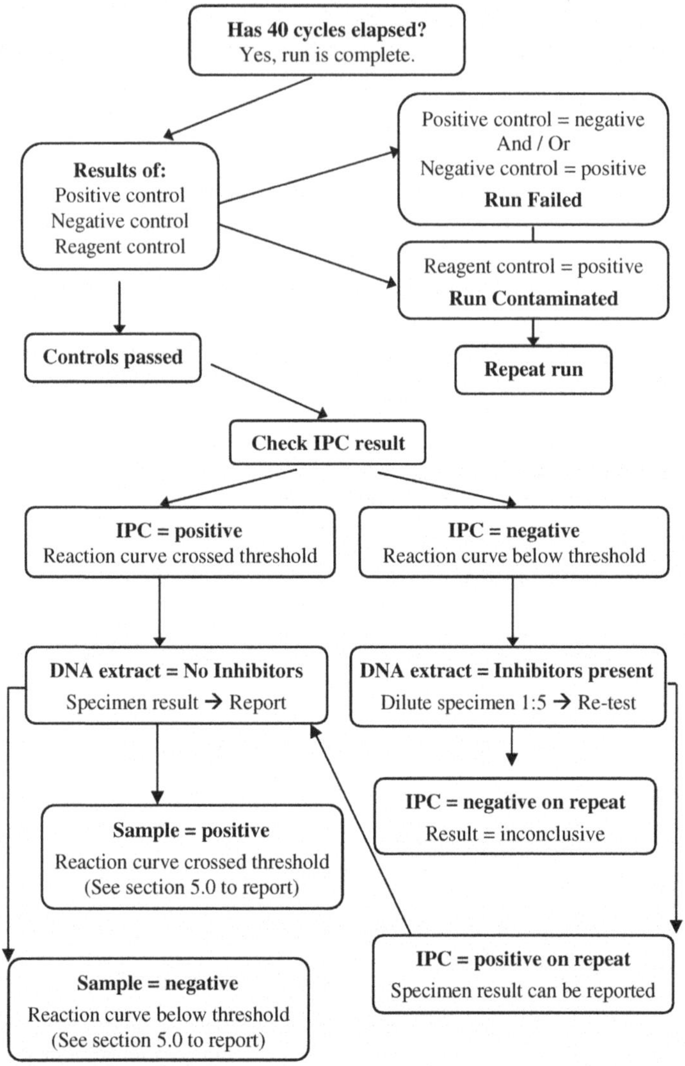

Fig. 18.3 Result interpretation

The online version of the original chapter can be found at
http://dx.doi.org/10.1007/978-90-481-9039-3_18

Index

M. Schuller et al. (eds.), *PCR for Clinical Microbiology*,
DOI 10.1007/978-90-481-9039-3, © Springer Science+Business Media B.V. 2010

The manufacturer's authorised representative in the EU is Springer Nature Customer Service Centre GmbH, Europaplatz 3, 69115 Heidelberg, Germany. If you have any concerns regarding our products, please contact ProductSafety@springernature.com

Printed and bound by CPI Group (UK) Ltd, Croydon, CR0 4YY
08/01/2026
02031440-0010